Plant Epigenetics

Annual Plant Reviews

A series for researchers and postgraduates in the plant sciences. Each volume in this series focuses on a theme of topical importance, and emphasis is placed on rapid publication.

Plant Epigenetics

Edited by

PETER MEYER
Professor of Plant Genetics
Centre for Plant Sciences
The University of Leeds
UK

© 2005 by Blackwell Publishing Ltd

Editorial Offices:
Blackwell Publishing Ltd, 9600 Garsington Road, Oxford OX4 2DQ, UK
    Tel: +44 (0)1865 776868
Blackwell Publishing Professional, 2121 State Avenue, Ames, Iowa 50014-8300, USA
    Tel: +1 515 292 0140
Blackwell Publishing Asia Pty Ltd, 550 Swanston Street, Carlton, Victoria 3053, Australia
    Tel: +61 (0)3 8359 1011

The right of the Author to be identified as the Author of this Work has been asserted in
accordance with the Copyright, Designs and Patents Act 1988.

First published 2005 by Blackwell Publishing Ltd

Library of Congress Cataloging-in-Publication Data
is available

ISBN-10: 1-4051-2977-8
ISBN-13: 978-14051-2977-0

A catalogue record for this title is available from the British Library

Set in 10/12 Times
by SPI Publisher Services, Pondicherry, India.
Printed and bound in India
by Replika Press, Pvt, Ltd, Kundli

The publisher's policy is to use permanent paper from mills that operate a sustainable
forestry policy, and which has been manufactured from pulp processed using acid-free and
elementary chlorine-free practices. Furthermore, the publisher ensures that the text paper
and cover board used have met acceptable environmental accreditation standards.

For further information on Blackwell Publishing, visit our website:
www.blackwellpublishing.com

Contents

9   MicroRNAs: micro-managing the plant genome   **244**
SANDRA K. FLOYD and JOHN L. BOWMAN

Contributors

Dr Werner Aufsatz

Gregor Mendel-Institut, GMI GmbH, Dr Ignaz Seipel-Platz 2, A-1010 Vienna, Austria

Professor John L. Bowman

Section of Plant Biology, One Shields Avenue, University of California Davis, Davis, CA 95616, USA

Dr Tamas Dalmay

School of Biological Sciences, University of East Anglia, Norwich NR4 7TJ, UK

Dr Lucia Daxinger

Gregor Mendel-Institut, GMI GmbH, Dr Ignaz Seipel-Platz 2, A-1010 Vienna, Austria

Professor Ann Depicker

Universiteit Gent, K.L. Ledeganckstraat 35, BE-9000 Gent, Belgium

Dr A. Fischer

Martin-Luther-Universität Halle-Wittenberg, Institut für Genetik, Weinbergweg 10, 06120 Halle, Germany

Dr Sandra K. Floyd

Section of Plant Biology, One Shields Avenue, University of California Davis, Davis, CA 95616, USA

Professor Ueli Grossniklaus

Institute of Plant Biology, University of Zürich, Zollikerstrasse 107, CH-8008 Zürich, Switzerland

Dr Max Haring

Swammerdam Institute for Life Sciences, Universiteit van Amsterdam, Kruislaan 318, 1098 SM Amsterdam, The Netherlands

Dr I. Hofmann

Martin-Luther-Universität Halle-Wittenberg, Institut für Genetik, Weinbergweg 10, 06120 Halle, Germany

Dr Bruno Huettel   Gregor Mendel-Institut, GMI GmbH, Dr Ignaz Seipel-Platz 2, A-1010 Vienna, Austria

Dr Estelle Jaligot   Gregor Mendel-Institut, GMI GmbH, Dr Ignaz Seipel-Platz 2, A-1010 Vienna, Austria

Dr Tatsuo Kanno   Gregor Mendel-Institut, GMI GmbH, Dr Ignaz Seipel-Platz 2, A-1010 Vienna, Austria

Dr Jan Kooter   Department of Genetics, Vrije Universiteit Amsterdam, De Boelelaan 1085, Kamer P544, 1081-HV Amsterdam, The Netherlands

Dr David P. Kreil   Department of Genetics/Inference Group, University of Cambridge, Cambridge CB2 3EH, UK

Dr Marieke Louwers   Swammerdam Institute for Life Sciences, Universiteit van Amsterdam, Kruislaan 318, 1098 SM Amsterdam, The Netherlands

Professor Marjori Matzke   Gregor Mendel-Institut, GMI GmbH, Dr Ignaz Seipel-Platz 2, A-1010 Vienna, Austria

Dr Antonius J. M. Matzke   Gregor Mendel-Institut, GMI GmbH, Dr Ignaz Seipel-Platz 2, A-1010 Vienna, Austria

Dr M. Florian Mette   IPK Gatersleben, Corrensstrasse 3, D-06466 Gatersleben, Germany

Professor Peter Meyer   Plant Genetics, Centre for Plant Sciences, University of Leeds, Leeds LS2 9JT, UK

Dr Nuno Neves   Faculdade de Ciências e Tecnologia, Universidade Nova de Lisboa, Monte da Caparica, 2859-516 Caparica, Portugal

Professor Craig S. Pikaard   Department of Biology, Washington University, Campus Box 1137, One Brookings Drive, St. Louis, MO 63130, USA

Professor Gunter Reuter   Martin-Luther-Universität Halle-Wittenberg, Institut für Genetik, Weinbergweg 10, 06120 Halle, Germany

Dr Philipp Rovina   Gregor Mendel-Institut, GMI GmbH, Dr Ignaz
                          Seipel-Platz 2, A-1010 Vienna, Austria

Dr Matthew Sanders   Universiteit Gent, K.L. Ledeganckstraat 35,
                          BE-9000 Gent, Belgium

Dr Maike Stam   Swammerdam Institute for Life Sciences,
                          Universiteit van Amsterdam, Kruislaan 318, 1098
                          SM Amsterdam, The Netherlands

Dr Wanda Viegas   Faculdade de Ciências e Tecnologia,
                          Universidade Nova de Lisboa, Monte da
                          Caparica, 2859-516 Caparica, Portugal

Preface

With the discovery of RNAi pathways and the histone code, epigenetics has become a popular and fast evolving research topic. Plant science has made a number of elementary contributions to this field, and the common elements of epigenetic systems have linked research groups interested in plant, fungal and animal systems.

This volume provides a comprehensive update on epigenetic mechanisms and biological processes in plants, illustrating the wider relevance of this research to work in other plant science areas and on non-plant systems. Directed at researchers and professionals, together with postgraduate students, it discusses recent advances in our knowledge of basic mechanisms and molecular components that control transcriptional and post-transcriptional silencing. An understanding of these mechanisms is essential for plant researchers who use transgenic lines for stable expression of a recombinant construct or for targeted inactivation of an endogenous gene. These aspects should be of special interest to the agricultural industry.

The volume also illustrates the relevance of epigenetic control systems to gene regulation and plant development, examining paramutation, genomic imprinting and microRNA-based gene regulation mechanisms. Finally, it demonstrates the significance of epigenetic systems to viral defence and genome organisation.

The depth and level of detail now attainable in individual research areas has encouraged a specialisation in the biological sciences that has often inhibited fruitful interaction across species or subject boundaries. By treating plant epigenetics as a group of basic molecular phenomena of wider relevance to the plant and non-plant research communities alike, it is hoped that this volume will help to establish new collaborations between research teams from different subject areas.

I would like to thank the authors who have contributed to this volume for their enthusiasm and commitment to the project.

P. Meyer

1 Transgene silencing

Ann Depicker, Matthew Sanders and Peter Meyer

1.1 Introduction: variation of transgene expression

Transgene technology is widely used for the development of novel crops and can be expected to provide a new area in modern agriculture that will see much more precise design and introduction of genetic traits into crop species, compared with the laborious and often limited options offered by classical breeding. By 2002, the global area of transgenic crop cultivation had risen to 58.7 million hectares (Konig et al., 2004). Expectations (Popelka et al., 2004) and concerns (Cellini et al., 2004) about the technology are equally high, and it is unrealistic to expect that scientific arguments alone will ultimately form the basis for decisions about the individual use of transgene technology in modern agriculture (Pohl Nielsen et al., 2001). There is, however, an obstacle to the long-term use of the technology that will require a scientific solution: the variation of transgene activity.

Reliable long-term activity of transgenes is an essential prerequisite for agronomic plant production; it also influences the value of transgenic plants in basic research for quantitative experiments. The high expression variability of recombinant constructs in individual transformants often compromises the comparability of different constructs, or at least requires extensive sample sizes to secure meaningful conclusions. This still hampers an efficient functional analysis of plant expression signals such as promoters, 5′ untranslated regions and terminators in transgenic plants. In retrospect, it is very likely that incorrect conclusions have been drawn from such experiments due to high variability of transgene expression. The situation is further complicated when transgenic lines that were produced in independent experiments or even via different transformation methods are compared.

Early observations of stable expression and transmission of transgenic traits (Budar et al., 1986) were soon followed by reports about a strong tendency of transgenes to become unstable, frequently in association with DNA methylation of promoter regions (John & Amasino, 1989). It also became obvious that expression instability was not exclusive for transgenes but could affect homologous endogenes (Napoli et al., 1990; Van der Krol et al., 1990), a phenomenon named cosuppression. For a number of years, transgene-silencing events were separated into two categories: transcriptional gene silencing (TGS) linked to an altered epigenetic state of the transgene and posttranscriptional gene silencing (PTGS) mediated by RNA turnover. This distinction has been blurred with the discovery that RNA molecules are essential for the induction of both transgene methylation (Wassenegger *et al.*, 1994) and transcriptional silencing events (Matzke *et al.*, 2004).

1.2   Molecular mechanisms of transgene silencing

Transgene silencing is the result of endogenous epigenetic mechanisms of plants. Apparently, transgenes and transgene loci can produce sequence-specific signals that are recognised by general gene expression surveillance mechanisms. The result is feedback silencing of the transgene either at the transcriptional level when promoter-specific signals are generated or at the posttranscriptional level when transcript-specific signals are produced.

  Below, we describe in short what is known for the two levels of silencing: transcriptional and posttranscriptional. A common aspect of many silencing events that fall into the two categories is that double-stranded RNA (dsRNA) acts as the core signalling molecule and that the downstream signals are short interfering RNAs (siRNAs). The determining factor for the establishment of TGS or PTGS effects at such loci is whether dsRNA is made from the transgene promoter or from the transcribed region. dsRNA formation can be constitutive or it can occur at different time points during development, resulting in constitutive or developmental silencing.

1.2.1   Transcriptional silencing

1.2.1.1   Chromatin remodelling
In recent years, we have seen a remarkable string of discoveries that highlight the importance of chromatin remodelling for the regulation of transcription. As the first level of chromatin organisation, nuclear DNA is wrapped around a histone octamer forming nucleosomes with a core histone fold centre and with protruding histone tails (Luger *et al.*, 1997). Binding of histone H1 to the linker regions that separate individual histones organises the chromatin into a zigzag-compacted 30-nm fibre that forms the next level of chromatin packaging (Bednar *et al.*, 1998). Structural and functional rearrangements of chromatin are mediated by the modification of histone tails, which can include acetylation (Kurdistani & Grunstein, 2003), ubiquitinylation (Sun & Allis, 2002) or sumoylation (Shiio & Eisenman, 2003) of lysines, methylation of lysines and arginines (Zhang & Reinberg, 2001), phosphorylation of serines and threonines (Cheung *et al.*, 2000), or ADP-ribosylation of glutamic acids (Garcia-Salcedo *et al.*, 2003). The resulting complexity of this histone code far exceeds the information potential of the genetic code (Jenuwein & Allis, 2001). In addition to the multiplicity of target sites for histone modifications,

the complexity is further extended by the potential variety of specific modifications. Methylation, for example, can lead to mono-, di- or trimethylated residues (Lachner *et al.*, 2003). Compared with animals and yeast, plants differ in the histone modification sites they use and in the enzymes involved, suggesting that there are distinct histone modification pathways in plants (Loidl, 2004).

1.2.1.2 DNA methylation

In plants and other species that contain DNA methylation functions, we find a close link between histone modifications and changes in the DNA methylation profile. In contrast to fungi and animals, which predominantly methylate cytosine residues located in a CpG context, plants have 5-methyl-cytidine at CpG, CpNpG and even asymmetric CpNpN sites.

Maintenance and *de novo* CpG methylation is regulated by MET1, a homologue of the mammalian DNA methyltransferase 1. In mutants with *met1*-null alleles, CpG methylation is completely erased, which compromises the maintenance of transcriptional silencing, at least for certain transgenes (Saze *et al.*, 2003). MET1 also plays a vital role in gametophytic imprinting (Kinoshita *et al.*, 2004), as maintenance of CpG methylation is essential for the inheritance of epigenetic states during gametogenesis.

Methylation of cytosines outside a CpG context is regulated by *CHROMO-METHYLASE3* (*CMT3*) and by two members of the domain-rearranged methyltransferase (DRM) family, *DRM1* and *DRM2*. *DRM* genes are required for the *de novo* methylation of cytosines in all known sequence contexts (Cao and Jacobsen, 2002b). *DRM2* is expressed at much higher levels than *DRM1* (Cao *et al.*, 2000) and is most likely the predominant *de novo* methylase in *Arabidopsis thaliana*. *DRM* genes are also required for maintenance of asymmetrical methylation but at some loci they act redundantly with *CMT3*. In *drm1/drm2* mutants, asymmetrical methylation is lost at different sites, but for the *SUPERMAN* (*SUP*) locus, this requires the additional mutation of *CMT3* (Cao & Jacobsen, 2002a). Mutation of *cmt3* alone leads not only to a genome-wide loss of CpNpG methylation but also to a depletion of asymmetrical methylation at some loci (Lindroth *et al.*, 2001). Mutations of *drm1/drm2* or *cmt3* do not produce significant phenotypic effects. In contrast, *drm1/drm2/cmt3* triple mutants show developmental abnormalities, which suggest that *DRM* and *CMT3* are part of partially redundant and locus-specific non-CpG methylation pathways (Cao & Jacobsen, 2002a).

Transgene-silencing events are also dependent on the methylation functions. The promoter of the *FWA* gene is normally methylated within two direct repeats. Transgene copies of *FWA* are a useful tool in *de novo* methylation assays, as they are efficiently methylated and silenced. *Drm1/drm2* double mutants, however, lack *de novo* methylation of the direct repeats of the *FWA* locus, and retain transgene activity, which leads to a late flowering phenotype

(Cao & Jacobsen, 2002b). Interestingly, *FWA* transgenes retain the hypo-methylation pattern even when wild-type *DRM* alleles are subsequently crossed in (Cao & Jacobsen, 2002b). This suggests, at least for some trans-genes, the presence of a specific window for *DRM*-specific methylation during the transformation or regeneration process, which might even be linked to environmental stress associated with the transformation. *Drm1/drm2* double mutants also prevent *de novo* methylation and silencing of an inverted-repeat *SUP* transgene, which is efficiently re-established when functional *DRM* alleles are crossed in (Cao & Jacobsen, 2002b). The *drm* mutants do not reactivate previously methylated and silenced *FWA* or *SUP* epigenetic alleles, highlighting their role in the initiation of silencing.

1.2.1.3   *Interactions between DNA and histone methylation functions*

Individual modifications have been associated with the transcriptional competence of the relevant genetic regions, and especially methylation of lysine 9 histone H3 (H3K9) seems to be a hallmark of silent chromatin. In *Arabidopsis*, dimethylated lysine 4 of histone H3 is associated with euchromatin, while dimethylated H3K9 is found in heterochromatin (Jasencakova *et al.*, 2003). In plant species with large genomes, we find a uniform distribution of dimethy-lated H3K9, which may be required for silencing of transposons and interspersed repeats (Houben *et al.*, 2003).

H3K9 methylation can act as a signal for DNA methylation, but can also be reinforced by DNA methylation. In *Arabidopsis*, CpNpG DNA methylation is controlled by H3K9 methylation, through interaction of CMT3 with methy-lated chromatin (Jackson *et al.*, 2002), and in *Neurospora crassa*, one of the roles of heterochromatin HP1 is to recognise trimethylated lysines on histone H3, directing DNA methylation functions to this region (Freitag *et al.*, 2004a). On the other hand, H3K9 methylation requires the presence of CpG methyla-tion (Soppe *et al.*, 2002). A central role has been proposed for the chromatin-remodelling factor: decrease in DNA methylation 1 (DDM1), which is required for H3K9 methylation and H4K16 deacetylation. It was suggested that after replication, newly formed nucleosomes with acetylated H4K16 are still accessible for MET1. The resulting DNA methylation, followed by H3K9 methylation, and the DDM1-mediated recruitment of a H4K16-specific deacetylase are essential to re-establish heterochromatic chromatin (Soppe *et al.*, 2002). While the cause or consequence aspects are still unclear, the liaison between H3K9 methylation and DNA methylation is obvious, and both are important hallmarks of transcriptional silencing.

1.2.1.4   *RNA signals for transcriptional silencing*

A surprising observation that blurred the distinction between TGS and PTGS was the discovery that the RNA interference (RNAi) pathway plays a critical role in chromatin modification (Stevenson & Jarvis, 2003). The first evidence for RNA-directed DNA methylation came from the observation that viroids can target *de novo* DNA methylation to homologous sequences (Wassenegger *et al.*, 1994). This RNA-directed DNA methylation mechanism affects all C residues and requires DRM activity (Cao *et al.*, 2003). Promoters can be specifically targeted for *de novo* methylation by homologous dsRNA homologues (Mette *et al.*, 2000), which also requires the activity of HDA6, a putative histone deacetylase (Aufsatz *et al.*, 2002). In yeast, inactivation of the RNAi machinery leads to transcriptional de-repression of transgenes integrated near the centromere and loss of H3K9 methylation (Volpe *et al.*, 2002). In *Arabidopsis*, the RNAi pathway is required for DRM-dependent *de novo* methylation of the direct-repeat elements in the 5′ region of a *FWA* transgene (Chan *et al.*, 2004).

1.2.1.5 RNA-independent chromatin modification

Many transcriptional silencing events can be attributed to chromatin remodelling, initiated by small RNAs, and associated with increased DNA methylation and H3K9 methylation. There are, however, exceptions to this scheme that highlight the potential diversity of transcriptional silencing pathways that plants can use. While the release of transcriptional silencing in *ddm1* mutants also affects DNA methylation levels, reactivation of silent transgenes in the *Morpheus' molecule 1* (*mom1*) mutant is methylation independent. This suggests that MOM1 is part of a transcriptional silencing pathway that is independent from but complementary to the DDM1 pathway. Consequently, *ddm1/ mom1* double mutants show severe developmental abnormalities (Mittelsten Scheid *et al.*, 2002).

It is also uncertain if all TGS events require RNAi signals. In *Neurospora*, DNA methylation and heterochromatin formation is independent of the RNA-silencing machinery, and the unchanged localisation of HP1 in RNA-silencing pathway mutants suggests that H3K9 methylation is also unaffected (Freitag *et al.*, 2004b). In yeast, an RNAi-independent pathway has been identified that regulates heterochromatin nucleation, involving two proteins Atf1 and Pcr1 that target H3K9 methylation (Jia *et al.*, 2004).

1.2.2 Posttranscriptional silencing with different RNA degradation pathways

Our present understanding of PTGS is mainly based on genetic studies in plants and in *Caenorhabditis elegans*, and on biochemical studies in *Drosophila melanogaster*. What can be deduced from these studies is that RNA silencing appears to consist of a series of general steps involving dsRNA synthesis, processing of the dsRNA trigger into siRNAs and targeting of homologous single-stranded RNAs (ssRNAs) for degradation. Processing of dsRNA and subsequent targeting of ssRNAs seem to be the most conserved steps among organisms and therefore we refer to these two steps as the core mechanism of RNA silencing. This basic pathway is presented in Figure 1.1.

1.2.2.1   Initiation

RNA silencing is initiated in response to dsRNA. Direct evidence for this was initially obtained through the discovery of RNAi in *C. elegans*, showing the potential of dsRNA as an effective elicitor of sequence-specific RNA silencing (Fire *et al.*, 1998). Consistently, transcription of inverted-repeat structures and simultaneous expression of sense and antisense transgenes induce PTGS in plants at a high frequency (De Buck *et al.*, 2001; Muskens *et al.*, 2000; Waterhouse *et al.*, 1999). Furthermore, RNA viruses that generate dsRNA during their infection cycle are potent inducers of PTGS in plants (Voinnet, 2001). Evidence for the silencing inducing role of dsRNA has been expanded through the demonstration of RNAi in *Dictyostelium discoideum* (Martens

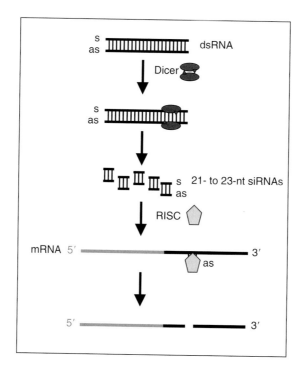

Figure 1.1   Initiation and RNA degradation of single-stranded target RNAs in RNA silencing. RNA silencing is induced upon synthesis of double-stranded RNA (dsRNA) trigger, which is processed by an RNase-III-like protein (termed Dicer in *Drosophila*) into 21- to 23-nt short interfering RNAs (siRNAs). These siRNAs are thought to guide a nuclease complex (termed RNA-induced silencing complex or RISC in *Drosophila*) to homologous (dark grey region) target RNA sequences that are subsequently cleaved near the centre of the region spanned by the siRNA molecule.

et al., 2002), *Trypanosoma brucei* (Ngo *et al.*, 1998), *Drosophila* (Elbashir *et al.*, 2001) and mammals (Wianny & Zernicka-Goetz, 2000).

The dsRNA-silencing trigger is processed to generate 21- to 23-nt RNAs. These short RNAs, referred to as siRNAs, were first discovered in plants exhibiting PTGS (Hamilton & Baulcombe, 1999). In a *Drosophila* cell-free system, addition of dsRNA led to rapid degradation of homologous mRNAs and accumulation of small sense and antisense 21- to 23-nt RNAs (Elbashir *et al.*, 2001). The coincidence of RNA silencing and siRNA accumulation has been established in many organisms and these small RNA molecules are now considered to be integral to RNA silencing.

Cleavage of the dsRNA trigger molecules into siRNAs is catalysed by the RNase-III-related protein, Dicer. The Dicer enzyme and its function were originally identified in *Drosophila* (Bernstein *et al.*, 2001). Dicer belongs to a specific RNase III family of proteins that have a distinctive structure containing dual catalytic domains, dsRNA-binding domains and helicase and PAZ motifs. This RNase III family of nucleases, which cleave dsRNAs, is evolutionarily conserved in worms, flies, plants, fungi and mammals. Homologues of Dicer are found in *Schizosaccharomyces pombe*, *C. elegans* (DCR-1), *Arabidopsis* (CARPEL FACTORY or CAF) and mammals (Bernstein *et al.*, 2001). Genetic evidence implicating the Dicer orthologues *dcr-1* and *carpel factory* in RNA silencing was obtained in *C. elegans* (Grishok *et al.*, 2001) and *Arabidopsis* (Park *et al.*, 2002). These findings strongly support a central role for Dicer and its orthologues in the initiation step of the RNA-silencing mechanism across species.

In summary, endogenous and foreign sequences have the capacity to generate dsRNA structures, thereby initiating an RNA-silencing mechanism. DsRNAs are processed by an RNase-III-related enzyme into 21- to 23-nt RNAs. These siRNAs are thought to mediate sequence-specific RNA degradation of homologous RNA targets.

1.2.2.2   Sequence-specific degradation of single-stranded target RNAs
The sequence-specific RNA degradation step of RNA silencing is catalysed by a multi-subunit nuclease, referred to as the RNA-induced silencing complex (RISC). RISC, originally discovered in *Drosophila* cell extracts, was shown to contain siRNAs, to be required for target RNA cleavage and to be separable from Dicer (Bernstein *et al.*, 2001; Hammond *et al.*, 2001). The siRNAs guide RISC to its target, which upon Watson–Crick base pairing, is cleaved. It is unknown whether RISC contains single-stranded or double-stranded siRNAs. In a *Drosophila* cell-free system, it was shown that target RNA cleavage is endonucleolytic and occurs only near the centre of the region spanned by the siRNAs, and that the ruler to define the position of cleavage is set by the 5′ end of the target-complementary siRNA strand (Elbashir *et al.*, 2001).

The configuration of the siRNAs appears to be crucial for incorporation into the RISC. Dicer produces double-stranded siRNAs with 2-nt-long 3' over-hangs and 5' phosphate and 3' hydroxyl ends (Elbashir *et al.*, 2001). This type of RNA is an effective inducer of RNAi, suggesting an efficient incorporation into the RISC. In contrast, synthetic duplex siRNAs longer than 30 bp or siRNAs with extensive 2'-deoxy modifications failed to mediate RNAi efficiently, probably by lack of or misincorporation into the RISC. The homology between the siRNAs and the target RNA is another critical requirement for cleavage by RISC. RNA degradation mediated by siRNAs is sensitive to sequence mismatches.

In conclusion, small 21- to 23-nt siRNAs, which are processed from a dsRNA trigger by an RNase-III-related protein, guide an RNA-degrading protein complex to homologous targets for endonucleolytic cleavage. This implies that the siRNAs are transferred from the (Dicer) initiation complex to the (RISC) RNA degradation complex.

1.2.2.3 *RNA-dependent RNA polymerases involved in signal generation and amplification*

The Dicer/RISC pathway accounts for the initiation of the silencing process and the homology-dependent selection and degradation of silencing targets. However, this pathway cannot explain all observed silencing phenomena, especially siRNA amplification and systemic spread of RNA silencing. This suggests the existence of additional processes involved in RNA silencing.

In general, sense transgenes are constructed to give high expression and are not expected to produce dsRNA from the transgene construct. Thus, when RNA silencing occurs, RNA that is transcribed from the transgene locus must have been converted in one way or another into dsRNA. It was therefore postulated that so far unidentified transgene RNA features would be recognised as aberrant by an RNA-dependent RNA polymerase (RdRP). The duplication of these RNAs would then result in the production of the required dsRNA from which primary siRNAs could be generated.

The hypothesis of an RdRP requirement for the triggering of RNA silencing was confirmed by several genetic studies revealing that homologues of the RdRP enzyme are essential for RNA silencing in *Neurospora* (Cogoni & Macino, 1999), *Arabidopsis* (Dalmay *et al.*, 2000; Mourrain *et al.*, 2000) and *C. elegans* (Smardon *et al.*, 2000). In *Arabidopsis*, RdRP is required to trigger silencing by a sense transgene, but not by a hairpin construct or by a virus (Dalmay *et al.*, 2000). This suggests that for PTGS induction from sense transgene expression, involvement of the RdRP for signal generation is essential. It indicates, however, that in the case of high expression of a hairpin RNA, amplification is not required, and that upon viral infection, the virus-encoded RNA polymerase produces enough dsRNA to induce gene silencing.

The impact of the RdRP enzyme activity on RNA silencing was recently illustrated by the observation that the incidence of highly expressing transformants shifts from 20% in wild type to 100% in a *sgs2* mutant RdRP-defective background, independent of the expressed transgene (Butaye *et al.*, 2004). Since the same drastic reduction in poorly expressing transgene transformants was observed in another PTGS line, sgs3, the authors concluded that the main cause of reduced transgene expression in an *Arabidopsis* wild-type background is linked to PTGS (Butaye *et al.*, 2004).

Besides the RdRP-mediated conversion of the primary silencing RNA signals into dsRNA, RdRP also synthesises dsRNA from target mRNAs that are marked by primary siRNAs. The resulting dsRNAs become the source for the production of secondary siRNAs, as illustrated in Figure 1.2. Evidence for a target-dependent amplification of the silencing signal can be found in many initial reports on transgene silencing in plants. Very often, a correlation is seen between the timing of induction of the endogene and the switch from a high to silenced expression of the transgene. This may now be interpreted as the need for sufficient target RNA in order to amplify the silencing signals.

In conclusion, amplification of the silencing signal occurs not only on the basis of the aberrant trigger RNA but also on the basis of its target RNA. This is required to obtain sufficient amounts of dsRNA and to achieve *cis* and *in trans* silencing. The amplification process is thus dependent on the combined activities of the RdRP and Dicer. Interestingly, the RdRP/Dicer pathway itself results in degradation of the template RNA and could take over RISC function when required (Figure 1.2). An siRNA amplification mechanism could be required for several aspects of RNA silencing. For instance, the amplification process could be a necessity for the production of the silencing signal molecules in amounts sufficient for systemic spread of RNA silencing. Amplification of siRNAs could also be required for efficient siRNA-guided target RNA degradation catalysed by RISC.

1.2.2.4 Transitive silencing

It is generally recognised that a silencing-inducing locus can efficiently reduce the expression of genes that produce transcripts partially homologous to those produced by the silencing-inducing locus (primary targets). Interestingly, the expression of genes that produce transcripts without homology to the silencing-inducing locus (secondary targets) can also be decreased dramatically via transitive RNA silencing. This process, referred to as transitive silencing, reflects the activity of RdRP mediating the amplification of siRNA-guided target RNAs, as described above.

Transitive silencing has been demonstrated for *C. elegans* (Sijen *et al.*, 2001) and for plants (Vaistij *et al.*, 2002; Van Houdt *et al.*, 2003). It is based on *de novo* synthesis of secondary siRNAs that were originally absent from the original pool of siRNAs. These secondary siRNAs, non-homologous to the

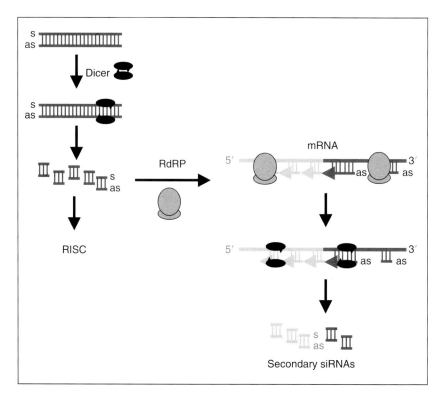

Figure 1.2   Amplification of short interfering RNAs (siRNAs) and transitive silencing. The initial double-stranded RNA (dsRNA) trigger is cleaved into 21- to 23-nt siRNAs by Dicer. Small trigger-derived antisense siRNAs can anneal to homologous (dark grey) RNAs. The resulting partial dsRNA–RNA hybrids can be recognised by the RNA-induced silencing complex (RISC) for direct degradation or by an RNA-dependent RNA polymerase (RdRP) for elongation. The dsRNA molecules produced through RdRP activity are further processed by a Dicer-like enzyme, resulting in amplification of siRNAs. This could also result in the accumulation of secondary siRNAs that do not correspond to the original trigger sequence (light grey) and that can direct degradation of secondary targets, a process referred to as transitive silencing.

silencing locus, correspond specifically to the adjacent upstream region of the original target sequence. This suggests an amplification cycle in which an RdRP can use siRNAs as primers and the target RNA as template to produce dsRNA (Figure 1.2). The process therefore leads to spreading of the RNA-silencing target spectrum to regions adjacent to the silenced target genes. The data suggest that in transgenic plants, targets of RNA silencing are involved in the expansion of the pool of functional siRNAs. Furthermore, methylation of target genes in sequences without homology to the initial silencing inducer indicates not only that RNA silencing can expand to adjacent target RNAs, but also that methylation can spread to adjacent chromosomal regions (Van Houdt *et al.*, 2003).

In plants, both transgenes and injected siRNAs have been reported to induce spreading of gene silencing along the target gene. Remarkably, the RNA targeting can spread to the 3′ and 5′ regions of the original target sequence, implying both primed and unprimed amplification cycles (Vaistij *et al.*, 2002; Voinnet *et al.*, 1998).

1.2.2.5 The role of DNA methylation and chromatin modification in RNA silencing

In plants, DNA methylation and RNA silencing have been linked through the observed capacity of dsRNA to mediate sequence-specific DNA methylation. RNA-directed DNA methylation was first described in studies of plants infected with viroids (Wassenegger *et al.*, 1994). Genomic targets with only 30 bp of sequence identity to the viroid RNA were shown to be methylated upon viroid infection (Pelissier & Wassenegger, 2000).

Methylation of the 3′ coding regions is a common feature of transgenes silenced *in trans* by PTGS (English *et al.*, 1996; Ingelbrecht *et al.*, 1994; Sijen *et al.*, 1996), whereas such correlation is not observed for endogenes. The first indications for a role of DNA methylation in RNA silencing came from genetic studies in *Arabidopsis*. Mutations in a predicted RdRP blocked both RNA silencing and methylation (Dalmay *et al.*, 2000). PTGS was also affected in certain *Arabidopsis* mutants that were deficient in DNA methylation (Morel *et al.*, 2000). Furthermore, drug-induced hypomethylation of cosuppressed and inverted transgenes results in partial reactivation of these genes (Kovařík *et al.*, 2000).

DNA methylation associated with RNA silencing could be a dispensable 'prelude' or a consequence of chromatin modification, which would be the essential process to ensure silencing. In essence, it is obvious that in plants, methylation, chromatin remodelling and silencing are linked, but the causal relationship is not completely understood. On the one hand, siRNA could direct DNA methylation, followed by chromatin remodelling, while on the other hand, chromatin remodelling may be the primary effect and DNA methylation serves to stabilise the altered chromatin structure. Yet a third option cannot be excluded, which is that chromatin modification and/or methylation are the very first events that set up the whole silencing process.

DNA methylation and chromatin remodelling might serve as a transitory mechanism between RNA silencing and TGS. In methylation-competent organisms, this could, for example, be achieved by spreading of DNA methylation from coding regions into promoter regions. Transition of PTGS into TGS could be beneficial to the organism for several reasons. PTGS is a 'wasteful' process since it involves synthesis and immediate degradation of target RNA whereas TGS is more 'economical', since no RNA is transcribed. On the other hand, this 'wasteful' RNA-silencing process could be an effective way to regulate genes encoding proteins involved in quick on and off responses, such as growth factors and other regulatory proteins. Therefore the transition of PTGS into TGS only makes sense from an 'economical' point of view for genes that need to be permanently shut off.

Evidence for this view comes from the analysis of DNA methylation kinetics in a transgenic line (Fojtova *et al.*, 2003). Parental silenced HeLo1 (hemizygous for locus 1) plants show posttranscriptional silencing of the residing neomycin phosphotransferase II (*nptII*) transgenes and cytosine methylation restricted to the 3′ end and the central part of the transcribed regions. With an increasing number of cell cycles during callus growth, DNA methylation changes gradually and is introduced into the promoter region. After 24 months of callus cultivation, an epigenetic variant, designated locus 1E, was obtained in which cytosine methylation of symmetrical sites was almost complete within the 5′ end of the *nptII*-transcribed region and the 35S promoter. The newly established epigenetic patterns were stably trans- mitted from calli into regenerated plants and their progeny. Nuclear run-on assays of locus 1E could not find any detectable amounts of primary tran- scripts along the *nptII* gene, indicating that the methylated promoter has become inactivated. Thus, a switch between PTGS and TGS could be a mechanism leading to an irrevocable shutdown of gene expression within a finite number of generations (Fojtova *et al.*, 2003).

In summary, DNA methylation and chromatin remodelling are clearly associated with RNA silencing, but their precise role in RNA silencing remains to be clarified. Over the years it has become more and more evident that RNA plays an important role in controlling DNA structure and chromatin status. It can be expected that this particular regulatory role of RNA will be an important topic for further research in the near future: how does RNA silencing influence genomic modification and vice versa?

1.3   Systemic silencing

RNA silencing in plants can produce a systemic silencing signal that is sequence-specific, moves long distances through the plant and is presumably amplified (Palauqui & Balzergue, 1999). Nevertheless, the factors involved in such amplification and/or transition from local to systemic silencing are not well understood.

Non-clonal patterns of transgene silencing and cosuppression were observed in tobacco plants transformed with the highly expressed 35S-driven S- adenosyl-L-methionine synthase (*SAM-S*) gene (Boerjan *et al.*, 1994) or with derivatives of 35S-driven nitrite (*NII*) or nitrate (*NIA*) reductase–encoding constructs (Palauqui *et al.*, 1997).

In the first case, the lower leaves of mature transgenic tobacco plants showed high *SAM-S* expression whereas younger fully expanded leaves showed an increasingly silenced phenotype. To explain this observation it was suggested that high levels of *SAM-S* expression in the older leaves induced silencing in the younger leaves because of a transported molecule, implicated in the negative regulation of the endogenous *SAM-S* genes (Boerjan *et al.*, 1994). A similar pattern was observed when cosuppression was initiated at an advanced state of tobacco plant development with an *NII* transgene. By contrast, with an *NIA* transgene construct, not all the young leaves but the leaves located at the same side of the plant as the first silenced leaves were the ones to become preferentially cosuppressed (Palauqui *et al.*, 1996). These results clearly suggest the existence of a non-metabolic and transgene-specific silencing signal transported in the plant. However, such a systemic silencing signal was only clearly demonstrated using grafting experiments.

Scions expressing *NIA* or *NII* reductase endogenes and transgenes became suppressed only when grafted onto stocks carrying the corresponding silenced transgene. Most probably, the signal is amplified in the scion, because the silencing state can persist a long time after the original source of silencing is eliminated, and a requirement for high expression of the target gene is observed (Palauqui & Balzergue, 1999). The systemic sequence-specific signal has not yet been identified. Candidates are small RNAs with a length of 24–26 nt, longer dsRNAs or aberrant RNAs, but evidence for either of them is lacking.

To gain insight into the factors involved in the spreading of PTGS through the plant, García-Pérez *et al.* (2004) combined the transitive and systemic silencing phenomena. They found that primary target sequences in combination with a silencing locus were able to promote the generation of systemic–transitive silencing signals. Sufficient production of the systemic signal was dosage-dependent for both the silencer and the primary target locus (García-Pérez *et al.*, 2004). These data suggest that the mobile signal contains an RNA amplification product. The data can also explain why antisense-mediated silencing does not produce a systemic silencing signal (Crété *et al.*, 2001), given that antisense transcripts are generally unstable. In order to find detectable amounts of siRNAs in the scion when grafted on a stock with silencer and primary target locus, the secondary target gene had to be present. This suggests that secondary target RNAs are essential for amplification, allowing in this way establishment of gene silencing in response to the mobile signal.

1.4 Silencing signals

Endogenes do not show the same degree of expression variation as transgenes. One approach to overcome transgene silencing is therefore to understand what makes transgenes so different from endogenes. We can define a series of factors that influence transgene silencing, and at least some of these aspects can be addressed when we want to design gene transfer strategies with reduced silencing potential.

As explained above, variation of transgene expression in *Arabidopsis* is mainly related to the induction of PTGS, which depends in most cases on the RdRP-mediated conversion of transgene RNA into dsRNA. It is unclear which features make the transgene RNA a substrate for RdRP-dependent conversion into dsRNA. This may be determined by the concentration of particular unprocessed or non-polyadenylated RNAs (Metzlaff *et al.*, 1997), by the structure of background-scanning transcripts that monitor the genome for invasive DNA elements, or it may be related to the association of RNAs with particular proteins upon transcription in a certain context. At least for RNA-silencing events, several correlations have been noted between the characteristics of a transgene locus and its propensity/tendency towards the induction of RNA silencing. The main factors are the expression level of the transgene and the locus structure, both of which are discussed below.

1.4.1   The transgene construct

The composition of the recombinant construct itself can significantly affect the susceptibility of a transgene to silencing.

Firstly, the promoter selection will determine at which frequency the transgene is transcribed and consequently silenced. By comparing the effects of strong and weak promoters that drive sense *CHALCONE SYNTHASE (CHS)* transgenes in large populations of independently transformed petunia plants, it was observed that a strong promoter was required for high-frequency cosuppression of the *chs* genes (Que *et al.*, 1997). Besides the frequency, the degree and pattern of cosuppression were also strongly modulated by the promoter strength, which was correlated with either one or four enhancer elements added to the minimal 35S promoter (Que *et al.*, 1997).

Secondly, transcript stability seems to be an important factor in determining the frequency and degree of RNA silencing. This was elegantly shown by introducing frameshift mutations in otherwise identical 35S-driven *CHS* transgene constructs. Each different frameshift mutation was found to reduce the frequency and the degree of cosuppression, presumably by altering the transcript stability. These results suggest that sense-transgene-induced cosuppression is a response to a high concentration of accumulated transgene mRNAs or a derivative therefrom (Que *et al.*, 1997).

Thirdly, features of the 3′ regions may affect the frequency at which transgenes trigger RNA silencing. This assumption is supported by the analysis of a 35S-driven transgene encoding the light chain of an antibody. When linked to the 3' region of the octopine synthase gene, the transgene was highly expressed in a hemizygous condition but reproducibly silenced in a homozygous condition (De Neve *et al.*, 1999), while this effect was not observed when the same recombinant gene construct was linked to the 3' region of the rubisco gene (Peeters *et al.*, 2001).

For single-transgene copies, we can conclude that the composition of the transgene and the level of transcript accumulation determine whether a single-copy transgene will trigger RNA silencing. The likelihood of silencing seems to be higher in lines that are homozygous for the transgene.

1.4.2 *The impact of the transgene locus structure*

While we have total control over the design of the recombinant construct, we cannot influence the final structure of the integrated transgene locus. Over time, we had to correct the initial concept that gene transfer leads to simple transgenic locus structures. Considering that the free ends of linear DNA stimulate non-homologous end-joining systems in plants that are error-prone and associated with a high degree of rearrangements (Gorbunova & Levy, 1999), it could be expected that biolistic transformation techniques or polyethylene glycol transfer methods that included linear carrier DNA fragments would result in a high degree of rearrangement at the transgene loci. But even transgenic loci in plants produced by *Agrobacterium*-based technology, often predicted to result in the clean transfer of a T-DNA fragment bracketed by the left and right border, frequently violate the textbook models, forming complex rearrangements. Transgene copy numbers can be highly variable and are probably dependent on numerous factors that have not been fully elucidated. A common feature of all transformation methods is that transgenes tend to integrate as multiple copies into one or a few insertion sites. This could be the result of extrachromosomal ligation of different duplicated T-strands being transferred from a single or multiple agrobacteria (De Buck *et al.*, 1998), but replication *in planta* before or during integration can also not be excluded (Jorgensen *et al.*, 1987).

There seems to be a correlation between the transformation method and the number of integrated T-DNA copies. Whereas co-cultivation of root explants in general yields about 50% of transformants with a single T-DNA copy, *in planta* transformation normally yields only 5–10% of transformants with a single T-DNA (De Buck *et al.*, 2004). Also the use of different *Agrobacterum* strains, explant material and co-cultivation conditions most likely affects the T-DNA integration pattern. For instance, Grevelding *et al.* (1993) found that transgenic plants derived from root transformation tended to have fewer inserts than plants derived from leaf disc transformation. For PEG-mediated gene transfer, the cell cycle stage of the protoplasts has been shown to influence the complexity of the transgene locus (Kartzke *et al.*, 1990).

The composition of T-DNA loci is further complicated by the presence of non-T-DNA sequences. Vector sequences located outside the T-DNA region are frequently integrated into the plant genome, linked or unlinked to T-DNA sequences (Kononov *et al.*, 1997).

The report of Hobbs *et al.* (1990) was a milestone in the awareness of the importance of the transgene locus structure. The intertransformant variability was found to be either high or low in a bimodal fashion without continuous variation in progeny plants of several transformants over two generations. Transformants having high expression all had similar expression, suggesting little position effect. These transformants with high transgene expression had single T-DNA insertions, while those with 100-fold lower expression had multiple T-DNA insertions at the same or different loci. Furthermore, the same research group demonstrated that loci encoding the low expression phenotype were acting epistatically on loci encoding the high expression phenotype (Hobbs *et al.*, 1993).

Thus, transformed plants with multicopy loci and with a dispersed number of transgene copies have a much higher tendency towards being silenced at the posttranscriptional level than transformed plants with a single-transgene copy. This probably reflects a higher probability for the production of threshold levels of transgene transcripts, and for inverted-repeat arrangements.

The presence of multiple T-DNA copies increases the likelihood of reaching the transcript threshold level at which the RdRP-mediated primary silencing signals are generated. Provided no transcriptional silencing affects the transgene activity, the dosage effect of the increasing number of transgenes results in a significant excess of transcript levels beyond the transgene-specific threshold at which RNA silencing is triggered, correlating with more pronounced silencing. Many transgene loci with several T-DNAs indeed show evidence for the threshold hypothesis, turning on RNA silencing and cosuppression only under homozygous conditions (de Carvalho *et al.*, 1992; Goodwin *et al.*, 1996; Kunz *et al.*, 1996). It should be noted, however, that in many multicopy transformants, RNA silencing is not triggered and that single-copy transformants are not always the highest expressers. Despite strong indications for a threshold at which RNA silencing is triggered, it remains a mystery why certain transformants with the same number of transgenes do not trigger silencing (De Buck *et al.*, 2004).

When two T-DNA copies are present in an inverted-repeat orientation, with transcription of these transgene copies proceeding in convergent orientation, a threshold-independent pathway seems to be responsible for triggering RNA silencing. Indeed, a strikingly strong silencing has been observed with a variety of transgenes when these are integrated as convergent inverted repeat, even in hemizygous condition (Cluster *et al.*, 1996; De Buck *et al.*, 2001; Depicker *et al.*, 1996; Jorgensen *et al.*, 1996; Que *et al.*, 1997). In order to analyse experimentally the correlation between transgene silencing and the presence of T-DNA-inverted repeats in transgenic *Arabidopsis* plants, expression of the β-glucuronidase (*gus*) gene was studied when present as a convergent transcribed inverted repeat or as a single copy in otherwise iso-genic lines (De Buck *et al.*, 2001). The results clearly show that convergent transcription of inverted-repeat transgenes triggers *trans*-acting silencing very efficiently, and that a spacer in between the inverted genes reduces the ef-ficiency of initiating and maintaining silencing (De Buck *et al.*, 2001).

Even promoterless constructs can induce RNA silencing and cosuppression when the constructs are integrated as inverted repeats, suggesting that in these cases a high transcription rate of the transgenes is not a prerequisite to induce this kind of RNA silencing (Stam *et al.*, 1997; Van Blokland *et al.*, 1994; Voinnet *et al.*, 1998). It is not clear why transgenes trigger silencing efficiently when integrated as inverted repeats. Possible explanations are that conver-gently transcribed repeats could form dsRNA by read-through transcription in conjunction with the fact that the palindromic centre may change transcription termination. In support of this model, disruption of the inverted repeat by a non-palindromic sequence significantly weakens the RNA-silencing strength of the locus (De Buck *et al.*, 2001).

In conclusion, many data support the assumption that the transgene locus structure has a significant impact on transgene expression. It is, however, important to distinguish among the effects of single-copy transgenes with high expression and multicopy tandem or inverted-repeat arrangements of transgenes.

1.4.3 RNA silencing induced by constructs carrying inverted repeats (sequence homology and repeats)

Constructs carrying transcribed sequences arranged as intramolecular inverted repeats trigger RNA gene silencing very efficiently. If dsRNA can be derived directly from transcription of a hairpin transgene, there is actually no need for the conversion of ssRNA into dsRNA by RdRP. In this way, powerful vectors for the cloning of coding sequences in a hairpin configuration between a plant promoter and terminator have been developed as effective tools for investi-gating plant gene function in a high-throughput, genome-wide manner (reviewed by Helliwell and Waterhouse, 2003; Wang and Waterhouse, 2002).

1.5 Position effects

The reliable expression of an endogenous gene not only depends on the control units immediately adjacent to the coding region, such as promoter or enhancer elements, but also on the wider nuclear environment (Alvarez *et al.*, 2003). The functionality of a gene can therefore be influenced by its chromosomal environment, which probably explains why the insertion of transgenes into different genomic positions often leads to unreliable transgene expression. The classical example of position effect variegation (PEV) refers to the translocation of a eukaryotic gene next to a heterochromatic block, which results in strong cell-to-cell variation in the activity of the gene. The initial mechanistic model for PEV suggested a *cis* spreading of condensed, heterochromatic chromatin that reduces or inhibits expression of the translocated gene. As some PEV effects, however, can act over considerable distances, a nuclear compartment model was proposed, suggesting that *trans*-interactions between heterochromatic regions determine the three-dimensional organisation of chromosomes in interphase loci, and that the miss-function of the displaced gene is the consequence of its association with a nuclear compartment that lacks sufficiently high concentrations of the required transcription factors (Wakimoto, 1998). It has been proposed that the only reliable techniques for analysing gene function will involve recombination technologies to either manipulate a gene at its natural chromosomal locus or alternatively select a 'neutral' locus for recombination where ectopic genes are least likely to be exposed to complex epigenetic factors (Jackson, 2000).

At present, it is still unclear how much position effects contribute to the expression stability of transgenes. Molecular and cytogenetic analysis of stably and unstably expressed transgene loci in tobacco showed that two stably expressed transgene loci had integrated as simple T-DNA arrangements near AT-rich regions that bind to nuclear matrices *in vitro*, which may also function as matrix attachment regions *in vivo*. Two unstably expressed loci, one a single-copy transgene and the other a multicopy rearranged transgene, were integrated at intercalary and paracentromeric locations (Iglesias *et al.*, 1997). The variety of factors that can influence transgene expression makes it difficult to determine if the variable expression of a transgene locus is due to its genomic position or to its composition. Rearrangements of transgenes are a frequent phenomenon, which can foster silencing mechanisms (Muskens *et al.*, 2000). A transgene locus that contains partially inverted sequences has the potential to form dsRNAs, which can cause RNA-induced DNA methylation (Mette *et al.*, 2000). The same applies for a transgene that harbours two transcribed units in head-to-head orientation, or to a transgene inserted near an endogenous transcription unit that might produce antisense transcripts of the transgene. For cosuppression of CHS in petunia, it has actually been demonstrated that the determining factors for PTGS were the repetitiveness and organisation pattern of the transgene, while the genomic sequence surrounding the integration site had little influence on the cosuppression (Jorgensen *et al.*, 1996).

An extensive study of 112 transgenes in *Arabidopsis* demonstrated that transgene insertion in heterochromatic regions is not necessarily associated with silencing effects (Forsbach *et al.*, 2003). In another study, 21 single-copy

T-DNA *Arabidopsis* transformants were characterised in detail for their insertion position and expression levels; 19 of these 21 lines showed comparable transgene expression levels independent of the orientation of the T-DNA or its integration into an intergenic or genic region, or into an exon or an intron (De Buck *et al.*, 2004). At least for *Arabidopsis,* these results strongly suggest that position effects play only a small, if any, role in transgene silencing, which seems to be predominantly based on PTGS effects.

In petunia, however, three single-copy transgenes, which did not contain any rearrangements, showed striking differences in their expression patterns. Each transgene contained the same two genes in tandem orientation, and both genes were inactivated in a line where the transgene had integrated into a highly repetitive and hypermethylated region. Two other lines, with transgenes inserted into unique regions, displayed stable transgene expression but differed in expression levels. A transgene with a relatively low expression level had integrated into a unique but methylated region, which imposed its hypermethylation pattern on the border regions of the transgene. The other line showed a much stronger transgene activity. In this line, the transgene was inserted into a unique and hypomethylated genomic environment (Pröls & Meyer, 1992). These results suggest that single-copy transgenes can come under the control of the integration region, and that classical position effects can play a role in transgene silencing, at least in certain plant species.

1.6   Environmental effects

Plants are usually exposed to highly variable environmental conditions and might have developed mechanisms of gene regulation that are distinct from those of animals.

There have been reports about transgene silencing specifically affecting transformants planted in the field and that could not be detected when the same lines were grown in the greenhouse. Silencing can affect the transgene and its homologous endogene, reminiscent of cosuppression (Brandle *et al.*, 1995), or it can affect transgenes, altering their DNA methylation state (Meyer *et al.*, 1992), and chromatin state as monitored by an altered susceptibility of the transgene region to enzymes (Van Blokland *et al.*, 1997). This implies that both transcriptional and posttranscriptional silencing mechanisms can come under environmental control.

The diversity of environmental effects and their combinations complicates the development of test programmes that could be applied to detect transgenes susceptible to such effects. A transgene may be reliably expressed for several generations before the specific environmental conditions develop that may alter its epigenetic state. Moreover, there appears to be variation in susceptibility in stress-induced transgene silencing, which suggests that 'competence gradients' are established in individual plants that determine how the next generation responds to environmental stress. This effect became apparent from a field trial with plants derived from a petunia transformant that contained a stably expressed *A1* transgene that was responsible for the production of a brick-red floral pigment (Meyer *et al.*, 1992). As petunia continuously produces new flowers, a large pool of progeny plants could be generated from one line by continuous pollination. Continuous production of flowers could also be used to monitor the activity of the *A1* transgene over a longer period in the next generation. As expected, all progeny plants initially displayed the same stable expression of the transgene as had been observed in the progenitor line. At later stages, however, A1 activity became reduced or completely silenced in the majority of the F1 generation. Stable transgene activity was, however, maintained in those plants that were derived from early pollination of the young flowers of the parental line. Thus, although the transgene had been active in all flowers of the parental lines, the older flowers had already imposed an epigenetic state onto the transgene that made it susceptible to become silenced in the next generation. In practical terms this observation has implications for breeding strategies as it implies that progeny from early pollinations have an increased ability to avoid being silenced in the next generation.

Although the actual environmental conditions that ultimately lead to transgene silencing are difficult to define, and may actually vary for individual transgenes, it would be useful to develop controlled stress tests that facilitate the identification of silencing-prone lines at an early stage. A phosphinothricin acetyltransferase *(PAT)* transgene-encoding herbicide resistance in a single-cell culture of *Medicago sativa* became silenced after a 10-day heat treatment at 37°C (Walter *et al.*, 1992). The effect seemed to be transgene-specific as the growth performance of the culture was not affected by the heat treatment. The line showed a moderate tendency for silencing as the transgene became PAT sensitive in 12% of the cells, but heat treatment enhanced this effect to 95%. Molecular analysis demonstrated that PAT sensitivity was not linked to a loss of transgene activity but was most likely due to an increased turnover of the PAT enzyme (Broer, 1996). A similar heat treatment of transgenic tobacco lines that contained a *PAT* transgene also resulted in a PAT-sensitive phenotype, which was restored when the heat treatment was terminated (Broer, 1996). Unaltered production of *PAT* transcripts in the PAT-sensitive lines suggested that protein production or stability may have been affected.

A more stable silencing effect was observed in transgenic petunia and tobacco lines that were cultured on propionic or butyric acid (ten Lohuis *et al.*, 1995). The treatment induced a high level of variegated or completely abolished expression, which persisted for several months after termination of the treatment. The effect was more pronounced in lines that were homozygous for the transgene compared with heterozygous lines, and correlated in all case with increased methylation of the transgene promoter region.

The behaviour of transgenes under environmental stress conditions is reminiscent of observations made for mobile elements. UV-B exposure and ozone depletion activate transposable elements (Walbot, 1999), and individual retrotransposons can become active in tissue culture (Hirochika *et al.*, 1996). Transposon activity appears to be higher in lines that have been placed under extreme evolutionary selection (Jiang *et al.*, 2003) and may reflect an increased epigenetic variation required for quick adaptation.

At first glance, there appears to be a difference as transposable elements are activated by environmental effects, while transgenes become silenced. This may, however, simply reflect the different assay systems that predominantly analyse the changing state of active transgenes and dormant transposable elements. The common feature of environmental stress applied to transposable elements and transgenes may therefore be an increased flexibility of epigenetic states that changes their expression profile. The activity of a very poorly expressed *A1* transgene could be enhanced when the transformant was grown in the field (Saedler *et al.*, 1992), suggesting that environmental effects can equally influence epigenetic states in active and silenced transgenes.

Certain transgenes and transposable elements may therefore share features that make them specific targets for epigenetic systems. Other genes are, however, also able to change their epigenetic state (Cubas *et al.*, 1999), and many such changes may go undetected if they do not result in a visible phenotype or if other copies of the targeted allele mask the effect. At least some of such epigenetic modifications are augmented under environmental stress. Mutagenic treatment of *Arabidopsis* produces epigenetic alleles of the *SUP* locus (Jacobsen & Meyerowitz, 1997), and the maize *P* locus shows a high frequency of DNA methylation–associated mutation in plants that were regenerated in tissue culture (Cocciolone *et al.*, 2001). Equally, an epiallele of the *bal* locus, isolated from an *Arabidopsis* mutant defective in DNA methylation (*DDM*), is very stable even in a wild-type background, but reverts at a high frequency upon genomic damage and stress (Stokes *et al.*, 2002).

1.7 Strategies for the prevention of transgene silencing

1.7.1 Selection of single-copy transgenes with no rearrangement

Transgenes that have the potential to produce dsRNA are prime candidates for silencing. It is therefore desirable to exclude transgenic lines that can form inverted repeats at an early stage, due to transgene rearrangements, or that have inserted near endogenous genes that might produce transgene-specific antisense transcripts via read-through transcription. Obviously, any head-to-head arrangement of transcription units within a transgene construct should be avoided in the first place, as many polyadenylation regions fail to control the efficient termination of transcription (Thompson & Myatt, 1997).

1.7.2   Selection of favourable integration regions

At least in some species, silencing of single-copy transgenes can correlate with the repetitiveness or hypermethylation state of the integration region (Pröls & Meyer, 1992). It may therefore be advisable to screen transgenic lines for the insertion of single-copy transgenes into unique and hypomethylated genomic regions. Especially for species for which comprehensive sequence information is available, such analysis should be relatively simple (Sallaud et al., 2004) and could in the future be complemented by an increased knowledge about the epigenetic states of genomic regions.

Alternatively, it may be useful to develop selection strategies to identify favourable insertion sites. One such approach is the transfer of constructs that combine two genes arranged in opposite orientation and under the control of the same promoter region (Akashi et al., 2002). Culture of transgenic material under high selection conditions for one of the two genes lead to the selection of cell lines that also displayed a high activity of the adjacent gene for at least 1.5 years. This co-selection experiment was, however, conducted in suspension culture lines, which allowed continuous selection, and which will obviously be difficult to maintain in soil-grown plants. For suspension cultures, co-selection does appear to be a promising strategy, which interestingly only works if a high selection pressure is applied from the very start. When lines that had lost the activity of both transgenes were subsequently cultured under high selection conditions, the selectable marker could be reactivated but the adjacent transgene remained silent. This demonstrates that reversion of silencing is relatively rare, and that reactivation is restricted to only a small region of the transgene.

1.7.3   Reactivation of silent transgenes

There have been numerous reports about the use of 5-azacytidine and other methylation-inhibiting drugs for the reactivation of silent transgenes (Emani et al., 2002; John & Amasino, 1989). Apart from uncertainties about the long-term stability of reactivated transgenes, it remains doubtful, however, if this approach will provide a solid solution to the problem. DNA methylation plays a key role in developmental processes, such as flowering and endosperm development (Finnegan et al., 2000). Azacytidine treatment may therefore lead to secondary effects that prevent the use of such material in agriculture.

1.7.4   Scaffold/matrix attachment regions

In animal systems, scaffold/matrix attachment regions (S/MARs) have been used very successfully to prevent transgene expression variegation. In response to torsional stress, unpaired structures are generated at core unwinding elements of S/MARs, which mediate the attachment of the region to the nuclear scaffold and thus constrain the topology of the DNA (Benham *et al.*, 1997). In mammalian genomes, S/MARs are conserved in intergenic regions preceding the 5′ ends of genes (Glazko *et al.*, 2003). They are required in a number of biological contexts, one being domain opening as a prerequisite for the formation of accessible chromatin (Kas *et al.*, 1993). As some plant genes are also flanked by S/MARs, which are most likely required for reliable expression (Van der Geest *et al.*, 1994), S/MARs could offer transgenes protection from silencing events that are based on chromatin condensation.

In plants, individual S/MARs have been shown to provide gene dosage-correlated expression levels (Schoffl *et al.*, 1993) or to enhance transformation frequencies (Galliano *et al.*, 1995). This probably reflects not only an influence of S/MARs on gene expression but also their specific participation in illegitimate recombination (Muller *et al.*, 1999). When a marker gene within a T-DNA was flanked by the chicken lysosyme MAR, the variability in transgene expression was greatly reduced, but not completely eliminated. The authors conclude that embedding a transgene into MARs reduces transgene expression variability to that caused by environmental factors, thus effectively eliminating position effect–specific variation (Mlynárová *et al.*, 1996). S/MARs could also have a beneficial effect for the prevention of read-through transcription from an endogenous promoter, located in the vicinity of the transgene, into the transgene region.

1.7.5 The use of silencing mutants

S/MARs have also been used very successfully in combination with PTGS mutants to obtain stable and high-level transgene expression in *Arabidopsis* (Butaye *et al.*, 2004). When 35S-promoter-driven transgenes were transferred into the PTGS mutants *sgs2* or *sgs3*, all transgenes displayed a stable expression level, which was only observed in about 20% of the wild-type transformants. Transgene stability was maintained in the T2 generation, in contrast to the variegated expression patterns observed in wild-type transformants. Constructs flanked by the MARs of the chicken lysosyme gene were expressed at significantly enhanced levels. These data suggest that PTGS effects are the major inducers of transgene silencing in *Arabidopsis,* and that enhanced transcription levels, which would frequently induce PTGS effects, can be achieved if the PTGS pathway is inactivated. At least in plants where PTGS is the dominant cause of transgene silencing, the exploitation of silencing mutants should be a promising route towards a stabilisation of transgene expression.

1.7.6 Targeted integration of transgenes

The common transformation protocols do not offer any control about the genomic region, into which a transgene integrates, nor can they influence the ultimate structure of an often-rearranged and deleted transgene locus. Targeted integration into preselected sites via site-specific or homologous recombination systems would improve the control over the final structure of the transgene, and may also assist in silencing prevention.

A number of site-specific recombination systems have been developed for plant transformation, one of the most prominent being the *Cre/lox* system, which can be used to transfer up to 230-kb regions into the plant genome (Choi *et al.*, 2000). Transfer of such large regions offers the opportunity to insert gene constructs embedded into a wider chromosomal context, which enhances the likelihood that all neighbouring genomic components that are required for a reliable gene expression are present. The system also allows selection of individual genomic regions that provide a favourable expression of any integrated transgene, provided the integration process itself does not influence the stable expression of the transgene. Apparently, not all genomic positions are equally suitable to prevent gene silencing. *Cre/lox*-mediated integration of the same transgene into different genomic positions of the tobacco genome supported the hypothesis that individual loci influence the expression potential of a transgene, as distinct chromosomal positions could have distinct effects on transgene expression levels (Day *et al.*, 2000). Surprisingly, however, the targeted transgenes were also subject to gene silencing, which was accompanied by DNA methylation islands within the transgene region. The authors suggested that the transgene may be subject to an imprinting mechanism that may be triggered by the transferred plasmid DNA, its transcripts, environmental stress associated with the cell culture or the integration process (Day *et al.*, 2000). It remains to be seen if the susceptibility of transgenes to such mechanisms can be avoided if appropriate insertion sites can be identified.

Homologous recombination may provide an even more reliable strategy to control the structure and activity of transgenes. Although homologous recombination events are very rare in somatic plant cells, gene-targeting events can be identified among a sufficiently large number of transformants. Among 750 *Arabidopsis* transformants targeted by a T-DNA binary vector that contained an *nptII* gene within the *AGL5* sequence, a line could be identified that showed targeted disruption of the *AGL5* MADS-box gene by homologous recombination (Kempin *et al.*, 1997).

While homologous recombination into the nuclear genome still faces a technical challenge, controlled targeting into the plastid genome has been widely used in functional genomics by performing gene knockouts and site-directed mutageneses of plastid genes. The same strategy has been successfully used to transfer transgene constructs, which provides high-level transgene expression and offers the additional advantage of an improved control over transgene dissemination due to the lack of pollen transmission of plastid in many species (Bock, 2001). With the extension of this technology to different species (Zubko *et al.*, 2004), it may become especially attractive for the use of plants as high-level expression systems in the field.

1.8 Conclusions

Although we can define strategies to reduce the susceptibility of transgenes to silencing, transgene silencing still remains a problem to be solved. The phenomenon is observed in all eukaryotes but it seems to be particularly prominent in plants. This may indicate that epigenetic patterns are much more flexible in plants, which enables them to cope with genome duplication and invasive DNA, while retaining the ability of stochastic epigenetic variation to develop new expression profiles via activation of silent copies under specific conditions. In contrast, epigenetic states in animals may be more stringently controlled to ensure a reliable performance of differentiated cell lines that reduces the risk of carcinogenesis. The consequence would be more stable maintenance of epigenetic states once a cell line has been established, which would reduce the probability that an active transgene will become silenced.

When the phenomenon of transgene silencing became apparent about 20 years ago, it was first considered an exceptional behaviour of a few lines that could be ignored, or at best, an enigmatic problem that would be sorted out relatively quickly. Very few colleagues recognised the potential of transgene-silencing research to understand how plants use epigenetic strategies for gene regulation, stress response, genome organisation and other vital biological aspects. This aspect has now become more apparent after transgene research has identified various epigenetic mechanisms as part of basic molecular strategies used in different eukaryotic systems. Plant epigenetic research has made substantial contributions, many of which have become relevant or even essential for discoveries in the animal field. The wider epigenetic community appreciates the powerful potential of plants as genetic tools, which has strengthened trans-kingdom research activities. Recently, environmental carcinogens and their effects on epigenetic patterns have become an important focus in medical research (Sutherland & Costa, 2003), and the value of plant research in this context is increasingly being recognised by non-plant research laboratories.

Paradoxically, although the problem itself has not been fully solved, transgene-silencing research in plants has been very successful. It is still difficult to predict when our understanding of epigenetic mechanisms will finally have advanced to a state that will allow us to overcome all barriers of transgene silencing. We can, however, be relatively certain that transgene research in plants will continue to provide tools for the discovery of new aspects of epigenetic gene regulation.

Acknowledgements

We would like to thank Ian Manfield for discussions and critical reading of the manuscript. We would also like to acknowledge the continuous support that the European Commission has provided to plant epigenetics in several network programmes (HRX-CT94-0530, BIO4-96-0253, QLKR-2000-00078, LSHG-CT-2004-503433).

References

Akashi, H., Kurata, H., Seki, M., Taira, K. & Furusaki, S. (2002) Screening for transgenic plant cells that highly express a target gene from genetically mixed cells. *Biochemical Engineering Journal*, **10**, 175–82.

Alvarez, M., Rhodes, S. J. & Bidwell, J. P. (2003) Context-dependent transcription: all politics is local. *Gene*, **313**, 43–57.

Aufsatz, W., Mette, M. F., Van der Winden, J., Matzke, M. & Matzke, A. J. M. (2002) HDA6, a putative histone deacetylase needed to enhance DNA methylation induced by double-stranded RNA. *EMBO Journal*, **21**, 6832–41.

Bednar, J., Horowitz, R. A., Grigoryev, S. A., Carruthers, L. M., Hansen, J. C., Koster, A. J. & Woodcock, C. L. (1998) Nucleosomes, linker DNA, and linker histone form a unique structural motif that directs the higher-order folding and compaction of chromatin. *Proceedings of the National Academy of Sciences of the United States of America*, **95**, 14173–8.

Benham, C., Kohwi-Shigematsu, T. & Bode, J. (1997) Stress-induced duplex DNA destabilization in scaffold/matrix attachment regions. *Journal of Molecular Biology*, **274**, 181–96.

Bernstein, E., Caudy, A. A., Hammond, S. M. & Hannon, G. J. (2001) Role for a bidentate ribonuclease in the initiation step of RNA interference. *Nature*, **409**, 295–6.

Bock, R. (2001) Transgenic plastids in basic research and plant biotechnology. *Journal of Molecular Biology*, **312**, 425–38.

Boerjan, W., Bauw, G., Van Montagu, M. & Inzé, D. (1994) Distinct phenotypes generated by overexpression and suppression of S-adenosyl-L-methionine synthetase reveal developmental patterns of gene silencing in tobacco. *Plant Cell*, **6**, 1401–14.

Brandle, J. E., McHugh, S. G., James, L., Labbé, H. & Miki, B. L. (1995) Instability of transgene expression in field grown tobacco carrying the *csr1-1* gene for sulfonylurea herbicide resistance. *Biotechnology*, **13**, 994–8.

Broer, I. (1996) Stress inactivation of foreign genes in transgenic plants. *Field Crops Research*, **45**, 19–25.

Budar, F., Thia-Toong, L., Van Montagu, M. & Hernalsteens, J.-P. (1986) Agrobacterium-mediated gene transfer results mainly in transgenic plants transmitting T-DNA as a single Mendelian factor. *Genetics*, **114**, 303–13.

Butaye, K. M. J., Goderis, I. J. W. M., Wouters, P. F. J., Pues, J. M. T. G., Delaure, S. L., Broekaert, W. F., Depicker, A., Cammue, B. P. A. & De Bolle, M. F. C. (2004) Stable high-level transgene expression in *Arabidopsis thaliana* using gene silencing mutants and matrix attachment regions. *Plant Journal*, **39**, 440–49.

Cao, X. F. & Jacobsen, S. E. (2002a) Locus-specific control of asymmetric and CpNpG methylation by the *DRM* and *CMT3* methyltransferase genes. *Proceedings of the National Academy of Sciences of the United States of America*, **99**, Suppl. 4, 16491–8.

Cao, X. F. & Jacobsen, S. E. (2002b) Role of the *Arabidopsis* DRM methyltransferases in *de novo* DNA methylation and gene silencing. *Current Biology*, **12**, 1138–44.

Cao, X., Springer, N. M., Muszynski, M. G., Phillips, R. L., Kaeppler, S. & Jacobsen, S. E. (2000) Conserved plant genes with similarity to mammalian *de novo* DNA methyltransferases. *Proceedings of the National Academy of Sciences of the United States of America*, **97**, 4979–84.

Cao, X. F., Aufsatz, W., Zilberman, D., Mette, M. F., Huang, M. S., Matzke, M. & Jacobsen, S. E. (2003) Role of the DRM and CMT3 methyltransferases in RNA-directed DNA methylation. *Current Biology*, **13**, 2212–17.

Cellini, F., Chesson, A., Colquhoun, I., Constable, A., Davies, H. V., Engel, K. H., Gatehouse, A. M. R., Karenlampi, S., Kok, E. J. & Leguay, J. J. (2004) Unintended effects and their detection in genetically modified crops. *Food and Chemical Toxicology*, **42**, 1089–125.

Chan, S. W. L., Zilberman, D., Xie, Z., Johansen, L. K., Carrington, J. C. & Jacobsen, S. E. (2004) RNA silencing genes control *de novo* DNA methylation. *Science*, **303**, 1336.

Cheung, P., Allis, C. D. & Sassone-Corsi, P. (2000) Signaling to chromatin through histone modifications. *Cell*, **103**, 263–71.

Choi, S., Begum, D., Koshinsky, H., Ow, D. W. & Wing, R. A. (2000) A new approach for the identification and cloning of genes: the pBACwich system using Cre/*lox* site-specific recombination. *Nucleic Acids Research*, **28**, e19.

Cluster, P. D., O'Dell, M., Metzlaff, M. & Flavell, R. B. (1996) Details of T-DNA structural organization from a transgenic petunia population exhibiting co-suppression. *Plant Molecular Biology*, **32**, 1197–203.

Cocciolone, S. M., Chopra, S., Flint-Garcia, S. A., McMullen, M. D. & Peterson, T. (2001) Tissue-specific patterns of a maize Myb transcription factor are epigenetically regulated. *Plant Journal*, **27**, 467–78.

Cogoni, C. & Macino, G. (1999) Gene silencing in *Neurospora crassa* requires a protein homologous to RNA-dependent RNA polymerase. *Nature*, **399**, 166–9.

Crété, P., Leuenberger, S., Iglesias, V. A., Suarez, V., Schöb, H., Holtorf, H., Van Eeden, S. & Meins, F. J. (2001) Graft transmission of induced and spontaneous post-transcriptional silencing of chitinase genes. *Plant Journal*, **28**, 493–501.

Cubas, P., Vincent, C. & Coen, E. (1999) An epigenetic mutation responsible for natural variation in floral symmetry. *Nature*, **401**, 157–61.

Dalmay, T., Hamilton, A., Rudd, S., Angell, S. & Baulcombe, D. C. (2000) An RNA-dependent RNA polymerase gene in *Arabidopsis* is required for posttranscriptional gene silencing mediated by a transgene but not by a virus. *Cell*, **101**, 543–53.

Day, C. D., Lee, E., Kobayashi, J., Holappa, L. D., Albert, H. & Ow, D. W. (2000) Transgene integration into the same chromosome location can produce alleles that express at a predictable level, or alleles that are differentially silenced. *Genes & Development*, **14**, 2869–80.

De Buck, S., Jacobs, A., Van Montagu, M. & Depicker, A. (1998) *Agrobacterium tumefaciens* transformation and cotransformation frequencies of *Arabidopsis thaliana* root explants and tobacco protoplasts. *Molecular Plant–Microbe Interactions*, **11**, 449–57.

De Buck, S., Van Montagu, M. & Depicker, A. (2001) Transgene silencing of invertedly repeated transgenes is released upon deletion of one of the transgenes involved. *Plant Molecular Biology*, **46**, 433–45.

De Buck, S., Windels, P., De Loose, M. & Depicker, A. (2004) Single-copy T-DNAs integrated at different positions in the *Arabidopsis* genome display uniform and comparable β-glucuronidase accumulation levels. *Cellular and Molecular Life Sciences*, **61**, 2632–45.

de Carvalho, F., Gheysen, G., Kushnir, S., Van Montagu, M., Inzé, D. & Castresana, C. (1992) Suppression of β-1,3-glucanase transgene expression in homozygous plants. *EMBO Journal*, **11**, 2595–602.

De Neve, M., De Buck, S., De Wilde, C., Van Houdt, H., Strobbe, I., Jacobs, A., Van Montagu, M. & Depicker, A. (1999) Gene silencing results in instability of antibody production in transgenic plants. *Molecular and General Genetics*, **260**, 582–92.

Depicker, A., Ingelbrecht, I., Van Houdt, H., De Loose, M. & Van Montagu, M. (1996) Post-transcriptional reporter transgene silencing in transgenic tobacco. In D. Grierson, G. W. Lycett & G. A. Tucker (eds) *Mechanisms and Applications of Gene Silencing*. Nottingham University Press, Nottingham, pp. 71–84.

Elbashir, S. M., Lendeckel, W. & Tuschl, T. (2001) RNA interference is mediated by 21- and 22-nucleotide RNAs. *Genes & Development*, **15**, 188–200.

Emani, C., Sunilkumar, G. & Rathore, K. S. (2002) Transgene silencing and reactivation in sorghum. *Plant Science*, **162**, 181–92.

English, J. J., Mueller, E. & Baulcombe, D. C. (1996) Suppression of virus accumulation in transgenic plants exhibiting silencing of nuclear genes. *Plant Cell*, **8**, 179–88.

Finnegan, E. J., Peacock, W. J. & Dennis, E. S. (2000) DNA methylation, a key regulator of plant development and other processes. *Current Opinion in Genetics and Development*, **10**, 217–23.

Fire, A., Xu, S., Montgomery, M. K., Kostas, S. A., Driver, S. E. & Mello, C. C. (1998) Potent and specific genetic interference by double-stranded RNA in *Caenorhabditis elegans*. *Nature*, **391**, 806–11.

Fojtova, M., Van Houdt, H., Depicker, A. & Kovarik, A. (2003) Epigenetic switch from posttranscriptional to transcriptional silencing is correlated with promoter hypermethylation. *Plant Physiology*, **133**, 1240–50.

Forsbach, A., Schubert, D., Lechtenberg, B., Gils, M. & Schmidt, R. (2003) A comprehensive characterization of single-copy T-DNA insertions in the *Arabidopsis thaliana* genome. *Plant Molecular Biology*, **52**, 161–76.

Freitag, M., Hickey, P. C., Khlafallah, T. K., Read, N. D. & Selker, E. U. (2004a) HP1 is essential for DNA methylation in *Neurospora*. *Molecular Cell*, **13**, 427–34.

Freitag, M., Lee, D. W., Kothe, G. O., Pratt, R. J., Aramayo, R. & Selker, E. U. (2004b) DNA methylation is independent of RNA interference in *Neurospora*. *Science*, **304**, 1939.

Galliano, H., Muller, A. E., Lucht, J. M. & Meyer, P. (1995) The transformation booster sequence from *Petunia hybrida* is a retrotransposon derivative that binds to the nuclear scaffold. *Molecular and General Genetics*, **247**, 614–22.

García-Pérez, R. D., Van Houdt, H. & Depicker, A. (2004) Spreading of posttranscriptional gene silencing along the target gene promotes systemic silencing. *Plant Journal*, **38**, 594–602.

Garcia-Salcedo, J. A., Gijon, P., Nolan, D. P., Tebabi, P. & Pays, E. (2003) A chromosomal SIR2 homologue with both histone NAD-dependent ADP-ribosyltransferase and deacetylase activities is involved in DNA repair in *Trypanosoma brucei*. *EMBO Journal*, **22**, 5851–62.

Glazko, G. V., Koonin, E. V., Rogozin, I. B. & Shabalina, S. A. (2003) A significant fraction of conserved noncoding DNA in human and mouse consists of predicted matrix attachment regions. *Trends in Genetics*, **19**, 119–24.

Goodwin, J., Chapman, K., Swaney, S., Parks, T. D., Wernsman, E. A. & Dougherty, W. G. (1996) Genetic and biochemical dissection of transgenic RNA mediated virus resistance. *Plant Cell*, **8**, 95–105.

Gorbunova, V. & Levy, A. A. (1999) How plants make ends meet: DNA double-strand break repair. *Trends in Plant Science*, **4**, 263–9.

Grevelding, C., Fantes, V., Kemper, E., Schell, J. & Masterson, R. (1993) Single copy T-DNA insertions in *Arabidopsis* are the predominant form of integration in root derived transgenics, whereas multiple insertions are found in leaf discs. *Plant Molecular Biology*, **23**, 847–60.

Grishok, A., Pasquinelli, A. E., Conte, D., Li, N., Parrish, S., Ha, I., Baillie, D. L., Fire, A., Ruvkun, G. & Mello, C. C. (2001) Genes and mechanisms related to RNA interference regulate expression of the small temporal RNAs that control *C. elegans* developmental timing. *Cell*, **106**, 23–34.

Hamilton, A. J. & Baulcombe, D. C. (1999) A species of small antisense RNA in posttranscriptional gene silencing in plants. *Science*, **286**, 950–52.

Hammond, S. M., Boettcher, S., Caudy, A. A., Kobayashi, R. & Hannon, G. J. (2001) Argonaute2, a link between genetic and biochemical analyses of RNAi. *Science*, **293**, 1146–50.

Helliwell, C. & Waterhouse, P. (2003) Constructs and methods for high throughput gene silencing in plants. *Methods*, **30**, 289–95.

Hirochika, H., Sugimoto, K., Otsuki, Y., Tsugawa, H. & Kanda, M. (1996) Retrotransposons of rice involved in mutations induced by tissue culture. *Proceedings of the National Academy of Sciences of the United States of America*, **93**, 7783–8.

Hobbs, S. L. A., Kpodar, P. & DeLong, C. M. O. (1990) The effect of T DNA copy number, position and methylation on reporter gene expression in tobacco transformants. *Plant Molecular Biology*, **15**, 851–64.

Hobbs, S. L. A., Warkentin, T. D. & DeLong, C. M. O. (1993) Transgene copy number can be positively or negatively associated with transgene expression. *Plant Molecular Biology*, **21**, 17–26.

Houben, A., Demidov, D., Gernand, D., Meister, A., Leach, C. R. & Schubert, I. (2003) Methylation of histone H3 in euchromatin of plant chromosomes depends on basic nuclear DNA content. *Plant Journal*, **33**, 967–73.

Iglesias, V. A., Moscone, E. A., Papp, I., Neuhuber, F., Michalowski, S., Phelan, T., Spiker, S., Matzke, M. & Matzke, A. J. (1997) Molecular and cytogenetic analyses of stably and unstably expressed transgene loci in tobacco. *Plant Cell*, **9**, 1251–64.

Ingelbrecht, I., Van Houdt, H., Van Montagu, M. & Depicker, A. (1994) Posttranslational silencing of reporter transgenes in tobacco correlates with DNA modification. *Proceedings of the National Academy of Sciences of the United States of America*, **91**, 10502–506.

Jackson, D. A. (2000) Features of nuclear architecture that influence gene expression in higher eukaryotes: confronting the enigma of epigenetics. *Journal of Cellular Biochemistry*, 69–77.

Jackson, J. P., Lindroth, A. M., Cao, X. F. & Jacobsen, S. E. (2002) Control of CpNpG DNA methylation by the KRYPTONITE histone H3 methyltransferase. *Nature*, **416**, 556–60.

Jacobsen, S. E. & Meyerowitz, E. M. (1997) Hypermethylated SUPERMAN epigenetic alleles in *Arabidopsis*. *Science*, **277**, 1100–103.

Jasencakova, Z., Soppe, W. J. J., Meister, A., Gernand, D., Turner, B. M. & Schubert, I. (2003) Histone modifications in *Arabidopsis* – high methylation of H3 lysine 9 is dispensable for constitutive heterochromatin. *Plant Journal*, **33**, 471–80.

Jenuwein, T. & Allis, C. D. (2001) Translating the histone code. *Science*, **293**, 1074–80.

Jia, S., Noma, K. & Grewal, S. I. S. (2004) RNAi-independent heterochromatin nucleation by the stress-activated ATF/CREB family proteins. *Science*, **304**, 1971–6.

Jiang, N., Bao, Z., Zhang, X., Hirochika, H., Eddy, S. R., McCouch, S. R. & Wessler, S. R. (2003) An active DNA transposon family in rice. *Nature*, **421**, 163–7.

John, M. C. & Amasino, R. M. (1989) Extensive changes in DNA methylation patterns accompany activation of a silent T-DNA *ipt* gene in *Agrobacterium tumefaciens*-transformed plant cells. *Molecular and Cellular Biology*, **9**, 4298–303.

Jorgensen, R. A., Snyder, C. & Jones, J. D. G. (1987) T-DNA is organized predominantly in inverted repeat structures in plants transformed with *Agrobacterium tumefaciens* C58 derivatives. *Molecular and General Genetics*, **207**, 471–7.

Jorgensen, R. A., Cluster, P. D., English, J., Que, Q. & Napoli, C. A. (1996) Chalcone synthase cosuppression phenotypes in petunia flowers: comparison of sense vs. antisense constructs and single-copy vs. complex T-DNA sequences. *Plant Molecular Biology*, **31**, 957–73.

Kartzke, S., Saedler, H. & Meyer, P. (1990) Molecular analysis of transgenic plants derived from transformation of protoplasts at various stages of the cell cycle. *Plant Science*, **67**, 63–72.

Kas, E., Poljak, L., Adachi, Y. & Laemmli, U. K. (1993) A model for chromatin opening: stimulation of topoisomerase II and restriction enzyme cleavage of chromatin by distamycin. *EMBO Journal*, **12**, 115–26.

Kempin, S. A., Liljegren, S. J., Block, L. M., Rounsley, S. D., Yanofsky, M. F. & Lam, E. (1997) Targeted disruption in *Arabidopsis*. *Nature*, **389**, 802–803.

Kinoshita, T., Miura, A., Choi, Y., Kinoshita, Y., Cao, X. F., Jacobsen, S. E., Fischer, R. L. & Kakutani, T. (2004) One-way control of FWA imprinting in *Arabidopsis* endosperm by DNA methylation. *Science*, **303**, 521–32.

Konig, A., Cockburn, A., Crevel, R. W. R., Debruyne, E., Grafstroem, R., Hammerling, U., Kimber, I., Knudsen, I., Kuiper, H. A. & Peijnenburg, A. A. C. M. (2004) Assessment of the safety of foods derived from genetically modified (GM) crops. *Food and Chemical Toxicology*, **42**, 1047–88.

Kononov, M. E., Bassuner, B. & Gelvin, S. B. (1997) Integration of T-DNA binary vector 'backbone' sequences into the tobacco genome: evidence for multiple complex patterns of integration. *Plant Journal*, **11**, 945–57.

Kovařík, A., Van Houdt, H., Holý, A. & Depicker, A. (2000) Drug-induced hypomethylation of a posttranscriptionally silenced transgene locus of tobacco leads to partial release of silencing. *FEBS Letters*, **467**, 47–51.

Kunz, C., Schöb, H., Stam, M., Kooter, J. M. & Meins, F. J. (1996) Developmentally regulated silencing and reactivation of tobacco chitinase transgene expression. *Plant Journal*, **10**, 437–50.

Kurdistani, S. K. & Grunstein, M. (2003) Histone acetylation and deacetylation in yeast. *Nature Reviews. Molecular Cell Biology*, **4**, 276–84.

Lachner, M., O'Sullivan, R. J. & Jenuwein, T. (2003) An epigenetic road map for histone lysine methylation. *Journal of Cell Science*, **116**, 2117–24.

Lindroth, A. M., Cao, X. F., Jackson, J. P., Zilberman, D., McCallum, C. M., Henikoff, S. & Jacobsen, S. E. (2001) Requirement of CHROMOMETHYLASE3 for maintenance of CpXpG methylation. *Science*, **292**, 2077–80.

Loidl, P. (2004) A plant dialect of the histone language. *Trends in Plant Science*, **9**, 84–90.

Luger, K., Maeder, A. W., Richmond, R. K., Sargent, D. F. & Richmond, T. J. (1997) X-ray structure of the nucleosome core particle at 2.8 Å resolution. *Nature*, 251–259.

Martens, H., Novotny, J., Oberstrass, J., Steck, T. L., Postlethwait, P. & Nellen, W. (2002) RNAi in *Dictyostelium*: the role of RNA-directed RNA polymerases and double-stranded RNase. *Molecular Biology of the Cell*, **13**, 445–53.

Matzke, M., Aufsatz, W., Kanno, T., Daxinger, L., Papp, I., Mette, M. F. & Matzke, A. J. M. (2004) Genetic analysis of RNA-mediated transcriptional gene silencing. *Biochimica et Biophysica Acta*, **1677**, 129–41.

Mette, M. F., Aufsatz, W., Van der Winden, J., Matzke, M. A. & Matzke, A. J. M. (2000) Transcriptional silencing and promoter methylation triggered by double-stranded RNA. *EMBO Journal*, **19**, 5194–201.

Metzlaff, M., O'Dell, M., Cluster, P. D. & Flavell, R. B. (1997) RNA-mediated RNA degradation and chalcone synthase A silencing in petunia. *Cell*, **88**, 845–54.

Meyer, P., Linn, F., Heidmann, I., Meyer, H., Niedenhof, I. & Saedler, H. (1992) Endogenous and environmental factors influence 35S promoter methylation of a maize A1 gene construct in transgenic petunia and its colour phenotype. *Molecular and General Genetics*, **231**, 345–52.

Mittelsten Scheid, O., Probst, A. V., Afsar, K. & Paszkowski, J. (2002) Two regulatory levels of transcriptional gene silencing in *Arabidopsis*. *Proceedings of the National Academy of Sciences of the United States of America*, **99**, 13659–62.

Mlynárová, L., Keizer, L. C. P., Stiekema, W. J. & Nap, J.-P. (1996) Approaching the lower limits of transgene variability. *Plant Cell*, **8**, 1589–99.

Morel, J.-B., Mourrain, P., Béclin, C. & Vaucheret, H. (2000) DNA methylation and chromatin structure affect transcriptional and post-transcriptional transgene silencing in *Arabidopsis*. *Current Biology*, **10**, 1591–4.

Mourrain, P., Béclin, C., Elmayan, T., Feuerbach, F., Godon, C., Morel, J.-B., Jouette, D., Lacombe, A.-M., Nikic, S., Picault, N., Rémoué, K., Sanial, M., Vo, T.-A. & Vaucheret, H. (2000) *Arabidopsis SGS2* and *SGS3* genes are required for posttranscriptional gene silencing and natural virus resistance. *Cell*, **101**, 533–42.

Müller, A. E., Kamisugi, Y., Grüneberg, R., Niedenhof, I., Hörold, R. J. & Meyer, P. (1999) Palindromic sequences and A+T-rich DNA elements promote illegitimate recombination in *Nicotiana tabacum*. *Journal of Molecular Biology*, **291**, 29–46.

Muskens, M. W. M., Vissers, A. P. A., Mol, J. N. M. & Kooter, J. M. (2000) Role of inverted DNA repeats in transcriptional and post-transcriptional gene silencing. *Plant Molecular Biology*, **43**, 243–60.

Napoli, C., Lemieux, C. & Jorgensen, R. (1990) Introduction of a chimeric chalcone synthase gene into petunia results in reversible co-suppression of homologous genes *in trans*. *Plant Cell*, **2**, 279–89.

Ngo, H., Tschudi, C., Gull, K. & Ullu, E. (1998) Double-stranded RNA induces mRNA degradation in *Trypanosoma brucei*. *Proceedings of the National Academy of Sciences of the United States of America*, **95**, 14687–92.

Palauqui, J.-C. & Balzergue, S. (1999) Activation of systemic acquired silencing by localised introduction of DNA. *Current Biology*, **9**, 59–66.

Palauqui, J.-C., Elmayan, T., Dorlhac de Borne, F., Crété, P., Charles, C. & Vaucheret, H. (1996) Frequencies, timing and spatial patterns of co-suppression of nitrate reductase and nitrite reductase in transgenic tobacco plants. *Plant Physiology*, **112**, 1447–56.

Palauqui, J.-C., Elmayan, T., Pollien, J.-M. & Vaucheret, H. (1997) Systemic acquired silencing: transgene-specific post-transcriptional silencing is transmitted by grafting from silenced stocks to non-silenced scions. *EMBO Journal*, **16**, 4738–45.

Park, W., Li, J., Song, R., Messing, J. & Chen, X. i. (2002) CARPEL FACTORY, a Dicer homolog, and HEN1, a novel protein, act in microRNA metabolism in *Arabidopsis thaliana*. *Current Biology*, **12**, 1484–95.

Peeters, K., De Wilde, C., De Jaeger, G., Angenon, G. & Depicker, A. (2001) Production of antibodies and antibody fragments in plants. *Vaccine*, **19**, 2756–61.

Pélissier, T. & Wassenegger, M. (2000) A DNA target of 30 bp is sufficient for RNA-directed DNA methylation. *RNA*, **6**, 55–65.

Pohl Nielsen, C., Robinson, S. & Thierfelder, K. (2001) Genetic engineering and trade: panacea or dilemma for developing countries. *World Development*, **29**, 1307–24.

Popelka, J. C., Terryn, N. & Higgins, T. J. V. (2004) Gene technology for grain legumes: can it contribute to the food challenge in developing countries? *Plant Science*, **167**, 195–206.

Pröls, F. & Meyer, P. (1992) The methylation patterns of chromosomal integration regions influence gene activity of transferred DNA in *Petunia hybrida*. *Plant Journal*, **2**, 465–75.

Que, Q., Wang, H.-Y., English, J. J. & Jorgensen, R. A. (1997) The frequency and degree of co-suppression by sense chalcone synthase transgenes are dependent on transgene promoter strength and are reduced by premature nonsense codons in the transgene coding sequence. *Plant Cell*, **9**, 1357–68.

Saedler, H., Giefers, W. & Meyer, P. (1992) A transgenic petunia line carrying the maize A1 gene not only features a new flower colour but is also field resistant against several fungal pathogens. In S. Lim (ed.) *The Development of Natural Resources and Environmental preservation*. Seoul, Korea, pp. 41–50.

Sallaud, C., Gay, C., Larmande, P., Bes, M., Piffanelli, P., Piegu, B., Droc, G., Regad, F., Bourgeois, E., Meynard, D., Perin, C., Sabau, X., Ghesquiere, A., Glaszmann, J. C., Delseny, M. & Guiderdoni, E. (2004) High throughput T-DNA insertion mutagenesis in rice: a first step towards *in silico* reverse genetics. *Plant Journal*, **39**, 450–64.

Saze, H., Mittelsten Scheid, O. & Paszkowski, J. (2003) Maintenance of CpG methylation is essential for epigenetic inheritance during plant gametogenesis. *Nature Genetics*, **34**, 65–9.

Schöffl, F., Schröder, G., Kliem, M. & Rieping, M. (1993) An SAR sequence containing 395 bp DNA fragment mediates enhanced, gene-dosage-correlated expression of a chimaeric heat shock gene in transgenic tobacco plants. *Transgenic Research*, **2**, 93–100.

Shiio, Y. & Eisenman, R. N. (2003) Histone sumoylation is associated with transcriptional repression. *Proceedings of the National Academy of Sciences of the United States of America*, **100**, 13225–30.

Sijen, T., Wellink, J., Hiriart, J. B. & Van Kammen, A. (1996) RNA-mediated virus resistance: role of repeated transgenes and delineation of targeted regions. *Plant Cell*, **8**, 2277–94.

Sijen, T., Fleenor, J., Simmer, F., Thijssen, K. L., Parrish, S., Timmons, L., Plasterk, R. H. A. & Fire, A. (2001) On the role of RNA amplification in dsRNA-triggered gene silencing. *Cell*, **107**, 465–76.

Smardon, A., Spoerke, J. M., Stacey, S. C., Klein, M. E., Mackin, N. & Maine, E. M. (2000) EGO-1 is related to RNA-directed RNA polymerase and functions in germ-line development and RNA interference in *C. elegans*. *Current Biology*, **10**, 169–78.

Soppe, W. J. J., Jasencakova, Z., Houben, A., Kakutani, T., Meister, A., Huang, M. l. S., Jacobsen, S. E., Schubert, I. & Fransz, P. F. (2002) DNA methylation controls histone H3 lysine 9 methylation and heterochromatin assembly in Arabidopsis. *EMBO Journal*, **21**, 6549–59.

Stam, M., de Bruin, R., Kenter, S., Van der Hoorn, R. A. L., Van Blokland, R., Mol, J. N. M. & Kooter, J. M. (1997) Post transcriptional silencing of chalcone synthase in petunia by inverted transgene repeats. *Plant Journal*, **12**, 63–82.

Stevenson, D. S. & Jarvis, P. (2003) Chromatin silencing: RNA in the driving seat. *Current Biology*, **13**, R13–5.

Stokes, T. L., Kunkel, B. N. & Richards, E. J. (2002) Epigenetic variation in *Arabidopsis* disease resistance. *Genes Development*, **16**, 171–82.

Sun, Z. -W. & Allis, C. D. (2002) Ubiquitination of histone H2B regulates H3 methylation and gene silencing in yeast. *Nature*, **418**, 104–108.

Sutherland, J. E. & Costa, M. A. X. (2003) Epigenetics and the environment. *Annals of the New York Academy of Sciences*, **983**, 151–60.

ten Lohuis, M., Galliano, H., Heidmann, I. & Meyer, P. (1995) Treatment with propionic and butyric acid enhances expression variegation and promoter methylation in plant transgenes. *Biological Chemistry Hoppe-Seyler*, **376**, 311–20.

Thompson, A. J. & Myatt, S. C. (1997) Tetracycline-dependent activation of an upstream promoter reveals transcriptional interference between tandem genes within T-DNA in tomato. *Plant Molecular Biology*, **34**, 687–92.

Vaistij, F. E., Jones, L. & Baulcombe, D. C. (2002) Spreading of RNA targeting and DNA methylation in RNA silencing requires transcription of the target gene and a putative RNA-dependent RNA polymerase. *Plant Cell*, **14**, 857–67.

Van Blokland, R., Van der Geest, N., Mol, J. N. M. & Kooter, J. M. (1994) Transgene mediated suppression of chalcone synthase expression in *Petunia hybrida* results from an increase in RNA turnover. *Plant Journal*, **6**, 861–77.

Van Blokland, R., ten Lohuis, M. & Meyer, P. (1997) Condensation of chromatin in transcriptional regions of an inactivated transgene: an indication for an active role of transcription in gene silencing. *Molecular and General Genetics*, **257**, 1–13.

Van der Geest, A. H. M., Hall, G. E. Jr, Spiker, S. & Hall, T. C. (1994) The β-phaseolin gene is flanked by matrix attachment regions. *Plant Journal*, **6**, 413–23.

Van der Krol, A. R., Mur, L. A., Beld, M., Mol, J. N. M. & Stuitje, A. R. (1990) Flavonoid genes in petunia – addition of a limited number of gene copies may lead to a suppression of gene-expression. *Plant Cell*, **2**, 291–9.

Van Houdt, H., Bleys, A. & Depicker, A. (2003) RNA target sequences promote spreading of RNA silencing. *Plant Physiology*, **131**, 245–53.

Voinnet, O. (2001) RNA silencing as a plant immune system against viruses *Trends in Genetics*, **17**, 449–59.

Voinnet, O., Vain, P., Angell, S. & Baulcombe, D. C. (1998) Systemic spread of sequence-specific transgene RNA degradation in plants is initiated by localized introduction of ectopic promoterless DNA. *Cell*, **95**, 177–87.

Volpe, T. A., Kidner, C., Hall, I. M., Teng, G., Grewal, S. I. S. & Martienssen, R. A. (2002) Regulation of heterochromatic silencing and histone H3 lysine-9 methylation by RNAi. *Science*, **297**, 1833–7.

Wakimoto, B. T. (1998) Beyond the nucleosome: epigenetic aspects of position-effect variegation in Drosophila. *Cell*, **93**, 321–4.

Walbot, V. (1999) UV-B damage amplified by transposons in maize. *Nature*, **397–399**, 398.

Walter, C., Broer, I., Hillemann, D. & Pühler, A. (1992) High frequency, heat treatment-induced inactivation of the phosphinothricin resistance gene in transgenic single cell suspension cultures of *Medicago sativa*. *Molecular and General Genetics*, **235**, 189–96.

Wang, M.-B. & Waterhouse, P. M. (2002) Application of gene silencing in plants. *Current Opinion in Plant Biology*, **5**, 146–50.

Wassenegger, M., Heimes, S., Riedel, L. & Sanger, H. L. (1994) RNA-directed *de novo* methylation of genomic sequences in plants. *Cell*, **76**, 567–76.

Waterhouse, P. M., Smith, N. A. & Wang, M. -B. (1999) Virus resistance and gene silencing: killing the messenger. *Trends in Plant Science*, **4**, 452–7.

Wianny, F. & Zernicka-Goetz, M. (2000) Specific interference with gene function by double-stranded RNA in early mouse development. *Nature Cell Biology*, **2**, 70–75.

Zhang, Y. & Reinberg, D. (2001) Transcription regulation by histone methylation: interplay between different covalent modifications of the core histone tails. *Genes & Development*, **15**, 2343–60.

Zubko, M., Zubko, E., Zuilen, K., Meyer, P. & Day, A. (2004) Stable transformation of petunia plastids. *Transgenic Research*, **13**, 523–30.

2 RNA interference: double-stranded RNAs and the processing machinery

Jan M. Kooter

2.1 Introduction

More than 3.5 billion years ago, 'life' may have been initiated by RNA molecules acting as primitive genomes and as catalysts (Joyce, 2002). The existence of this 'RNA world' will be hard to prove but what we know for sure is that RNAs still perform essential tasks in present-day life forms. We are familiar with the well-known properties of mRNAs, tRNAs, rRNAs, U-RNAs, and the like, but the turn of the twentieth century will be remembered as the time when the impact of double-stranded RNA (dsRNA) and short RNAs became apparent. Small RNAs are involved in a variety of cellular processes, in the cytoplasm and in the nucleus, ranging from RNA degradation to signalling epigenetic modifications of DNA and histones. Over the years, people came up with several names for the phenomena in which small RNAs are implicated, illustrating the evolution of the field: it started with transgene-induced cosuppression in petunia (Napoli *et al.*, 1990; Van der Krol *et al.*, 1990), followed by posttranscriptional gene silencing (PTGS), RNA silencing, RNA-directed DNA methylation, transcriptional gene silencing (TGS), RNA-induced silencing, homology-dependent gene silencing and genetic interference. Nowadays the most commonly used name for the process in which small RNAs target mRNAs for degradation is 'RNA interference' or RNAi (Tuschl *et al.*, 1999). Step by step, details of the mechanisms are being uncovered, but two key findings speeded up the field – in particular, the discovery by Fire *et al.* (1998) that dsRNA strongly induced RNA degradation in *Caenorhabditis elegans* and the discovery by Hamilton and Baulcombe (1999) of small RNAs, 20–25 nt in size, that are specifically associated with RNA silencing in plants. Short RNAs were soon found in all other systems where dsRNA was able to induce gene silencing. These tiny RNAs targeting complementary mRNAs for degradation are now usually called small interfering RNAs (siRNAs). But there is a whole collection of short RNAs performing different functions. They not only induce mRNA degradation but also guide other processes, including inhibition of translation, cytosine methylation in plants and heterochromatin formation in *Schizosaccharomyces pombe* (Hall *et al.*, 2002; Verdel *et al.*, 2004; Volpe *et al.*, 2003) and plants (Lippman & Martienssen, 2004; Lippman *et al.*, 2004).

The mechanism of RNAi and the biological function(s) of dsRNAs and the various types of short RNAs have led to the discovery of new regulatory pathways. This in itself is a fascinating area of biology at the moment. However, the greatest impact RNAi appears to have is its application as an experimental tool to knock down gene expression in cells and intact organisms. It can be achieved in a fairly simple manner and has become a very important genomics tool. Genetic tricks commonly used in model systems became available for examining the function of genes and pathways in more complex systems, including human cells, on both a small and a large scale (Berns *et al.*, 2004). It is therefore understandable that the term 'revolution' was used in a historical account of the major findings in the RNAi field (Matzke & Matzke, 2004) and the impact it has on present-day molecular genetics of eukaryotes and the way we view biological processes and evolution.

It appears that in plants all processes known to be controlled by RNAi occur side by side, including synthesis of dsRNA from DNA or RNA, cleavage of dsRNAs by distinct RNases in the nucleus and in the cytoplasm and degradation of mRNAs by siRNAs. Also viral RNAs in an infected plant fall victim to the RNAi machinery in order to keep the viral infection under control. Furthermore, mRNA translation is inhibited by a class of siRNAs, called microRNAs (miRNAs). In the nucleus, RNAi guides not only DNA methylation by a process called RNA-directed DNA methylation, but also the assembly of particular DNA sequences into silent heterochromatin. Some of these functions can be considered regulatory whereas others can be viewed as defence against exogenous pathogens, like viruses, and endogenous 'parasites', such as transposable elements (TEs) and repetitive DNA. In any case, it shows that small RNAs control many different processes in a cell and that they have an enormous impact on an organism's performance.

This chapter mainly reviews the RNAi mechanisms responsible for PTGS. The various types of RNA involved in RNAi, the effector RNA molecules and how they are generated, and where possible, applications are discussed. Despite the fact that gene-silencing phenomena were first identified in plants, most detailed information about the mechanism of RNAi stems from animal systems. RNAi in animal and plant systems are discussed side by side, with an emphasis on RNAi in plants. For further reading, reference to the many excellent reviews cited throughout the chapter is helpful.

2.2 Mechanism of RNA interference

Biochemical studies have revealed a detailed picture of the different steps of RNAi and the protein factors involved (Figure 2.1). Initial experiments were done with extracts from *Drosophila* embryos (Hammond *et al.*, 2000; Tuschl *et al.*, 1999), followed by *C. elegans* (Ketting *et al.*, 2001) and mouse cells

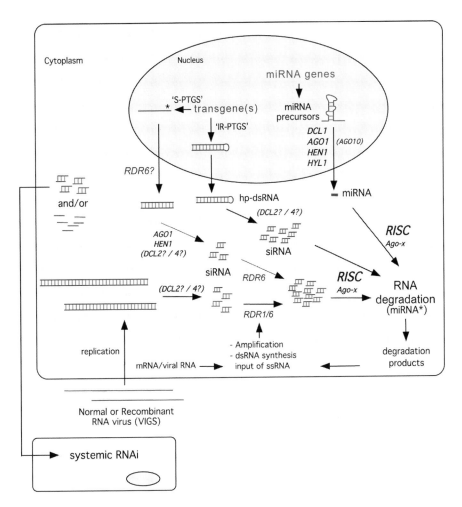

Figure 2.1 RNA interference (RNAi)/posttranscriptional gene silencing (PTGS) pathways in the nucleus and cytoplasm of plants. Endogenous sequences, such as microRNA (miRNA) genes and also viruses, are indicated in green; transgene sequences are indicated in red. The enzymes involved in the different pathways are depicted in blue. Asterisk on the RNA at the S-PTGS pathway indicates an aberrant species. Asterisk on the miRNA indicates that only a limited number of miRNAs in plants inhibit translation. Ago-x in the RISC denotes the putative PAZ–PIWI protein with nuclease activity, which remains to be identified. The numerous steps, the proteins involved and uncertainties are described in the main text.

(Billy *et al.*, 2001). Some reactions of the RNAi pathway are also possible with wheat germ and cauliflower extracts (Tang & Zamore, 2004; Tang *et al.*, 2003). This section briefly describes the key steps of RNAi and the following sections discuss them in greater detail. dsRNA is cleaved by an RNase-III-like endonuclease, called Dicer (Bernstein *et al.*, 2001) or Dicer-like (DCL), generating siRNA duplexes of 21–25 nucleotides depending on the species, with two nucleotides overhanging at the 3′ end. The strands carry a 5′ phosphate and a 3′ hydroxyl, which is typical for products generated by RNase III enzymes. Each of the siRNA strands become incorporated into the RNA-induced silencing complex (RISC), thereby targeting the complex to a complementary or partially complementary RNA sequence (Martinez *et al.*, 2002). It then cleaves the target RNA in the centre of the double-stranded region. Cleavage yields a 5′ phosphate and 3′ hydroxyl terminus (Martinez & Tuschl, 2004), after which the unprotected RNAs are degraded. These steps are the core events of RNAi in all species examined thus far.

Some species have the ability to synthesise additional dsRNAs. These so-called secondary dsRNAs are made by an RNA-dependent RNA polymerase (RDR) that selects target mRNA or RISC cleavage products as templates (see below). The involvement of this enzyme in gene silencing/RNAi came out of genetic screens in *Neurospora crassa* (Cogoni & Macino, 1999), *Arabidopsis* (Dalmay *et al.*, 2000b; Mourrain *et al.*, 2000), *C. elegans* (Sijen *et al.*, 2001a; Smardon *et al.*, 2000) and *Dictyostelium* (Martens *et al.*, 2002).

In *S. pombe*, RNAi controls the formation of heterochromatin in the nucleus (Grewal & Moazed, 2003) and mutations in the RDR homologue *rdp1* result in loss of heterochromatin and increased expression of centromeric sequences, among others (Volpe *et al.*, 2002). It is thought that the heterochromatin-associated *rdp1* synthesises complementary RNA on RNAs transcribed from the expressed heterochromatic sequences. The dsRNAs formed are then cleaved into ~21-nt siRNA. Plants appear to have a very similar system-controlling histone modification and DNA methylation (Xie *et al.*, 2004). It is not known yet if all eukaryotes possess this nuclear RNAi pathway involved in epigenetics because vertebrates and *Drosophila melanogaster* seem to lack any obvious RDR gene (Schwarz *et al.*, 2002; Stein *et al.*, 2003).

siRNAs capable of triggering the degradation of complementary mRNAs can be synthesised *in vitro* and added directly to cells (Caplen *et al.*, 2001; Elbashir *et al.*, 2001). The double-stranded siRNAs work efficiently in cultured mammalian cells where long dsRNA are not tolerated as they activate the interferon pathway and cell death. From a mechanistic point of view, it shows that the siRNAs can be incorporated into the RISC that is responsible for selecting complementary mRNAs and their degradation. This approach has been used mainly for animal cells. There are only a few reports describing the direct application of siRNAs to plant cells (Klahre *et al.*, 2002; Vanitharani *et al.*, 2003).

2.3 Sources of dsRNA

The small RNAs are the real guiding molecules in RNAi and related processes. As they are processed from dsRNA, an important question is: what are the various sources of dsRNA? One can divide them into 'artificial' dsRNAs from transgenes or viruses that have been modified *in vitro* and natural or endogenous dsRNAs that are encoded by endogenous plant genes and other sequences. The replication intermediates of RNA viruses can also be considered 'endogenous'. In addition, plant cells contain several RDRs that can synthesise dsRNA either in the cytoplasm or in the nucleus.

2.3.1 Transgene-encoded dsRNA

The 'man-made' transgene dsRNAs are in most cases fully complementary and relatively long (>100–200 bp up to >1 kb). They are deliberately produced by expressing a gene composed of coding and/or UTR cDNA sequences, which are arranged as inverted repeats (IRs). These so-called hairpin (hp)RNA constructs are expressed from constitutive or inducible promoters (Waterhouse *et al.*, 1998; Wesley *et al.*, 2001) and are very efficient in triggering silencing. The self-complementary RNAs are assumed to readily form an hpRNA and transported to the cytoplasm where they are processed into siRNAs. Waterhouse and Helliwell (2003) improved the hpRNA system considerably. An interesting observation was the enhanced silencing by hpRNA constructs containing an intron in between the complementary sequences (Smith *et al.*, 2000; Stoutjesdijk *et al.*, 2002). The reason for this enhanced silencing is not entirely clear. Splicing may increase hairpin formation by bringing the complementary sequences in close proximity during the splicing process, the hpRNA might be more stable and less sensitive to nucleases because of the relatively small loop and splicing may facilitate transport of dsRNA to the cytoplasm, where it is further processed (Stoutjesdijk *et al.*, 2002). However, dsRNAs containing a relative large loop that is not spliced out can also be very effective (e.g. Chuang & Meyerowitz, 2000; Sijen *et al.*, 2001b). It is therefore perhaps more important to reach the highest possible level of dsRNA in the cytoplasm, which can be obtained either by a strong promoter or by including an intron. In any case, several of the commercially available hpRNA vectors are based on the intron-dsRNA principle and they work indeed very efficiently (Helliwell & Waterhouse, 2003). The system has been adapted for high-throughput cloning of cDNA sequences in generic hpRNA vectors and silencing of the corresponding genes (Wesley *et al.*, 2004).

2.3.2 Fortuitous synthesis of transgene dsRNA

Silencing has been observed in transgenic plants with various kinds of constructs and genes (reviewed in Vaucheret *et al.*, 1998), before dsRNA was identified as a trigger. Obviously, gene silencing in these cases occurred unintentionally. Generating transformants with a silenced phenotype was a matter of trial and error. In the beginning it was puzzling how silencing was induced, because it occurred with single-copy transgenes, repetitive transgene loci and even with promoterless genes. Many of the repetitive transgene loci associated with silencing consist of two or more transgene copies integrated into the genome as IRs (Muskens *et al.*, 2000). Read-through transcription of such IRs may yield hpRNAs that are cleaved by a DCL into siRNAs. Although little is known about aberrant transcription of inversely repeated transgenes, analysis of transcripts from such loci involved in triggering silencing of chalcone synthase in petunia indeed suggests that they are transcribed into self-complementary RNAs (R. Van Blokland, J. M. Kooter, unpublished data). Transcription of the IR loci and the level of dsRNA are usually low (Stam *et al.*, 1998), but in conjunction with the RDR-mediated amplification step (Sijen *et al.*, 2001a), by which secondary dsRNA is made from regular mRNAs, sufficient siRNAs could be produced to induce a strong silencing phenotype. This implies that single-stranded mRNAs play an important role in the efficiency of silencing (Sanders *et al.*, 2002). This scenario may also explain silencing by IRs consisting of promoterless transgenes. Such transgene sequences can yield low levels of dsRNA as a result of read-through transcription initiated at up- or downstream genes, depending on the genomic integration site (Stam *et al.*, 1997a).

Another possibility is that single-stranded transcripts from repetitive transgene loci are directly converted into dsRNA by an RDR, with (Makeyev & Bamford, 2002) or without (Tang & Zamore, 2004) siRNA as primer. This alternative is based on studies in *S. pombe*, showing a connection between RNAi and heterochromatin formation (Schramke & Allshire, 2003; Volpe *et al.*, 2002). Heterochromatinisation of centromeres and transcriptionally silent mating type loci requires *rdp1*, an RDR that is associated with heterochromatin (Hall *et al.*, 2002; Volpe *et al.*, 2002). It is proposed that RDR generates dsRNA from transcription-derived RNAs (Grewal & Moazed, 2003; Schramke & Allshire, 2003) and that these dsRNAs are cleaved into siRNA by *dcr1*. The siRNAs are then incorporated into the RNA-induced initiation of transcriptional silencing (RITS) complex that directs the methylation of histone H3 lysine 9. Dcr1, Ago1 and rdp1 are also required for regular RNAi in *S. pombe* (Sigova *et al.*, 2004), indicating that the complexes involved in transcriptional and posttranscriptional silencing share common factors and that the complexes operate in two distinct compartments. Such detailed experiments have not yet been done with plants but given the effect of mutations in RNAi factors on the accumulation of siRNAs derived from retro-elements and pericentromeric sequences (Xie *et al.*, 2004), plants may possess the same epigenetic RNAi pathway. Thus it is conceivable that single-stranded RNAs (ssRNAs) from repetitive transgene loci could produce dsRNA with the help of a nuclear RDR (Martienssen, 2003).

In addition to IR transgene loci, fortuitous silencing in plants has also been observed with single-copy transgenes (S-PTGS) (Beclin *et al.*, 2002;

Vaucheret *et al.*, 1998) and it is interesting to examine how these 'old' observations (Stam *et al.*, 1997b) can be explained by current RNAi models. Also, because the single-copy transgenes able to induce RNAi may resemble endogenous genes or other native sequences that are subjected and controlled by the same process, silencing is associated with siRNA production and thus likely induced by dsRNA. But how is dsRNA synthesised? To answer this question, a few features associated with single-copy-induced RNAi might be relevant: the transgenes are often highly transcribed and strong promoters increase the frequency and extent of RNAi (Que *et al.*, 1997); silencing can be induced by increasing the number of transgenes in the genome, for example, by making plants homozygous (de Carvalho *et al.*, 1992; Vaucheret *et al.*, 1998); and furthermore, silencing is triggered or enhanced when the endogenous genes are co-expressed, for example, in a tissue-specific or developmental manner (Kunz *et al.*, 1996; Vaucheret *et al.*, 1997). Another interesting observation is that RNAi by single-copy transgenes requires the sgs2/sde1 gene, RDR6, whereas RNAi by hpRNA expressing constructs does not (Beclin *et al.*, 2002). These findings point to the importance of dsRNA synthesis other than transcription and to the importance of mRNA concentration. Indeed wheat germ extracts are able to 'transcribe' ssRNA into dsRNA (Tang *et al.*, 2003), and also tomato contains an RDR that synthesises dsRNA (Schiebel *et al.*, 1993). Initiation of dsRNA synthesis does not require a primer, so basically any ssRNA can be used as template. However, this does not happen *in vivo*, because only the transgene RNAs and native homologues appear to be selected and ultimately degraded. How are these transcripts selected? There are two aspects that seem crucial to the importance of RNA dosage for inducing silencing: a small number of abnormal RNAs that can be converted into dsRNA and the large number of normal mRNAs that can participate in the amplification of RNAi signals. The normal mRNAs can be used for the synthesis of secondary dsRNA and siRNAs. Each step requires an RDR activity, possibly explaining the need of RDR6 in silencing that is induced by single-copy transgenes. If there is a continuous supply of siRNAs derived from the hpRNAs transcribed from the IR transgene (Beclin *et al.*, 2002) or from a virus (Dalmay *et al.*, 2000b), the need for an RDR is much less. A weak point in the model is the 'abnormal' RNA that is proposed to be selected by RDR and converted into dsRNA. How does such an RNA look? It may be a degradation intermediate lacking, for example, a 5′ CAP (Gazzani *et al.*, 2004). In addition to such RNAs, the concentration of target mRNA also seems important. Namely, *in vitro* studies with wheat germ extracts show that the production of the ∼24-nt siRNAs occurs mainly at a high mRNA concentration (Tang *et al.*, 2003). This mRNA dependence is in line with the observation that in plants PTGS depends on the number of active transgenes (Stam *et al.*, 1997b). This issue will be covered later when the various RDRs in plants are discussed.

2.3.3 Regulated and inducible RNAi

hpRNA transgenes controlled by a strong promoter, such as the CaMV-35S promoter, are now routinely used to knock down genes expression (Wesley *et al.*, 2001, 2004). But in addition to the 35S or nopaline synthase promoter, which are active in most plant cells, it would be convenient to have tissue- and cell-type-specific promoters to express hpRNA and induce RNAi in some cells but not in others. This would allow much more sophisticated applications of RNAi and studies of temporal and spatial gene knockdowns in plants. Seed-specific silencing is already possible by using a napin promoter (Wesley *et al.*, 2001).

One of the major problems with tissue- and cell-type-specific RNAi might be the systemic nature of RNAi in plants. This has been observed with virus-induced and transgene-induced RNAi (Himber *et al.*, 2003; Palauqui *et al.*, 1997; Voinnet & Baulcombe, 1997) and endogenous genes are also affected (Yoo *et al.*, 2004). One or more components of the RNAi process are mobile and able to move to other cells and tissues where they trigger silencing in the absence of the primary trigger (see also below). The moving silencing signal has not been fully characterised but is most likely a 21-nt siRNA, both for short- and long-distance movement (Baulcombe, 2004; Himber *et al.*, 2003). Most studies on systemic silencing have been done with *Nicotiana benthamiana* and *Arabidopsis* but differences in silencing between plant species, and also organs and cell types may exist. For example, suppression of the floral pigmentation gene chalcone synthase in petunia petals often result in variegated patterns (Napoli *et al.*, 1990; Van Blokland *et al.*, 1994; Van der Krol *et al.*, 1990), indicating that RNAi is more efficient in some parts than in others. If silencing in petals is systemic one would always expect uniformly white petals, which is not the case. Also in *Arabidopsis* and *Medicago truncatula*, silencing does not spread in the root epidermis and only inefficiently in shoots (Limpens *et al.*, 2004). Therefore, in some cases cell- and tissue-specific silencing might be feasible. An alternative would be to use mutants in which factors specifically involved in spreading are defective or absent.

Constitutive silencing of genes involved in fundamental cell processes and embryonic developmental pathways may be lethal. Therefore, inducible promoters controlling the expression of hpRNA transgenes provide an alternative approach for examining the exact functions of such genes. A few attempts have been made in this direction. In *Arabidopsis thaliana* and *N. benthamiana*, a chemically inducible Cre/loxP recombination system was used to trigger the expression of an hpRNA transgene (Guo *et al.*, 2003). Cre expression is controlled by a chimeric transcription factor, XVE, whose activity is regulated by the estrogens (Zuo *et al.*, 2000, 2001). Upon addition of 17β-estradiol, Cre is expressed and catalyses the *loxP*-mediated excision of a cassette that separates the hpRNA transgene and the constitutive G10-90 promoter (Zuo *et al.*, 2001). When the hpRNA transgene is placed directly downstream of the G10-90 promoter, RNAi is induced. In this way a green fluorescent protein (GFP) gene and the endogenous gene phytoene desaturase were silenced. By applying 17β-estradiol at different time points, one could induce silencing at different developmental stages. Because of the recombination events, which may not affect all *Cre/loxP* transgenes in a plant, one can generate chimeras that are useful for certain applications. Also the local application of 17β-estradiol induces local RNAi, which can be used to study systemic silencing.

In addition to inducing RNAi by a recombination event, which basically is irreversible, one could also use the XVE-17β-estradiol-dependent promoter to directly express the hpRNA transgene (Zuo *et al.*, 2000). In this case, constitutive RNAi requires repeated applications of 17β-estradiol, which under some circumstances can be a disadvantage (Zuo *et al.*, 2001). But for addressing questions concerning the amplification of RNAi and spreading, it would be convenient to have a system that can be switched on and off.

Heat shock promoters in *C. elegans* (Tavernarakis *et al.*, 2000) and *Arabidopsis* (Masclaux *et al.*, 2004) have also been used to regulate dsRNA expression from transgenes. By raising the temperature to 37–38°C, the *Arabidopsis* HSP18.2 heat shock promoter is strongly activated in all organs of the plant, except seeds. Targeting the phytoene desaturase gene resulted in photobleached leaves (Masclaux *et al.*, 2004) that were developing at the time of the heat shock. Leaves that emerged later were green, clearly indicating the transient nature of the induction of RNAi. The approach is simple and does not require additional factors like the estrogen-inducible system does. It remains to be seen how generally the system can be applied because the promoter is probably active at a basal level and inducible by other abiotic stresses. If the conditions can be controlled, the HSP promoter is a suitable alternative for studying genes whose inactivation is lethal or disturbs normal development (Masclaux *et al.*, 2004).

2.3.4   *Viral dsRNA and virus-induced gene silencing*

The majority of viruses infecting plants have ssRNA genomes. Upon entering cells new viral genomes are made by the viral replicase, giving rise to dsRNA intermediates. The presence of viral siRNAs in infected plants (Hamilton & Baulcombe, 1999) indicates that viral dsRNA is processed by one of the DCLs in plants. The siRNAs are assumed to attack and degrade the viral RNAs. Thus, viruses are RNAi inducers and targets at the same time, suggesting that one of the main functions of RNAi is to fight off viral infections. This is further supported by the molecular evolution of viruses. Due to the continuous arms race between host and pathogen, viruses have evolved ways to block the RNAi defence reaction. The first suppressor of RNAi encoded by a virus was

P1/Hc-Pro from potyviruses (Anandalakshmi *et al.*, 1998; Brigneti *et al.*, 1998; Kasschau & Carrington, 1998). In the absence of the virus, ectopic expression of Hc-Pro inhibits RNAi transgene–induced RNAi. Over the years several other suppressors encoded by both RNA and DNA viruses have been identified (Goldbach *et al.*, 2003; Moissiard & Voinnet, 2004). The proteins are different, suggesting that they evolved independently, and interfere with the RNAi pathway at different steps (Hamilton *et al.*, 2002; Llave *et al.*, 2000; Mallory *et al.*, 2001; Voinnet *et al.*, 2000). A common characteristic is that they reduce the accumulation of siRNAs (Hamilton *et al.*, 2002). The tombusvirus p19 protein, for example, binds double-stranded siRNAs *in vitro* (Silhavy *et al.*, 2002) and *in vivo* (Lakatos *et al.*, 2004), suggesting that p19 prevents systemic RNA silencing by sequestering siRNAs. A detailed discussion of viral suppressors and the way they act can be found elsewhere (Goldbach *et al.*, 2003; Moissiard & Voinnet, 2004).

The extent to which viral 'suppressors' inhibit RNAi varies (Robertson, 2004). Viruses encoding weak suppressors can be used as vectors to induce RNAi of endogenous genes when the recombinant virus contains a sequence from the gene to be silenced (Figure 2.1). This coordinated virus–gene silencing system, called virus-induced gene silencing (VIGS), has been developed into sophisticated applications to study gene function (Baulcombe, 1999; Ratcliff *et al.*, 2001; Waterhouse & Helliwell, 2003). Various DNA vectors based on potato virus X (PVX) (Baulcombe, 1999), tobacco rattle virus (TRV) (Ratcliff *et al.*, 2001), and satellite tobacco mosaic virus (STMV) (Gossele *et al.*, 2002) have been generated (Robertson, 2004). Either *in vitro* synthesised viral RNA is used to infect a plant (VIGS) or the DNA version of the chimeric viral genome is introduced into the plant after which the viral transcript is made *in planta* from a promoter that is active in plants, such as the CaMV-35S promoter. The latter is known as transgenic VIGS (Baulcombe, 1999). It acts as an 'amplicon' because the virus directs its own replication, moves to other organs and induces silencing of the target gene. When using an amplicon, not all viral genes have to be included, because of which viral symptoms often associated with a viral infection are avoided (Angell & Baulcombe, 1997).

A powerful tool to suppress gene expression is the local delivery of viral silencing constructs to leaves by *Agrobacterium tumefaciens*, known as Agro-VIGS or Agro infiltration. *A. tumefaciens* is able to introduce a transgene version of the virus when the genome sequence is cloned between the LB and RB boundaries of a T DNA plasmid to plant cells. The introduced transgene is transcribed, delivering the RNAs able to induce RNAi, and as the virus spreads together with the systemic properties, silencing is induced throughout the plant. By this approach the function of several plant genes and the pathways they are operating in have been examined (Ekengren *et al.*, 2003; Liu *et al.*, 2002, 2004b, c).

Advantages of VIGS are that it is relatively fast, transgenic plants do not have to be made and it is easy to implement. However, VIGS and Agro-VIGS have some limitations. One is the narrow host range of the virus. With the virus vectors that are available most experiments are limited to *N. benthamiana*. But this may change rapidly as vectors have become available for *Nicotiana tabacum*, tomato (Dinesh-Kumar *et al.*, 2003), *Arabidopsis* (Dalmay *et al.*, 2000a, Muangsan *et al.*, 2004), potato (Brigneti *et al.*, 2004) and even for legumes (Constantin *et al.*, 2004). Another disadvantage is that certain tissues or cells, e.g. meristems, may not be reached. A complete overview of the various types of VIGS, vectors and their applications can be found in Robertson (2004), Lu *et al.* (2003) and Waterhouse and Helliwell (2003).

2.3.5 Endogenous dsRNAs

An important class of endogenous dsRNAs is the collection of miRNA precursors. These RNAs are transcribed from non-protein-coding genes, are self-complementary and fold into a stem-loop structure (Park *et al.*, 2002; Reinhart *et al.*, 2002). In *Arabidopsis*, part of the molecule is recognised by DCL1/CAF that cleaves out the single-stranded miRNA (Finnegan *et al.*, 2003; Reinhart *et al.*, 2002). In contrast, siRNAs are cleaved out of usually fully base-paired dsRNAs by another DCL, and they are double-stranded with a 2-nucleotide 3′ overhang. Otherwise, miRNAs and siRNAs are similar in that they are ~21–24 nt in length, have a 3′-OH and a 5′ phosphate, and most plant miRNAs guide mRNA degradation like siRNAs do. This is a striking difference with animal miRNAs, which mostly control the translation of the target mRNA (Bartel, 2004). In plants, many miRNAs control development by regulating the abundance of mRNAs, of which the proteins are involved in developmental programmes (Carrington & Ambros, 2003; Kidner & Martienssen, 2004; Palatnik *et al.*, 2003). What is the origin of miRNA genes? The fact that they encode partially dsRNAs suggests that they have originated from IR genes, and indeed a recent study shows that several genes encoding miRNAs and other small RNA-generating loci possess hallmarks of inverted duplications, each encoding an arm of the stem-loop precursor RNA (Allen *et al.*, 2004).

TEs and remnants of these elements may contain inverted duplicated sequences and transcription of such sequences may yield dsRNAs (Sijen & Plasterk, 2003). A DCL enzyme may cleave these dsRNAs into short RNAs. By cloning short RNAs from *Arabidopsis* and performing a BLAST search of the genome, Mette *et al.* (2002) were indeed able to detect short RNAs derived from intergenic regions with hallmarks of TEs. The sequences can be folded into partial dsRNA structures, which include the short RNA sequence. It is thus likely that these short RNAs are cleaved out of the dsRNA. TE-derived siRNAs have also been reported for *C. elegans* (Sijen & Plasterk, 2003). The function of these RNAs, if any, is unclear. They may be involved in transposon silencing either by directing RNA degradation or by epigenetic modification of TE–DNA sequences. It is even conceivable that some of the TE–IR sequences may evolve or have evolved into miRNA genes.

Based on the identification of siRNAs that correspond to both the sense and antisense strands of a non-coding RNA in *Arabidopsis*, there is yet another class of endogenous dsRNAs whose production depends on a RDR (Vazquez *et al.*, 2004b). The non-coding ssRNA is most likely converted into dsRNA and then processed in 21-nt siRNAs. The production of these siRNAs depends on factors of the sense transgene silencing pathway (S-PTGS) (Beclin *et al.*, 2002), including RDR6 and SGS3 and components of the miRNA pathway, AGO1, DCL1, HEN1, and HYL1 (Vazquez *et al.*, 2004b), proteins that are discussed in greater detail below. These siRNAs behave like miRNAs by guiding mRNA degradation. How common this mode of dsRNA/siRNA production actually is remains to be determined. It would be interesting to learn why the non-coding RNA is converted into dsRNA whereas so many other transcripts remain single-stranded.

The third class of dsRNA also depends on RDR2 and is produced in the nucleus. RNAs transcribed from pericentromeric sequences and other repetitive sequences, such as TEs, seem to be converted into dsRNA (Lippman & Martienssen, 2004) by RDR2. The siRNAs originating from these dsRNAs are involved in maintaining the heterochromatic structure and high cytosine methylation status of the repetitive DNA (Xie *et al.*, 2004) (Figure 2.2).

2.4   The protein machinery of RNAi

2.4.1   *Double-stranded RNA-processing enzymes: the DCLs*

The dramatic suppressive effect of dsRNA on gene expression, in contrast to single-stranded sense or antisense RNAs (Fire *et al.*, 1998), implies that the double-stranded nature of the RNA is crucial. At the time this discovery was made, an enzyme known to cleave dsRNAs was the dsRNA-specific ribonuclease III (RNase III) from *Escherichia coli* (Robertson *et al.*, 1967, 1968). Although this activity was considered to be unique to bacteria, several years later, nucleases with similar specificity were found in eukaryotes (Libonati & Sorrentino, 1992). Based on the conserved dsRNA-binding domain and the catalytic domain of these proteins, the RNase III enzymes form a superfamily. In bacteria they are involved in various steps of RNA processing, including maturation of rRNA, tRNA, precursor mRNA and mRNA degradation. It is also implicated in antisense RNA-mediated control of plasmid replication (Nicholson, 1999). There are also species-specific processing pathways, illustrating the versatile character of RNase III enzymes. As long as the dsRNA region is 20 bp or more, the enzymes recognise it (Nicholson, 1999).

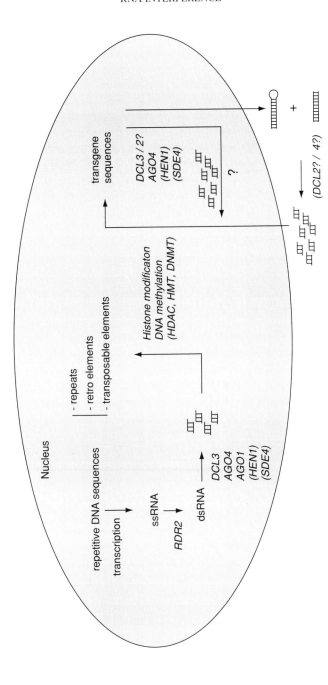

Figure 2.2 RNA interference (RNAi)/epigenetic pathways in the nucleus of plants. The various steps and proteins involved are described in the main text.

Bernstein *et al.* (2001) identified an activity in *Drosophila* embryo extracts that was able to cleave relatively long dsRNA into segments of 21–22 nt in length, the siRNAs. The enzyme was aptly called Dicer. The siRNAs resembled the small RNAs identified by Hamilton and Baulcome (1999) in plants containing posttranscriptionally silenced transgenes and endogenous genes, and in plants infected with viruses. So the connection between RNA degradation, the inducer (dsRNA) and the effector (siRNA) molecules was quickly made. Dicer proteins have been identified in various eukaryotes and several of them contain more than one dicer. Humans, *C. elegans* and *S. pombe* contain just one Dicer, *Neurospora crassa* two redundant DCR genes (Catalanotto *et al.*, 2004), *Drosophila* two (DCR-1 and DCR-2) (Lee *et al.*, 2004), and *Arabidopsis* contains four DCL proteins (DCL1–DCL4) (Schauer *et al.*, 2002). Dicers are multidomain proteins. They contain two RNase-III-like catalytic motifs and a C-terminaldsRNA-binding domain. In addition, some contain a PAZ domain based on the conserved amino acid sequence found in other proteins involved in RNAi: PIWI, Argonaute (AGO), Zwille (Cerutti *et al.*, 2000; Meister & Tuschl, 2004) and helicase. The *Drosophila* DCR1 lacks a functional helicase domain whereas DCR-2 lacks the PAZ domain (Lee *et al.*, 2004), indicating their functional divergence. Biochemical and genetic experiments of *Drosophila* DCR enzymes provided detailed insight into the activities, substrate specificities and biological functions of DCRs (Lee *et al.*, 2004; Meister & Tuschl, 2004; Pham *et al.*, 2004), which helps us understand the function of the four DCLs in plants. *Drosophila* DCR-1 preferentially cleaves the partially double-stranded miRNA precursors while DCR-2 cleaves long dsRNA into siRNAs (Lee *et al.*, 2004). So there are distinct pathways in which the DCRs operate, depending on the type of dsRNA they recognise. But they also function downstream of dsRNA cleavage because biochemical experiments with embryo lysates indicate that both DCRs are needed for the assembly of the siRNA–RISC (Lee *et al.*, 2004). In fact both proteins reside in the complex (Pham *et al.*, 2004). The nuclease activity of DCR is at this stage not required anymore because lysates containing RNase-III-defective versions of DCR-2 exhibit normal mRNA cleavage *in vitro* (Lee *et al.*, 2004). Apparently, the siRNA–mRNA duplex in RISC is cleaved by another enzyme of the complex, most likely a member of the AGO family of proteins of the RISC. In mammals, it is believed to be AGO2 because mutations in the cryptic RNase H domain of AGO-2 block siRNA-mediated cleavage by RISC (Liu *et al.*, 2004a).

Native gel electrophoresis of *Drosophila* embryo extracts suggests that there are several siRNP complexes that assemble with siRNAs, R1, R2, R3 (Pham *et al.*, 2004). The picture that is emerging is that R3 is formed from pre-existing complexes starting with the R1 complex containing a DCR-cleaving dsRNA. In subsequent steps factors are added, ultimately resulting in an activated 'holo-RISC' capable of cleaving the siRNA–mRNA duplex. Experiments suggest that 'holo-RISC' is associated with ribosomes, which is in line

with earlier observations in *Drosophila* (Hammond *et al.*, 2001) and in tryp-
anosomes (Djikeng *et al.*, 2003).

2.4.1.1   *What is known about plant DCLs?*
As mentioned, *Arabidopsis* contains four DCLs (Schauer *et al.*, 2002) and as in
Drosophila, they appear to have distinct functions.

2.4.1.1.1   *DCL1*

The first DCL was identified by characterising the developmental *Arabidopsis*
mutants Carpel factory (CAF), short integuments (SIN1), suspensor1 (SUS1)
and embryonic defective76 (EMB76) (Schauer *et al.*, 2002). The phenotypes
are caused by mutations in the same gene (Golden *et al.*, 2002; Jacobsen *et al.*,
1999), encoding a protein with an N-terminal DExH/DEAD-box type RNA
helicase domain, RNase III domains, two dsRBD domains and a PAZ domain.
It has all the features of the *Drosophila* Dicer (Bernstein *et al.*, 2001) and
was therefore renamed dicer-like 1 or DCL1. It also contains two functional
nuclear localisation signals and when fused to GFP, the hybrid protein is
transported to the nucleus (Papp *et al.*, 2003).

DCL1 is involved in the processing of miRNA precursors (Park *et al.*, 2002;
Xie *et al.*, 2004) (Figure 2.1) and in the cleavage of some other nuclear
dsRNAs (Vazquez *et al.*, 2004b). Thus in the CARPEL FACTORY–DCL1
mutants, RNAi induced by hpRNA constructs normally proceeds whereas
miRNAs fail to accumulate (Finnegan *et al.*, 2003). In addition to DCL1,
HEN1 (Park *et al.*, 2002), a protein with unknown function but which could be
a dsRNA methylase (Anantharaman *et al.*, 2002) and the dsRNA-binding
protein HYL1 are also needed for miRNA production (Han *et al.*, 2004;
Vazquez *et al.*, 2004a). Interestingly, HEN1 is also required for the production
of siRNAs derived from sense transgenes that are able to induce PTGS
(S-PTGS) (Boutet *et al.*, 2003). It is not required for the production of siRNAs
processed from long dsRNAs. The reason for this difference is not well
understood but might be related to the requirement of dsRNA synthesis by
RDR6 from aberrant RNAs transcribed from S-PTGS genes. Another striking
observation is that the HEN1 requirement for miRNA and siRNA accumulation can be uncoupled by a single-point mutation, indicating that the two
pathways are different. This again might be related to the need for S-PTGS
to first synthesise dsRNA from an aberrant RNA before siRNAs can be
generated (Dalmay *et al.*, 2000b; Mourrain *et al.*, 2000).

The involvement of DCL1 in miRNA production explains the developmental
defects in mutants because many of the miRNAs in plants generated by DCL1
appear to control or degrade mRNAs encoding proteins involved in development.
Strikingly, many of these are transcription factors (Reinhart *et al.*, 2002; Rhoades
et al., 2002), including *Arabidopsis* TCP involved in leaf morphogenesis

(Palatnik *et al.*, 2003), SCARECROW-like (Llave *et al.*, 2002), PHABULOSA and PHAVOLUTA (Rhoades *et al.*, 2002) involved in leaf polarity, the NAC domain proteins CUPSHAPED COTELYDONS-1 and -2, which are involved in embryonic, vegetative and floral development (Laufs *et al.*, 2004; Mallory *et al.*, 2004), and APETALA 2 (AP2) and homologues, involved in flowering time (Aukerman & Sakai, 2003). AP2 is interesting because unlike most other miRNA-controlled mRNAs, the AP2 mRNA is one of the few whose translation is inhibited by an miRNA (Aukerman & Sakai, 2003; Chen, 2004). However, in contrast to most animal miRNAs, the majority of plant miRNAs trigger mRNA degradation. The examples mentioned above indicate that complete complementarity between miRNA and its mRNA target sequence is not required. The consequence is that a single miRNA is able to control the expression of several closely related gene family members, which indeed has been observed for the AP2-like genes (Aukerman & Sakai, 2003), SCARECROW-like genes (Llave *et al.*, 2002) and TCP genes (Palatnik *et al.*, 2003).

It is interesting to note that DCL1 itself is regulated by miRNA-mediated mRNA decay (Xie *et al.*, 2003). Analysis of DCL1 transcripts revealed that wild-type plants contain various forms of DCL1 mRNA, including truncated transcripts derived from aberrant pre–mRNA processing (Xie *et al.*, 2003). The levels of DCL1 mRNA and truncated RNAs are relatively low but in a DCL1 mutant, the levels are several-fold higher than in DCL1 wild-type plants. The DCL1 RNA levels are also higher in the miRNA-defective HEN1 mutant and in plants expressing the RNAi suppressor P1/HC-Pro. These results point to negative feedback regulation of DCL1 expression. Consistent with this is the presence of a target sequence for miR162 in the middle of the DCL1 mRNA and also the presence of corresponding mRNA cleavage products. This negative feedback mechanism of DCL1 expression suggests that the activity of DCL1 must be tightly controlled.

2.4.1.1.2   *DCL2*

DCL2, like DCL1, is primarily a nuclear protein (Xie *et al.*, 2004). Some experiments suggest that DCL2 functions in the antiviral defence pathway (Xie *et al.*, 2004). But the differences between viruses and the fact that a dcl2 mutant infected with TCV exhibits only a delayed accumulation of siRNAs indicate that DCL2 is not the only one involved in viral defence, if at all. Also the nuclear localisation of DCL2 would argue against a virus-specific DCL as most viruses replicate in the cytoplasm. It is more likely that DCL2 has a preference for dsRNAs that are produced in the nucleus.

2.4.1.1.3   *DCL3*

DCL3 is also a nuclear protein, although not exclusively, and is involved in chromatin-associated events, together with RDR2, AGO4 and HEN1 (Xie *et al.*, 2004) (Figure 2.2). It acts on endogenous RDR2-produced dsRNAs derived from retroelements and other repetitive elements (Lippman & Martienssen, 2004). The siRNAs generated by DCL3 may direct histone H3-K9 methylation and DNA methylation by a mechanism that is poorly understood. This plant chromatin modification system is similar to that in *S. pombe* (Grewal & Rice, 2004). The *S. pombe dcr1* and *Arabidopsis* DCL3 are likely orthologous proteins.

With three DCLs primarily found in the nucleus, it is still unclear which DCL generates the siRNAs from 'long' hpRNAs derived from IR transgenes. It is very unlikely that it is DCL1 as it has been tested (Finnegan *et al.*, 2003). DCL3 is also unlikely because of its nuclear activities. This leaves DCL2, based on its effect on cytoplasmic viral RNAs despite its nuclear location, and DCL4, whose function remains to be determined, as possible candidates. It could be that there is considerable redundancy and that multiple DCLs are involved. This issue can only be addressed by making double, triple or even quadruple DCL mutants.

Another intriguing issue is the temperature dependence of siRNA accumulation as observed in *N. benthamiana* (Szittya *et al.*, 2003). At 18°C, siRNAs are clearly detectable but at 15°C, siRNAs from transgenes and viruses are not produced. Production of miRNAs at lower temperature is not affected. As the two types of short RNAs are derived from different dsRNA classes, it could indicate that DCL1 is not affected whereas the siRNA-generating DCL, which is not known, does not work at 15°C. However, it is equally possible that it is one of the other factors involved in siRNA production, for example RDR1, RDR6 or an AGO protein. Whatever the outcome is, this differential responsiveness to temperature can very helpful in unravelling the different RNAi pathways in plants.

2.4.2   DCL activities and the production of different size classes of siRNA

The length of siRNAs varies between roughly 20 and 26 nt (Hamilton & Baulcombe, 1999; Tang *et al.*, 2003; Xie *et al.*, 2004). In *N. benthamiana*, however, two major size classes of siRNAs have been identified, one of 21–22 nt and the other of 24–26 nt (Hamilton *et al.*, 2002). GFP transgene silencing, induced by Agro-infiltration of a plant containing an expressed GFP gene with a 35S-GFP construct, is associated with both size classes whereas siRNAs derived from an endogenous retroelement are only of the long class. The function of the two size classes is not well understood and it is also not clear how they are generated. Based on transient studies with viral suppressors, which inhibit the production of the longer class, it is proposed that the 21- to 22-nt siRNAs class is responsible for mRNA degradation whereas the longer species is involved in systemic silencing and DNA methylation (Hamilton *et al.*, 2002). Although these elegant experiments suggest a functional difference between siRNAs, additional experiments are needed to confirm it, because short- and long-range systemic silencing seems to involve siRNAs of the short class (Himber *et al.*, 2003).

For a large collection of *Arabidopsis* endogenous siRNAs, including miR-NAs, the length varies between 19 and 25, with a minor peak at 21 and a major one at 24 nt (Tang *et al.*, 2003). Xie *et al.* (2004) found siRNAs of all sizes (20–25 nt), with the 24-nt most abundant. In any case, Tang showed that each class has a distinct sequence bias with a 5′ adenosine predominating the longer class; whereas miRNAs in plants of the short class start with a 5′ uridine (Reinhart *et al.*, 2002; Tang *et al.*, 2003). The different siRNAs may therefore be generated by different DCLs, have a different origin or both. *In vitro* experiments with the wheat germ lysate have provided some of the answers. Tang *et al.* (2003) showed that a single dsRNA species is processed in the lysate into both classes of siRNAs, although the 24- to 25-nt class was about four-fold more abundant. This figure is similar to the size distribution of siRNAs in *Arabidopsis*. Inhibition experiments with synthetic siRNAs of 21- and 25-nt indicate that the two classes are generated by different DCL ortho-logues and in an ATP-dependent manner. Another interesting observation is that the longer class is coupled to the production of dsRNA by an RDR in the extract, suggesting that the DCL generating long siRNAs is linked to the RDR converting ssRNA into dsRNA. It may imply that the majority of endogenous siRNAs in plants is derived from RDR-derived dsRNAs, which does not exclude other possibilities, such as differences in siRNA stability.

2.4.3   *Argonaute proteins/PAZ and PIWI domain (PPD) proteins*

Among the many proteins involved in RNAi, the requirement of members of the so-called AGO family seems universal. The AGO story started with the *Neurospora crassa* QDE-2 mutant that was defective in quelling, a phenom-enon similar to RNAi (Catalanotto *et al.*, 2000; Cogoni & Macino, 1997), the *C. elegans* RDE-1 mutant (Tabara *et al.*, 1999), and a developmental *Arabi-dopsis* mutant (Bohmert *et al.*, 1998). The *Arabidopsis* mutant had unex-panded pointed cotyledons and very narrow rosette leaves. Due to this unusual appearance, which reminded the authors of a small squid, the mutant was called *argonaute* (Bohmert *et al.*, 1998)! The AGO1 gene that was defective in the mutant encoded a 115-kDa protein (Bohmert *et al.*, 1998) and turned out to resemble QDE-2 and RDE-1. It was soon shown that AGO1 was also required for PTGS or RNAi in plants (Fagard *et al.*, 2000). This clearly connected RNAi with plant development.

AGO genes are found in almost all eukaryotes but it is striking that the number of AGO genes greatly varies: *S. pombe* and *Neurospora crassa* contain only 1 AGO gene, *Drosophila* at least 5, humans 8, *Arabidopsis* 10, whereas *C. elegans* contains 27 AGO genes. In case there are multiple members, studies of mutants show that not all AGO proteins are involved in RNAi (Meister & Tuschl, 2004). The first feature of the AGO proteins noticed was the similarity with a rabbit eIF2C factor, which is involved in the initiation of translation. However, AGO proteins are characterised by more defined domains: PAZ and PIWI (Cerutti *et al.*, 2000). Because of these conserved motifs, they are also known as PPD proteins.

2.4.3.1   The PAZ domain

The PAZ domain, named after a conserved region in the proteins PIWI, AGO and Zwille, is found not only in AGO proteins but also in RNase III Dicer proteins (Bernstein *et al.*, 2001), including CAF. Since the *Drosophila* Dicer and AGO2 interact, it was initially proposed that the PAZ domain may function as a protein–protein interaction motif (Hammond *et al.*, 2001). However, structural and biochemical studies of AGO2 indicate that the PAZ domain binds the 2-nucleotide 3′ overhang of an siRNA or miRNA duplex (Lingel *et al.*, 2004; Ma *et al.*, 2004). The affinity for blunt-end siRNA duplexes is 30-fold lower, suggesting that it is selected for binding dicer-generated siRNA products, which have the specific 2-nucleotide 3′ overhang. The presence of an siRNA/miRNA in the PAZ domain of an AGO protein may guide the RISC/miRNP complex to the appropriate target. The function of the PAZ domain in Dicer is less clear. It may help docking the enzyme onto the ends of dsRNAs and miRNA precursors (Zhang *et al.*, 2002). It may also transiently hold the siRNA/miRNA before it is transferred to the PAZ domain of an AGO protein. Another interesting question is that if there are multiple Dicers, such as in plants and *Drosophila*, and if a Dicer does not contain a clear PAZ motif (Liu *et al.*, 2003), what is their specificity and exact mode of action? It is obvious that biochemical experiments are needed to address these issues. Thus far, most information has been obtained from *Drosophila* (Haley *et al.*, 2003), *C. elegans* and mammalian RNAi systems (Tahbaz *et al.*, 2004), but recently the biochemical analysis has been extended to the plant-silencing system using a standard wheat germ extract and a cauliflower extract (Tang & Zamore, 2004; Tang *et al.*, 2003).

2.4.3.2   The PIWI domain

The second interesting domain in AGO proteins is the PIWI motif named after the *Drosophila* PIWI protein, which is required for stem cell division (Cox *et al.*, 1998, 2000). It appears to be an evolutionary very conserved domain as it is found even in a few prokaryotes, including archaeabacteria (Song *et al.*, 2004). The crystal structure of the archeal AGO protein revealed a crescent-shaped base made up of the N-terminal, middle and PIWI domains. Strikingly, the tertiary structure of the PIWI domain resembles that of ribonucleaseH (RNaseH), which cleaves the RNA strand in a DNA/RNA hybrid. In addition, it contains the conserved active site aspartate–aspartate-glutamate amino acids thought to bind the divalent metal ion that is required for catalysis. The catalytic site is adjacent to a positively charged region that extends into the PAZ and that may provide the binding site for the siRNA/mRNA duplex. These structural data suggest that the PIWI domain of the AGO protein in the RISC performs the siRNA-guided cleavage (Song *et al.*, 2004). Functional assays indicate that mutations in the RNase H/PIWI domain of the human AGO2 protein indeed inactivate the RISC activity, supporting the notion that the AGO2 protein is the slicer subunit of the RISC (Liu *et al.*, 2004a; Sontheimer & Carthew, 2004). Also the tight association of this cleavage activity with AGO2 protein, as demonstrated by stringent washes of immuno-precipitates, shows that it resides in the protein itself and that it is not due to a protein tightly bound to it.

Other human AGO proteins (1, 3 and 4) were also tested for cleavage activity and although all proteins were able to bind siRNAs and miRNAs, these AGOs did not cleave mRNA (Liu *et al.*, 2004a; Meister *et al.*, 2004). This shows that the other AGO proteins have a different function in silencing despite the strong sequence conservation of the PIWI domains. It is likely that the cleavage ability of AGO2 is determined by amino acids in the N-terminal part or in the spacer region separating the PAZ and the PIWI domain (Meister *et al.*, 2004). Also posttranslational modifications specific for AGO2 have been suggested as playing a role either in activating AGO2 as an endonuclease or in regulating the association of AGO2 with a putative endonuclease protein. That even the same AGO protein may act in different siRNA-directed silencing pathways is shown by AGO1 of *S. pombe*. Like *Neurospora crassa*, *S. pombe* contains a single AGO protein and is able to mediate both transcriptional and posttranscriptional silencing (RNAi) (Sigova *et al.*, 2004). But given the numerous AGOs in other eukaryotes, an intriguing question is whether 'all these AGOs require or contain a nuclease activity as part of their normal activity' (Sontheimer & Carthew, 2004). Biochemical experiments on recombinant AGO proteins will yield some insights into the mechanism of small RNA-induced gene silencing.

One of the critical factors whether or not the *Drosophila* AGO2 cleaves an mRNA might be the base pairing between the si/miRNA and the target. Mammalian miRNAs are usually not fully complementary to their targets and act predominantly as translational inhibitors (Bartel, 2004). It is conceivable that AGO2/RISC has a dual function in mediating cleavage and translational inhibition, which depends in part on the extent of base pairing. Meister *et al.* (2004) suggested: 'depending on their RNA targets, some miRNAs may predominantly function as translational repressors since they may not have been under evolutionary selection for guiding effective mRNA cleavage, while others may have evolved to also use the degradative mechanism'. One could also argue the opposite order of events in that initially the degradation pathway was predominant and that later in evolution the translational repressive pathway became more dominant, because one should ask, why do all AGO proteins have an RNase-specific PIWI domain? This suggests that the degradation pathway came first. The other issue is the origin of miRNAs and miRNA genes. Allen *et al.* (2004) provide evidence for an attractive scenario. They propose that miRNA genes are derived from inverted duplications of gene segments. This implies that miRNAs were processed from initially fully complementary dsRNAs and that the miRNAs may have been fully complementary to their target mRNA. This most likely resulted in mRNA degradation. It is likely that mutations in the miRNA genes accumulated and that at some point in evolution the miRNAs only partially base-paired with their target, which may have reduced the degradation level. Also the miRNA target genes are likely to have accumulated mutations. However, to reach the same gene-suppressing effect, the miRNA/AGO complex, not able to efficiently degrade the target mRNA anymore, may have started to interfere with translation and reduced protein levels in this way. It is important to address evolutionary issues like these as it may provide insights into the origin and development of the various RNAi/silencing pathways and their selective advantages.

In comparison with *Drosophila* and mammalian AGO proteins, little is known about the biochemical properties of plant proteins. But assuming that the *Arabidopsis* AtAGO1 is an orthologue of the human huAGO2, it appears that the AtAGO1 is less stringent than its human counterpart in its requirements for full complementarity between the miRNA and its mRNA target to allow cleavage. Despite the few mismatches, most miRNAs in plants guide their target mRNA for degradation instead of inhibiting its translation (Bartel, 2004; Carrington & Ambros, 2003). It will be interesting to examine whether these differences between systems are due to different properties of the 'slicing' AGO proteins or to differences in other proteins of the RISC.

2.4.4 More about plant Argonautes

Arabidopsis contains ten AGO genes (Morel *et al.*, 2002) although it is not known whether they are all functional. Little is known about the number of AGO genes in other plant species. The presence of multiple genes suggests that at least some of them have distinct functions, which are as follows:

AGO1 is clearly involved in miRNA-mediated RNA degradation and since miRNAs control several developmental pathways, mutations in the AGO1 gene give rise to a range of different phenotypes, depending on the strength of the mutation (Bohmert *et al.*, 1998; Fagard *et al.*, 2000; Kidner & Martienssen, 2004). That the activity of AGO1 must be tightly controlled is suggested by the observation that the AGO1 mRNA level itself is controlled by an miRNA (Vaucheret *et al.*, 2004), like DCL1. It is now thought that AGO1 acts within the miRNA–RISC pathway, where it allows the miRNAs to direct the cleavage of mRNA targets. The AGO1 is also involved in transgene silencing but only when it is induced by single-copy transgenes, the so-called S-PTGS (Figure 2.1). Curiously enough, silencing induced by IR transgenes (IR-PTGS) (Beclin *et al.*, 2002; Boutet *et al.*, 2003), whose transcription yield dsRNA, does not require AGO1. Although these results suggest that different RNAi pathways are operating, the molecular basis is unclear. It might be that there are different RISCs, characterised by containing distinct AGO proteins. The loading of a particular AGO protein with an siRNA or an miRNA may depend on the type of dsRNA. It is conceivable that long hairpin dsRNAs derived from transcription, dsRNAs synthesised by an RDR, viral dsRNAs and a partially double-stranded miRNA precursor are differentially recognised and processed.

The differential involvement of AGO1 in some RNAi pathways (e.g. S-PTGS, miRNA) and not in others (e.g. IR-PTGS) is also noticed at the epigenetic level, in the nucleus. PTGS/RNAi in plants causes cytosine methylation of those DNA sequences that are identical to the sequences of the dsRNAs/siRNAs (Matzke *et al.*, 2004). Although many details of the RNA-directed DNA methylation mechanism remain to be determined, it seems likely that siRNAs are the guiding molecules in this sequence-specific methylation. In an ago1 mutant, in which silencing by S-PTGS does not occur, RNA-directed DNA methylation does not take place (Fagard *et al.*, 2000). In contrast, DNA methylation triggered by IR-PTGS appears to be independent of AGO1 (Beclin *et al.*, 2002). This suggests that the RISC involved in mRNA degradation and the machinery responsible for DNA methylation are similar or at least share common factors. Furthermore, AGO1 also seems involved in the suppression of a subset of TEs in *Arabidopsis* (Lippman *et al.*, 2003), indicating that it acts in a nuclear epigenetic pathway.

AGO10, the new name for the PINHEAD/ZWILLE protein, is also involved in plant development (Lynn *et al.*, 1999). It is required for the formation of primary and axillary shoot apical meristems. The PNH/ZLL mutant has also defects in floral organ number and shape, as well as aberrant embryo and ovule development. Mutant embryos that lack AGO1 and AGO10 do not develop; they fail to progress to bilateral symmetry and do not accumulate the shoot meristemless protein. The severity of the phenotype depends on whether the AGO-null mutations are homozygous or heterozygous, suggesting that they act together or have partially overlapping functions in allowing normal growth and gene expression during embryogenesis (Lynn *et al.*, 1999). It is not unlikely that other AGO proteins may also be partially redundant in their function (Vaucheret *et al.*, 2004) because single AGO mutants are viable, whereas a null mutation of DCL1, the Dicer that generates the miRNAs from their precursors, is lethal.

The biological role of AGO7 was identified genetically by examining *ZIPPY* (ZIP) mutants, which show a premature expression of vegetative traits

(Hunter *et al.*, 2003). AGO7 mutants do not have the severe developmental abnormalities of the AGO1 mutants, suggesting that its function is different and not overlapping with AGO1 or AGO10. In contrast to AGO1, AGO7 does not play a role in gene silencing induced by transgenes (S-PTGS). Furthermore, a phylogenetic analysis of the *Arabidopsis* AGO proteins places AGO7 outside the AGO1 group, indicating that the AGO1-like proteins, including AGO10 and AGO7, have diverged relatively early (Hunter *et al.*, 2003). The molecular function and its specificity remains to be determined but as AGO7 contains a clear PAZ and a PIWI domain it is likely to be involved in one or more miRNA-controlled processes.

AGO4 was obtained from a screen for mutants that repress silencing of the *Arabidopsis* SUPERMAN (SUP) gene (Zilberman *et al.*, 2003) and turns out to be a component of the RdDM pathway. AGO4 is not implicated in RNAi-directed mRNA degradation (Zilberman *et al.*, 2004). The ago4 mutant shows decreased cytosine methylation at CG, CNG and asymmetric sites, as well as reduced histone H3 lysine 9 methylation (Zilberman *et al.*, 2003, 2004). This was found at the SUP gene and at the SINE retroelement AtSN1. However, the role of AGO4 in DNA methylation is locus-specific, since the methylation pattern of the 180-bp centromeric repeat (*CEN*), the *Ta3* retrotransposon, and the *FWA* gene in the ago4 mutant was the same as in wild type. The reason for this difference is unclear but could be related to a preference of AGO4 to control DNA methylation for dsRNA as a trigger, for example, dsRNAs generated by IR genes (Zilberman *et al.*, 2004). Interestingly, in the case of IR-induced methylation, AGO4 seems mainly required for maintenance not for establishment of methylation. This is different for other types of loci, such as the FWA where it is involved in both pathways (Chan *et al.*, 2004). This suggests that different pathways control establishment and maintenance methylation by RNA at direct and IR sequences. Again, reasons for these differences are likely to include the type of dsRNA and how it is generated (RDR or transcription), the cellular compartment in which the dsRNAs are processed into siRNAs (nucleus vs cytoplasm) and by which AGO proteins these siRNAs are 'brought' to their target DNA (Zilberman *et al.*, 2004).

2.4.5   *RNA-dependent RNA polymerases*

dsRNA can be generated by RDR using an ssRNA, for example an mRNA, as template. The first RDRs were found in RNA viruses (Blumenthal & Carmichael, 1979) where it is required for transcription and replication of the viral genome but many eukaryotes contain RDRs as well. Although not found (yet) in human and *Drosophila*, RDRs are found in *Neurospora crassa* (Cogoni & Macino, 1999), *C elegans* (Sijen *et al.*, 2001a; Smardon *et al.*, 2000), plants (Dalmay *et al.*, 2000b; Mourrain *et al.*, 2000; Schiebel *et al.*, 1998), *Dictyostelium* (Martens *et al.*, 2002), and *S. pombe* (Volpe *et al.*, 2002). They have been identified by sequence comparison with the tomato RDR (Schiebel *et al.*, 1998) and by characterising RNAi mutants. They comprise an evolutionary ancient class of enzymes as they are found in such distant organisms, like many other proteins involved in RNAi. RDRs contain a conserved motif that resembles the catalytic domain of DNA-dependent RNA polymerase (Iyer *et al.*, 2003). What is the role of RDRs in gene silencing, in TGS and in PTGS? In addressing this question we mainly focus on RDRs from plants. Most information is derived from *Arabidopsis* and its RDR mutants (Dalmay *et al.*, 2000b; Mourrain *et al.*, 2000), and from elegant *in vitro* RNAi experiments using a wheat germ extract (Tang & Zamore, 2004; Tang *et al.*, 2003), which is best known for its use as an *in vitro* translation system.

2.4.5.1   RDR1 and RDR6: virus-induced RNAi and S-PTGS

The *Arabidopsis* genome contains six putative RDR genes and the RDR mutants examined clearly indicate that they participate in different silencing pathways, like several other proteins that have been discussed. RDR6, originally called SGS2/SDE1, was the first identified and was required for transgene-induced silencing (Dalmay *et al.*, 2000b; Mourrain *et al.*, 2000). However, it is only needed for S-PTGS since IR-PTGS normally occurs in an rdr6 mutant (Beclin *et al.*, 2002). One explanation is that non-IR transgenes (S-PTGS) trigger RNAi only if additional dsRNA from aberrant RNAs or target mRNA templates is made, while the amounts of dsRNA produced by IR transgenes might already be sufficient to initiate and maintain RNAi. However, it cannot be ruled out that one of the other five RDRs is involved in IR-PTGS.

Differences also exist in the susceptibility of rdr6 mutants to certain viruses. An Arabidopisis rdr6 mutant is much more susceptible to cucumber mosaic virus than the wild type (Mourrain *et al.*, 2000), but strikingly not to TRV and to tobacco mosaic virus (TMV) (Dalmay *et al.*, 2001). The reason for this difference is unclear. Besides RDR6, RDR1 of *Arabidopsis* is also involved in viral defence because an rdr1 mutant accumulates higher levels of tobamovirus and tobravirus than the wild type (Yu *et al.*, 2003). Furthermore, RDR1 expression and its activity is induced by a viral infection as well as by salicylic acid. The putative tobacco RDR1 orthologue is also involved in reducing TMV and PVX infections (Xie *et al.*, 2001) and although these results clearly indicate that RDRs are important in the viral defence pathway, it is not well understood how. But it indicates that in addition to the viral dsRNA replication intermediates, which apparently are not sufficient to fully induce an antiviral response, additional production of dsRNA by an RDR is needed to achieve this. The differences between viruses with regard to response to RDR are puzzling, as well as the selection of viral RNAs as apposed to regular mRNAs by RDRs. Given the fact that S-PTGS resembles viral RNA in several ways, examining the transcripts derived from the S transgenes may provide a good experimental system to examine the nature of RNAs that are selected by RDRs.

One of the first clues to the possible nature of such aberrant transcripts came from a screen for suppressors of an abnormal developmental phenotype caused by the expression of a single-copy transgene involved in meristem development (Gazzani *et al.*, 2004). One suppressor turned out to have a mutation in XRN4, a 5′–3′ exonuclease degrading decapped mRNAs that play a role in mRNA turnover. XRN4 also degrades the 3′ fragment of miRNA-cleaved mRNAs (Souret *et al.*, 2004). It is proposed that XRN4 prevents RNAi by degrading decapped mRNAs that otherwise could be used by RDR as template for dsRNA production. This would also agree with RNAi triggered in tobacco cells by direct introduction of single-stranded, uncapped RNA (Klahre *et al.*, 2002). However, XNR4 is only part of the story because the XRN4 mutant looks normal (Gazzani *et al.*, 2004), indicating that (not all) decapped endogenous mRNAs are converted into dsRNA by an RDR.

2.4.5.2   *RDR2: a role in epigenetics*
RDR2 functions in an endogenous pathway in which short RNAs are generated and modification of chromatin proteins occurs (Xie *et al.*, 2004). By examining different siRNA classes derived from different regions of the *Arabidopsis* genome, the authors showed in an rdr2 mutant that certain siRNAs were absent. These siRNAs were derived from a SINE retroelement and other repetitive sequences, including the spacer of 5S rDNA repeats. Methylation of the AtSN1 element in the rdr2 mutant was reduced as well as histone H3 lysine 9 methylation. Since the same was found in dcl3 and ago4 mutants, it suggests that the three proteins act in the same pathway. Considering the epigenetic changes in the mutants, RDR2 is most likely a nuclear protein that is associated with chromatin, like rpd1 of *S. pombe* (Volpe *et al.*, 2002), where it converts ssRNAs derived from repetitive sequences into dsRNA. The picture that emerges is that DCL3, AGO4 and RDR2, possibly with additional proteins (SDE4 and HEN1), perform the same functions in the nucleus as the *dcr1, ago1, rdp1* proteins of *S. pombe* (Grewal & Moazed, 2003; Hall *et al.*, 2002; Volpe *et al.*, 2002) (Figure 2.2). The next step would be the identification of the plant equivalent of the RITS complex (Verdel *et al.*, 2004) that directs the methylation of histone H3 K9, and together with plant DNMTs (DRM1 and DRM2), the methylation of DNA (Cao *et al.*, 2003; Lippman & Martienssen, 2004).

2.4.5.3   *Biochemical properties of RDRs*
To fully understand the function of each RDR in relation to the cellular process it is involved in, it is essential to know about biochemical properties like template specificity, dsRNA-synthesising properties, primer dependence and/or independence, cellular localisation and associated proteins. The first *in vitro*

experiments were done with a tomato RDR, which synthesises short RNAs on
ssRNA templates, with and without a primer (Schiebel *et al.*, 1993). A wheat
germ extract also contains RDR activity (Tang *et al.*, 2003). ssRNA templates
are transcribed into full-length complementary RNAs, which are initiated at
the very 3' end. This synthesis does not require a primer, although the extract is
able to extend a primer base-paired to the template. This primer-extension
activity was not found in lysates from *Drosophila* embryos, contrary to earlier
reports (Lipardi *et al.*, 2001) but in line with the absence of an obvious RDR
gene in the *Drosophila* genome. Which RDRs in the wheat germ extract are
active is yet unknown. An interesting observation is that the class of RNAs
produced in the extract depends on the concentration of the ssRNA template.
At high concentrations, RNAs ~24 nt long are *de novo* synthesised; RNAs
~21 nt long are not produced under these conditions (Tang *et al.*, 2003). When
dsRNA is incubated in the extract the fraction of 21-nt RNAs is much higher.
The authors concluded therefore that the production of dsRNA by an RDR
may be coupled to production of the 24-nt RNAs. The long class of siRNAs
are genuine cleavage products of a DCL, which appears to be distinct from the
DCL that cleaves a dsRNA produced in another way. It has been proposed that
the RDR converting ssRNA into dsRNA is physically linked to a DCL that
cleaves dsRNA into ~24-nt RNAs (Tang *et al.*, 2003). The DCL mainly
generating the short class and 'directly' selecting dsRNA may not be associated with an RDR.

Purified QDE-1, the RDR of the fungus *Neurospora crassa*, performs two
different reactions on ssRNA templates *in vitro*: it is able to synthesise full-
length complementary RNAs and small RNAs of 9–21 nt scattered along the
entire template (Makeyev & Bamford, 2002). QDE-1 supports both *de novo*
and primer-dependent initiation. These properties resemble those of the wheat
germ RDR(s). It is not known which mode of synthesis is the predominant one
in vivo, but it is conceivable that they all occur side by side, perhaps depending
on the types of RNAs the RDRs encounter and the availability of free primers
that can help initiate dsRNA synthesis.

2.4.5.4   *RDR activity: amplification and transitive RNAi*
RDR activity results in synthesis of new dsRNAs and siRNAs, and is therefore
responsible for amplification of the silencing 'signal'. A limited number of
aberrant transcripts and siRNAs could be sufficient to trigger silencing and due
to the presence of ssRNA templates and primer-dependent RNA synthesis,
silencing can be maintained and even further enhanced. As a viral defence
strategy, this amplification may be required to keep pace with the replication
of viruses during an infection. Amplification could also be responsible for
the systemic properties of RNAi in plants and in nematodes. The sequence-
specific signal, most likely a short siRNA species (Himber *et al.*, 2003),
moves to neighbouring cells through plasmodesmata and through the vascular

system. In *Arabidopsis*, movement over a short distance does not require RDR6 and SDE3, a helicase, whereas long-distance movement does (Himber *et al.*, 2003). This suggests a kind of relay amplification of the signal, ensuring the continuous production of 'new' silencing-inducing siRNAs. In fact, there are quite a number of endogenous small RNAs that are transported through the phloem to other parts of the plant (Yoo *et al.*, 2004). Whether they all trigger amplification and the production of secondary small RNAs is unknown. For a more detailed discussion of systemic silencing and the general movement of RNAs in plants, other reviews and research articles can be referred to (Baulcombe, 2004; Hamilton *et al.*, 2002; Himber *et al.*, 2003; Jorgensen, 2002; Lucas & Lee, 2004; Lucas *et al.*, 2001; Yoo *et al.*, 2004).

Spreading of silencing signals along the length of an mRNA occurs in plants and *C. elegans* and is termed transitive RNAi (Sijen *et al.*, 2001a). Thus, if an mRNA is targeted for degradation by a short double-stranded hpRNA expressed from a transgene, siRNAs, called secondary siRNAs, are produced from RNA sequences upstream of the target. In plants, they can also be generated from the region 3′ of the target (García-Pérez *et al.*, 2004), although not with every mRNA target. The secondary siRNAs are functional in that they trigger the degradation of mRNAs that do not contain the primary target sequence (García-Pérez *et al.*, 2004; Sijen *et al.*, 2001a; Vaistij *et al.*, 2002; Van Houdt *et al.*, 2003). Because of this effect the phenomenon was called transitive RNAi. It requires an RDR and in *C. elegans* it is primer-dependent (Sijen *et al.*, 2001a). 5′ spreading can be explained by primer-initiated dsRNA synthesis of an mRNA by an RDR, but 3′ spreading, as observed in plants, must occur differently. One possibility is that RDR copies the 3′ RNA cleavage product of RISC by initiating RNA synthesis at the 3′ end, as has been observed in wheat germ extracts (Tang *et al.*, 2003). The dsRNA is then cleaved by a DCL to produce the siRNAs. However, it remains to be seen how general 3′ transitive RNAi is because it has not been observed with all RNAi-targeted mRNAs and VIGS-mediated silencing (Vaistij *et al.*, 2002).

Although 5′ spreading has even been observed with a single siRNA (Klahre *et al.*, 2002), spreading has not been found with RNAs targeted and cleaved by miRNA–RISCs (McConnell *et al.*, 2001), even when the miRNA was fully complementary to the mRNA (Llave *et al.*, 2002). A potential risk of transitive RNAi is that secondary siRNAs could result in degradation of non-target mRNAs. This may occur by knockdown expression of individual members of a gene family. Even when a 3′ gene-'specific' fragment is used in the RNAi construct, secondary siRNAs derived from more conserved sequences of the mRNA upstream of targeted region may guide RISC to related RNAs. This is further enforced by the fact that a few mismatches in the siRNA–mRNA hybrid are allowed for cleavage. Whether this is indeed a serious problem in genomics and practical applications of RNAi remains to be seen, because several genes have been silenced in a specific manner (Waterhouse & Helliwell, 2003).

Acknowledgements

I thank my colleagues in the laboratory for their input and discussions, and Peter Meyer for reading the manuscript. Research has been supported by grants from The Netherlands Organisation for the advancement of Research (NWO-STW and NWO-ALW) and from the European Union.

References

Allen, E., Xie, Z., Gustafson, A. M., Sung, G. H., Spatafora, J. W. & Carrington, J. C. (2004) Evolution of microRNA genes by inverted duplication of target gene sequences in *Arabidopsis thaliana*. *Nature Genetics*, **36**, 1282–90.

Anandalakshmi, R., Pruss, G. J., Ge, X., Marathe, R., Mallory, A. C., Smith, T. H. & Vance, V. B. (1998) A viral suppressor of gene silencing in plants. *Proceedings of the National Academy of Sciences of the United States of America*, **95**, 13079–84.

Anantharaman, V., Koonin, E. V. & Aravind, L. (2002) SPOUT: a class of methyltransferases that includes spoU and trmD RNA methylase superfamilies, and novel superfamilies of predicted prokaryotic RNA methylases. *Journal of Molecular Microbiology and Biotechnology*, **4**, 71–5.

Angell, S. M. & Baulcombe, D. C. (1997) Consistent gene silencing in transgenic plants expressing a replicating potato virus X RNA. *EMBO Journal*, **16**, 3675–84.

Aukerman, M. J. & Sakai, H. (2003) Regulation of flowering time and floral organ identity by a microRNA and its APETALA2-like target genes. *Plant Cell*, **15**, 2730–41.

Bartel, D. P. (2004) MicroRNAs: genomics, biogenesis, mechanism, and function. *Cell*, **116**, 281–97.

Baulcombe, D. C. (1999) Fast forward genetics based on virus-induced gene silencing. *Current Opinion in Plant Biology*, **2**, 109–13.

Baulcombe, D.C. (2004) RNA silencing in plants. *Nature*, **431**, 356–63.

Beclin, C., Boutet, S., Waterhouse, P. & Vaucheret, H. (2002) A branched pathway for transgene-induced RNA silencing in plants. *Current Biology*, **12**, 684–8.

Berns, K., Hijmans, E. M., Mullenders, J., Brummelkamp, T. R., Velds, A., Heimerikx, M., Kerkhoven, R. M., Madiredjo, M., Nijkamp, W., Weigelt, B., Agami, R., Ge, W., Cavet, G., Linsley, P. S., Beijersbergen, R. L. & Bernards, R. (2004) A large-scale RNAi screen in human cells identifies new components of the p53 pathway. *Nature*, **428**, 431–7.

Bernstein, E., Caudy, A. A., Hammond, S. M. & Hannon, G. J. (2001) Role for a bidentate ribonuclease in the initiation step of RNA interference. *Nature*, **409**, 363–6.

Billy, E., Brondani, V., Zhang, H., Muller, U. & Filipowicz, W. (2001) Specific interference with gene expression induced by long, double-stranded RNA in mouse embryonal teratocarcinoma cell lines. *Proceedings of the National Academy of Sciences of the United States of America*, **98**, 14428–33.

Blumenthal, T. & Carmichael, G. G. (1979) RNA replication: function and structure of Qbeta-replicase. *Annual Review of Biochemistry*, **48**, 525–48.

Bohmert, K., Camus, I., Bellini, C., Bouchez, D., Caboche, M. & Benning, C. (1998) AGO1 defines a novel locus of *Arabidopsis* controlling leaf development. *EMBO Journal*, **17**, 170–80.

Boutet, S., Vazquez, F., Liu, J., Beclin, C., Fagard, M., Gratias, A., Morel, J. B., Crete, P., Chen, X. & Vaucheret, H. (2003) *Arabidopsis* HEN1: a genetic link between endogenous miRNA controlling development and siRNA controlling transgene silencing and virus resistance. *Current Biology*, **13**, 843–8.

Brigneti, G., Voinnet, O., Li, W. X., Ji, L. H., Ding, S. W. & Baulcombe, D. C. (1998) Viral pathogenicity determinants are suppressors of transgene silencing in *Nicotiana benthamiana*. *EMBO Journal*, **17**, 6739–46.

Brigneti, G., Martin-Hernandez, A. M., Jin, H., Chen, J., Baulcombe, D. C., Baker, B. & Jones, J. D. (2004) Virus-induced gene silencing in Solanum species. *Plant Journal*, **39**, 264–72.

Cao, X., Aufsatz, W., Zilberman, D., Mette, M. F., Huang, M. S., Matzke, M. & Jacobsen, S. E. (2003) Role of the DRM and CMT3 methyltransferases in RNA-directed DNA methylation. *Current Biology*, **13**, 2212–17.

Caplen, N. J., Parrish, S., Imani, F., Fire, A. & Morgan, R. A. (2001) Specific inhibition of gene expression by small double-stranded RNAs in invertebrate and vertebrate systems. *Proceedings of the National Academy of Sciences of the United States of America*, **98**, 9742–7.

Carrington, J. C. & Ambros, V. (2003) Role of microRNAs in plant and animal development. *Science*, **301**, 336–8.

Catalanotto, C., Azzalin, G., Macino, G. & Cogoni, C. (2000) Gene silencing in worms and fungi. *Nature*, **404**, 245.

Catalanotto, C., Pallotta, M., ReFalo, P., Sachs, M. S., Vayssie, L., Macino, G. & Cogoni, C. (2004) Redundancy of the two dicer genes in transgene-induced posttranscriptional gene silencing in *Neurospora crassa*. *Molecular Cell Biology*, **24**, 2536–45.

Cerutti, L., Mian, N. & Bateman, A. (2000) Domains in gene silencing and cell differentiation proteins: the novel PAZ domain and redefinition of the piwi domain. *Trends in Biochemical Sciences*, **25**, 481–2.

Chan, S. W., Zilberman, D., Xie, Z., Johansen, L. K., Carrington, J. C. & Jacobsen, S. E. (2004) RNA silencing genes control *de novo* DNA methylation. *Science*, **303**, 1336.

Chen, X. (2004) A microRNA as a translational repressor of APETALA2 in *Arabidopsis* flower development. *Science*, **303**, 2022–5.

Chuang, C. F. & Meyerowitz, E. M. (2000) Specific and heritable genetic interference by double-stranded RNA in *Arabidopsis thaliana*. *Proceedings of the National Academy of Sciences of the United States of America*, **97**, 4985–90.

Cogoni, C. & Macino, G. (1997) Isolation of quelling-defective (qde) mutants impaired in posttranscriptional transgene-induced gene silencing in *Neurospora crassa*. *Proceedings of the National Academy of Sciences of the United States of America*, **94**, 10233–8.

Cogoni, C. & Macino, G. (1999) Gene silencing in *Neurospora crassa* requires a protein homologous to RNA-dependent RNA polymerase. *Nature*, **399**, 166–9.

Constantin, G. D., Krath, B. N., Macfarlane, S. A., Nicolaisen, M., Elisabeth Johansen, I. & Lund, O. S. (2004) Virus-induced gene silencing as a tool for functional genomics in a legume species. *Plant Journal*, **40**, 622–31.

Cox, D. N., Chao, A., Baker, J., Chang, L., Qiao, D. & Lin, H. (1998) A novel class of evolutionarily conserved genes defined by piwi are essential for stem cell self-renewal. *Genes & Development*, **12**, 3715–27.

Cox, D. N., Chao, A. & Lin, H. (2000) piwi encodes a nucleoplasmic factor whose activity modulates the number and division rate of germline stem cells. *Development*, **127**, 503–14.

Dalmay, T., Hamilton, A., Mueller, E. & Baulcombe, D. C. (2000a) Potato virus X amplicons in *Arabidopsis* mediate genetic and epigenetic gene silencing. *Plant Cell*, **12**, 369–79.

Dalmay, T., Hamilton, A., Rudd, S., Angell, S. & Baulcombe, D. C. (2000b) An RNA-dependent RNA polymerase gene in *Arabidopsis* is required for posttranscriptional gene silencing mediated by a transgene but not by a virus. *Cell*, **101**, 543–53.

Dalmay, T., Horsefield, R., Braunstein, T. H. & Baulcombe, D. C. (2001) SDE3 encodes an RNA helicase required for post-transcriptional gene silencing in *Arabidopsis*. *EMBO Journal*, **20**, 2069–78.

de Carvalho, F., Gheysen, G., Kushnir, S., Van Montagu, M., Inze, D. & Castresana, C. (1992) Suppression of beta-1,3-glucanase transgene expression in homozygous plants. *EMBO Journal*, **11**, 2595–602.

Dinesh-Kumar, S. P., Anandalakshmi, R., Marathe, R., Schiff, M. & Liu, Y. (2003) Virus-induced gene silencing. *Methods in Molecular Biology*, **236**, 287–94.

Djikeng, A., Shi, H., Tschudi, C., Shen, S. & Ullu, E. (2003) An siRNA ribonucleoprotein is found associated with polyribosomes in *Trypanosoma brucei*. *RNA*, **9**, 802–808.

Ekengren, S. K., Liu, Y., Schiff, M., Dinesh-Kumar, S. P. & Martin, G. B. (2003) Two MAPK cascades, NPR1, and TGA transcription factors play a role in Pto-mediated disease resistance in tomato. *Plant Journal*, **36**, 905–17.

Elbashir, S. M., Harborth, J., Lendeckel, W., Yalcin, A., Weber, K. & Tuschl, T. (2001) Duplexes of 21-nucleotide RNAs mediate RNA interference in cultured mammalian cells. *Nature*, **411**, 494–8.

Fagard, M., Boutet, S., Morel, J. B., Bellini, C. & Vaucheret, H. (2000) AGO1, QDE-2, and RDE-1 are related proteins required for post-transcriptional gene silencing in plants, quelling in fungi, and RNA interference in animals. *Proceedings of the National Academy of Sciences of the United States of America*, **97**, 11650–54.

Finnegan, E. J., Margis, R. & Waterhouse, P. M. (2003) Posttranscriptional gene silencing is not compromised in the *Arabidopsis* CARPEL FACTORY (DICER-LIKE1) mutant, a homolog of Dicer-1 from Drosophila. *Current Biology*, **13**, 236–40.

Fire, A., Xu, S. Q., Montgomery, M. K., Kostas, S. A., Driver, S. E. & Mello, C. C. (1998) Potent and specific genetic interference by double-stranded RNA in *Caenorhabditis elegans*. *Nature*, **391**, 806–11.

García-Pérez, R. D., Houdt, H. V. & Depicker, A. (2004) Spreading of post-transcriptional gene silencing along the target gene promotes systemic silencing. *Plant Journal*, **38**, 594–602.

Gazzani, S., Lawrenson, T., Woodward, C., Headon, D. & Sablowski, R. (2004) A link between mRNA turnover and RNA interference in *Arabidopsis*. *Science*, **306**, 1046–8.

Goldbach, R., Bucher, E. & Prins, M. (2003) Resistance mechanisms to plant viruses: an overview. *Virus Research*, **92**, 207–12.

Golden, T. A., Schauer, S. E., Lang, J. D., Pien, S., Mushegian, A. R., Grossniklaus, U., Meinke, D. W. & Ray, A. (2002) SHORT INTEGUMENTS1/SUSPENSOR1/CARPEL FACTORY, a Dicer homolog, is a maternal effect gene required for embryo development in *Arabidopsis*. *Plant Physiology*, **130**, 808–22.

Gossele, V., Fache, I., Meulewaeter, F., Cornelissen, M. & Metzlaff, M. (2002) SVISS – a novel transient gene silencing system for gene function discovery and validation in tobacco plants. *Plant Journal*, **32**, 859–66.

Grewal, S. I. & Moazed, D. (2003) Heterochromatin and epigenetic control of gene expression. *Science*, **301**, 798–802.

Grewal, S. I. & Rice, J. C. (2004) Regulation of heterochromatin by histone methylation and small RNAs. *Current Opinion in Cell Biology*, **16**, 230–38.

Guo, H. S., Fei, J. F., Xie, Q. & Chua, N. H. (2003) A chemical-regulated inducible RNAi system in plants. *Plant Journal*, **34**, 383–92.

Haley, B., Tang, G. & Zamore, P. D. (2003) *In vitro* analysis of RNA interference in *Drosophila melanogaster*. *Methods*, **30**, 330–36.

Hall, I. M., Shankaranarayana, G. D., Noma, K. -I., Ayoub, N., Cohen, A. & Grewal, S. I. S. (2002) Establishment and maintenance of a heterochromatin domain. *Science*, **297**, 2232–7.

Hamilton, A. J. & Baulcombe, D. C. (1999) A species of small antisense RNA in posttranscriptional gene silencing in plants. *Science*, **286**, 950–52.

Hamilton, A. J., Voinnet, O., Chappell, L. & Baulcombe, D. (2002) Two classes of short interfering RNA in RNA silencing. *EMBO Journal*, **21**, 4671–9.

Hammond, S. M., Bernstein, E., Beach, D. & Hannon, G. J. (2000) An RNA-directed nuclease mediates post-transcriptional gene silencing in Drosophila cells. *Nature*, **404**, 293–6.

Hammond, S. M., Boettcher, S., Caudy, A. A., Kobayashi, R. & Hannon, G. J. (2001) Argonaute2, a link between genetic and biochemical analyses of RNAi. *Science*, **293**, 1146–50.

Han, M. H., Goud, S., Song, L. & Fedoroff, N. (2004) The Arabidopsis double-stranded RNA-binding protein HYL1 plays a role in microRNA-mediated gene regulation. *Proceedings of the National Academy of Sciences of the United States of America*, **101**, 1093–8.

Helliwell, C. & Waterhouse, P. (2003) Constructs and methods for high-throughput gene silencing in plants. *Methods*, **30**, 289–95.

Himber, C., Dunoyer, P., Moissiard, G., Ritzenthaler, C. & Voinnet, O. (2003) Transitivity-dependent and -independent cell-to-cell movement of RNA silencing. *EMBO Journal*, **22**, 4523–33.

Hunter, C., Sun, H. & Poethig, R. S. (2003) The *Arabidopsis* heterochronic gene ZIPPY is an ARGONAUTE family member. *Current Biology*, **13**, 1734–9.

Iyer, L. M., Koonin, E. V. & Aravind, L. (2003) Evolutionary connection between the catalytic subunits of DNA-dependent RNA polymerases and eukaryotic RNA-dependent RNA polymerases and the origin of RNA polymerases. *BMC Structural Biology*, **3**, 1.

Jacobsen, S., Running, M. & Meyerowitz, E. (1999) Disruption of an RNA helicase/RNAse III gene in *Arabidopsis* causes unregulated cell division in floral meristems. *Development*, **126**, 5231–43.

Jorgensen, R. A. (2002) RNA traffics information systemically in plants. *PNAS*, **99**, 11561–63.

Joyce, G. F. (2002) The antiquity of RNA-based evolution. *Nature*, **418**, 214–21.

Kasschau, K. D. & Carrington, J. C. (1998) A counterdefensive strategy of plant viruses: suppression of posttranscriptional gene silencing. *Cell*, **95**, 461–70.

Ketting, R. F., Fischer, S. E., Bernstein, E., Sijen, T., Hannon, G. J. & Plasterk, R. H. (2001) Dicer functions in RNA interference and in synthesis of small RNA involved in developmental timing in *C. elegans*. *Genes & Development*, **15**, 2654–9.

Kidner, C. A. & Martienssen, R. A. (2004) Spatially restricted microRNA directs leaf polarity through ARGONAUTE1. *Nature*, **428**, 81–4.

Klahre, U., Crete, P., Leuenberger, S. A., Iglesias, V. A. & Meins, F., Jr (2002) High molecular weight RNAs and small interfering RNAs induce systemic posttranscriptional gene silencing in plants. *Proceedings of the National Academy of Sciences of the United States of America*, **99**, 11981–6.

Kunz, C., Schöb, H., Stam, M., Kooter, J. M. & Meins, F. (1996) Developmentally regulated silencing and reactivation of tobacco chitinase genes. *Plant Journal*, **10**, 437–50.

Lakatos, L., Szittya, G., Silhavy, D. & Burgyan, J. (2004) Molecular mechanism of RNA silencing suppression mediated by p19 protein of tombusviruses. *EMBO Journal*, **23**, 876–84.

Laufs, P., Peaucelle, A., Morin, H. & Traas, J. (2004) MicroRNA regulation of the CUC genes is required for boundary size control in *Arabidopsis* meristems. *Development*, **131**, 4311–22.

Lee, Y. S., Nakahara, K., Pham, J. W., Kim, K., He, Z., Sontheimer, E. J. & Carthew, R. W. (2004) Distinct roles for Drosophila Dicer-1 and Dicer-2 in the siRNA/miRNA silencing pathways. *Cell*, **117**, 69–81.

Libonati, M. & Sorrentino, S. (1992) Revisiting the action of bovine ribonuclease-A and pancreatic-type ribonucleases on double-stranded RNA. *Molecular and Cellular Biochemistry*, **117**, 139–51.

Limpens, E., Ramos, J., Franken, C., Raz, V., Compaan, B., Franssen, H., Bisseling, T. & Geurts, R. (2004) RNA interference in *Agrobacterium rhizogenes*-transformed roots of *Arabidopsis* and *Medicago truncatula*. *Journal of Experimental Botany*, **55**, 983–92.

Lingel, A., Simon, B., Izaurralde, E. & Sattler, M. (2004) Nucleic acid 3′-end recognition by the Argonaute2 PAZ domain. *Nature Structural and Molecular Biology*, **11**, 576–7.

Lipardi, C., Wei, Q. & Paterson, B. M. (2001) RNAi as random degradative PCR: siRNA primers convert mRNA into dsRNAs that are degraded to generate new siRNAs. *Cell*, **107**, 297–307.

Lippman, Z. & Martienssen, R. (2004) The role of RNA interference in heterochromatic silencing. *Nature*, **431**, 364–70.

Lippman, Z., May, B., Yordan, C., Singer, T. & Martienssen, R. (2003) Distinct mechanisms determine transposon inheritance and methylation via small interfering RNA and histone modification. *PLoS Biology*, **1**, E67.

Lippman, Z., Gendrel, A. V., Black, M., Vaughn, M. W., Dedhia, N., McCombie, W. R., Lavine, K., Mittal, V., May, B., Kasschau, K. D., Carrington, J. C., Doerge, R. W., Colot, V. & Martienssen, R. (2004) Role of transposable elements in heterochromatin and epigenetic control. *Nature*, **430**, 471–6.

Liu, J., Carmell, M. A., Rivas, F. V., Marsden, C. G., Thomson, J. M., Song, J. J., Hammond, S. M., Joshua-Tor, L. & Hannon, G. J. (2004a) Argonaute2 is the catalytic engine of mammalian RNAi. *Science*, **305**, 1437–41.

Liu, Q., Rand, T. A., Kalidas, S., Du, F., Kim, H. E., Smith, D. P. & Wang, X. (2003) R2D2, a bridge between the initiation and effector steps of the Drosophila RNAi pathway. *Science*, **301**, 1921–5.

Liu, Y., Schiff, M., Marathe, R. & Dinesh-Kumar, S. P. (2002) Tobacco Rar1, EDS1 and NPR1/NIM1-like genes are required for N-mediated resistance to tobacco mosaic virus. *Plant Journal*, **30**, 415–29.

Liu, Y., Nakayama, N., Schiff, M., Litt, A., Irish, V. F. & Dinesh-Kumar, S. P. (2004b) Virus induced gene silencing of a DEFICIENS ortholog in *Nicotiana benthamiana*. *Plant Molecular Biology*, **54**, 701–11.

Liu, Y., Schiff, M. & Dinesh-Kumar, S. P. (2004c) Involvement of MEK1 MAPKK, NTF6 MAPK, WRKY/MYB transcription factors, COI1 and CTR1 in N-mediated resistance to tobacco mosaic virus. *Plant Journal*, **38**, 800–809.

Llave, C., Kasschau, K. D. & Carrington, J. C. (2000) Virus-encoded suppressor of posttranscriptional gene silencing targets a maintenance step in the silencing pathway. *Proceedings of the National Academy of Sciences of the United States of America*, **97**, 13401–406.

Llave, C., Xie, Z., Kasschau, K. D. & Carrington, J. C. (2002) Cleavage of Scarecrow-like mRNA targets directed by a class of *Arabidopsis* miRNA. *Science*, **297**, 2053–6.

Lu, R., Martin-Hernandez, A. M., Peart, J. R., Malcuit, I. & Baulcombe, D. C. (2003) Virus-induced gene silencing in plants. *Methods*, **30**, 296–303.

Lucas, W. J. & Lee, J. Y. (2004) Plasmodesmata as a supracellular control network in plants. *Nature Reviews. Molecular Cell Biology*, **5**, 712–26.

Lucas, W. J., Yoo, B. C. & Kragler, F. (2001) RNA as a long-distance information macromolecule in plants. *Nature Reviews. Molecular Cell Biology*, **2**, 849–57.

Lynn, K., Fernandez, A., Aida, M., Sedbrook, J., Tasaka, M., Masson, P. & Barton, M. K. (1999) The PINHEAD/ZWILLE gene acts pleiotropically in *Arabidopsis* development and has overlapping functions with the ARGONAUTE1 gene. *Development*, **126**, 469–81.

Ma, J. B., Ye, K. & Patel, D. J. (2004) Structural basis for overhang-specific small interfering RNA recognition by the PAZ domain. *Nature*, **429**, 318–22.

Makeyev, E. V. & Bamford, D. H. (2002) Cellular RNA-dependent RNA polymerase involved in posttranscriptional gene silencing has two distinct activity modes. *Molecular Cell*, **10**, 1417–27.

Mallory, A. C., Ely, L., Smith, T. H., Marathe, R., Anandalakshmi, R., Fagard, M., Vaucheret, H., Pruss, G., Bowman, L. & Vance, V. B. (2001) HC-Pro suppression of transgene silencing eliminates the small RNAs but not transgene methylation or the mobile signal. *Plant Cell*, **13**, 571–83.

Mallory, A. C., Dugas, D. V., Bartel, D. P. & Bartel, B. (2004) MicroRNA regulation of NAC-domain targets is required for proper formation and separation of adjacent embryonic, vegetative, and floral organs. *Current Biology*, **14**, 1035–46.

Martens, H., Novotny, J., Oberstrass, J., Steck, T. L., Postlethwait, P. & Nellen, W. (2002) RNAi in *Dictyostelium*: the role of RNA-directed RNA polymerases and double-stranded RNase. *Molecular Biology of the Cell*, **13**, 445–53.

Martienssen, R. A. (2003) Maintenance of heterochromatin by RNA interference of tandem repeats. *Nature Genetics*, **35**, 213–14.

Martinez, J. & Tuschl, T. (2004) RISC is a 5′ phosphomonoester-producing RNA endonuclease. *Genes & Development*, **18**, 975–80.

Martinez, J., Patkaniowska, A., Urlaub, H., Luhrmann, R. & Tuschl, T. (2002) Single-stranded antisense siRNAs guide target RNA cleavage in RNAi. *Cell*, **110**, 563–74.

Masclaux, F., Charpenteau, M., Takahashi, T., Pont-Lezica, R. & Galaud, J. P. (2004) Gene silencing using a heat-inducible RNAi system in *Arabidopsis*. *Biochemical and Biophysical Research Communications*, **321**, 364–9.

Matzke, M. A. & Matzke, A. J. (2004) Planting the seeds of a new paradigm. *PLoS Biology*, **2**, E133.

Matzke, M., Aufsatz, W., Kanno, T., Daxinger, L., Papp, I., Mette, M. F. & Matzke, A. J. (2004) Genetic analysis of RNA-mediated transcriptional gene silencing. *Biochimica et Biophysica Acta*, **1677**, 129–41.

McConnell, J. R., Emery, J., Eshed, Y., Bao, N., Bowman, J. & Barton, M. K. (2001) Role of PHABULOSA and PHAVOLUTA in determining radial patterning in shoots. *Nature*, **411**, 709–13.

Meister, G. & Tuschl, T. (2004) Mechanisms of gene silencing by double-stranded RNA. *Nature*, **431**, 343–9.

Meister, G., Landthaler, M., Patkaniowska, A., Dorsett, Y., Teng, G. & Tuschl, T. (2004) Human Argonaute2 mediates RNA cleavage targeted by miRNAs and siRNAs. *Molecular Cell*, **15**, 185–97.

Mette, M. F., Van der Winden, J., Matzke, M. & Matzke, A. J. (2002) Short RNAs can identify new candidate transposable element families in *Arabidopsis*. *Plant Physiology*, **130**, 6–9.

Moissiard, G. & Voinnet, O. (2004) Viral suppressors of RNA silencing in plants. *Molecular Plant Pathology*, **5**, 71–82.

Morel, J. B., Godon, C., Mourrain, P., Beclin, C., Boutet, S., Feuerbach, F., Proux, F. & Vaucheret, H. (2002) Fertile hypomorphic ARGONAUTE (ago1) mutants impaired in post-transcriptional gene silencing and virus resistance. *Plant Cell*, **14**, 629–39.

Mourrain, P., Beclin, C., Elmayan, T., Feuerbach, F., Godon, C., Morel, J. B., Jouette, D., Lacombe, A. M., Nikic, S., Picault, N., Remoue, K., Sanial, M., Vo, T. A. & Vaucheret, H. (2000) Arabidopsis SGS2 and SGS3 genes are required for posttranscriptional gene silencing and natural virus resistance. *Cell*, **101**, 533–42.

Muangsan, N., Beclin, C., Vaucheret, H. & Robertson, D. (2004) Geminivirus VIGS of endogenous genes requires SGS2/SDE1 and SGS3 and defines a new branch in the genetic pathway for silencing in plants. *Plant Journal*, **38**, 1004–14.

Muskens, M. W. M., Vissers, A. P. A., Mol, J. N. M. & Kooter, J. M. (2000) Role of inverted DNA repeats in transcriptional and posttranscriptional gene silencing. *Plant Molecular Biology*, **43**, 243–60.

Napoli, C., Lemieux, C. & Jorgensen, R. (1990) Introduction of a chimeric chalcone synthase gene into petunia results in reversible co-suppression of homologous genes *in trans*. *Plant Cell*, **2**, 279–89.

Nicholson, A. W. (1999) Function, mechanism and regulation of bacterial ribonucleases. *FEMS Microbiological Reviews*, **23**, 371–90.

Palatnik, J. F., Allen, E., Wu, X., Schommer, C., Schwab, R., Carrington, J. C. & Weigel, D. (2003) Control of leaf morphogenesis by microRNAs. *Nature*, **425**, 257–63.

Palauqui, J. C., Elmayan, T., Pollien, J. M. & Vaucheret, H. (1997) Systemic acquired silencing: transgene-specific post-transcriptional silencing is transmitted by grafting from silenced stocks to non-silenced scions. *EMBO Journal*, **16**, 4738–45.

Papp, I., Mette, M. F., Aufsatz, W., Daxinger, L., Schauer, S. E., Ray, A., van der Winden, J., Matzke, M. & Matzke, A. J. (2003) Evidence for nuclear processing of plant micro RNA and short interfering RNA precursors. *Plant Physiology*, **132**, 1382–90.

Park, W., Li, J., Song, R., Messing, J. & Chen, X. (2002) CARPEL FACTORY, a Dicer homolog, and HEN1, a novel protein, act in microRNA metabolism in *Arabidopsis thaliana*. *Current Biology*, **12**, 1484–95.

Pham, J. W., Pellino, J. L., Lee, Y. S., Carthew, R. W. & Sontheimer, E. J. (2004) A Dicer-2-dependent 80s complex cleaves targeted mRNAs during RNAi in Drosophila. *Cell*, **117**, 83–94.

Que, Q. D., Wang, H. Y., English, J. J. & Jorgensen, R. A. (1997) The frequency and degree of cosuppression by sense chalcone synthase transgenes are dependent on transgene promoter strength and are reduced by premature nonsense codons in the transgene coding sequence. *Plant Cell*, **9**, 1357–68.

Ratcliff, F., Martin-Hernandez, A. & Baulcombe, D. (2001) Technical advance: tobacco rattle virus as a vector for analysis of gene function by silencing. *Plant Journal*, **25**, 237–45.

Reinhart, B. J., Weinstein, E. G., Rhoades, M. W., Bartel, B. & Bartel, D. P. (2002) MicroRNAs in plants. *Genes & Development*, **16**, 1616–26.

Rhoades, M. W., Reinhart, B. J., Lim, L. P., Burge, C. B., Bartel, B. & Bartel, D. P. (2002) Prediction of plant microRNA targets. *Cell*, **110**, 513–20.

Robertson, D. (2004) VIGS vectors for gene silencing: many targets, many tools. *Annual Review of Plant Biology*, **55**, 495–519.

Robertson, H. D., Webster, R. E. & Zinder, N. D. (1967) A nuclease specific for double-stranded RNA. *Virology*, **12**, 718–19.

Robertson, H. D., Webster, R. E. & Zinder, N. D. (1968) Purification and properties of ribonuclease III from *Escherichia coli*. *Journal of Biological Chemistry*, **243**, 82–91.

Sanders, M., Maddelein, W., Depicker, A., Van Montagu, M., Cornelissen, M. & Jacobs, J. (2002) An active role for endogenous β-1,3-glucanase genes in transgene-mediated co-suppression in tobacco. *EMBO Journal*, **21**, 5824–32.

Schauer, S. E., Jacobsen, S. E., Meinke, D. W. & Ray, A. (2002) DICER-LIKE1: blind men and elephants in *Arabidopsis* development. *Trends in Plant Science*, **7**, 487–91.

Schiebel, W., Haas, B., Marinković, S., Klanner, A. & Sänger, H. L. (1993) RNA-directed RNA polymerase from tomato leaves. II Catalytic *in vitro* properties. *Journal of Biological Chemistry*, **268**, 11858–67.

Schiebel, W., Pelissier, T., Riedel, L., Thalmeir, S., Schiebel, R., Kempe, D., Lottspeich, F., Sänger, H. L. & Wassenegger, M. (1998) Isolation of an RNA-directed RNA polymerase-specific cDNA clone from tomato. *Plant Cell*, **10**, 1–16.

Schramke, V. & Allshire, R. (2003) Hairpin RNAs and retrotransposon LTRs effect RNAi and chromatin-based gene silencing. *Science*, **301**, 1069–74.

Schwarz, D. S., Hutvagner, G., Haley, B. & Zamore, P. D. (2002) Evidence that siRNAs function as guides, not primers, in the Drosophila and human RNAi pathways. *Molecular Cell*, **10**, 537–48.

Sigova, A., Rhind, N. & Zamore, P. D. (2004) A single Argonaute protein mediates both transcriptional and posttranscriptional silencing in *Schizosaccharomyces pombe. Genes & Development*, **18**, 2359–67.

Sijen, T. & Plasterk, R. H. (2003) Transposon silencing in the *Caenorhabditis elegans* germ line by natural RNAi. *Nature*, **426**, 310–14.

Sijen, T., Fleenor, J., Simmer, F., Thijssen, K. L., Parrish, S., Timmons, L., Plasterk, R. H. & Fire, A. (2001a) On the role of RNA amplification in dsRNA-triggered gene silencing. *Cell*, **107**, 465–76.

Sijen, T., Vijn, I., Rebocho, A., Van Blokland, R., Roelofs, D., Mol, J. N. M. & Kooter, J. M. (2001b) Transcriptional and posttranscriptional silencing are mechanistically related. *Current Biology*, **11**, 436–40.

Silhavy, D., Molnar, A., Lucioli, A., Szittya, G., Hornyik, C., Tavazza, M. & Burgyan, J. (2002) A viral protein suppresses RNA silencing and binds silencing-generated, 21- to 25-nucleotide double-stranded RNAs. *EMBO Journal.*, **21**, 3070–80.

Smardon, A., Spoerke, J. M., Stacey, S. C., Klein, M. E., Mackin, N. & Maine, E. M. (2000) EGO-1 is related to RNA-directed RNA polymerase and functions in germ-line development and RNA interference in *C. elegans. Current Biology*, **10**, 169–78.

Smith, N. A., Singh, S. P., Wang, M. B., Stoutjesdijk, P. A., Graan, A. G. & Waterhouse, P. M. (2000) Total silencing by intron-spliced hairpin RNAs. *Nature*, **407**, 319–320.

Song, J. J., Smith, S. K., Hannon, G. J. & Joshua-Tor, L. (2004) Crystal structure of Argonaute and its implications for RISC slicer activity. *Science*, **305**, 1434–7.

Sontheimer, E. J. & Carthew, R. W. (2004) Molecular biology. Argonaute journeys into the heart of RISC. *Science*, **305**, 1409–410.

Souret, F. F., Kastenmayer, J. P. & Green, P. J. (2004) AtXRN4 degrades mRNA in *Arabidopsis* and its substrates include selected miRNA targets. *Molecular Cell*, **15**, 173–83.

Stam, M., De Bruin, R., Kenter, S., Van der Hoorn, R. A. L., Van Blokland, R., Mol, J. N. M. & Kooter, J. M. (1997a) Post-transcriptional silencing of chalcone synthase in *Petunia* by inverted transgene repeats. *Plant Journal*, **12**, 63–82.

Stam, M., Mol, J. N. M. & Kooter, J. M. (1997b) The silence of genes in transgenic plants. *Annual Review of Botany*, **79**, 3–12.

Stam, M., Viterbo, A., Mol, J. N. M. & Kooter, J. M. (1998) Position-dependent methylation and transcriptional silencing of transgenes in inverted T-DNA repeats: implications for posttranscriptional silencing of homologous host genes in plants. *Molecular and Cellular Biology*, **18**, 6165–77.

Stein, P., Svoboda, P., Anger, M. & Schultz, R. M. (2003) RNAi: mammalian oocytes do it without RNA-dependent RNA polymerase. *RNA*, **9**, 187–92.

Stoutjesdijk, P. A., Singh, S. P., Liu, Q., Hurlstone, C. J., Waterhouse, P. A. & Green, A. G. (2002) hpRNA-mediated targeting of the *Arabidopsis* FAD2 gene gives highly efficient and stable silencing. *Plant Physiology*, **129**, 1723–31.

Szittya, G., Silhavy, D., Molnar, A., Havelda, Z., Lovas, A., Lakatos, L., Banfalvi, Z. & Burgyan, J. (2003) Low temperature inhibits RNA silencing-mediated defence by the control of siRNA generation. *EMBO J.*, **22**, 633–40.

Tabara, H., Sarkissian, M., Kelly, W. G., Fleenor, J., Grishok, A., Timmons, L., Fire, A. & Mello, C. C. (1999) The rde-1 gene, RNA interference, and transposon silencing in *C. elegans. Cell*, **99**, 123–32.

Tahbaz, N., Kolb, F. A., Zhang, H., Jaronczyk, K., Filipowicz, W. & Hobman, T. C. (2004) Characterization of the interactions between mammalian PAZ PIWI domain proteins and Dicer. *EMBO Reports*, **5**, 189–94.

Tang, G. & Zamore, P. D. (2004) Biochemical dissection of RNA silencing in plants. *Methods in Molecular Biology*, **257**, 223–44.

Tang, G., Reinhart, B. J., Bartel, D. P. & Zamore, P. D. (2003) A biochemical framework for RNA silencing in plants. *Genes & Development*, **17**, 49–63.

Tavernarakis, N., Wang, S. L., Dorovkov, M., Ryazanov, A. & Driscoll, M. (2000) Heritable and inducible genetic interference by double-stranded RNA encoded by transgenes. *Nature Genetics*, **24**, 180–83.

Tuschl, T., Zamore, P. D., Lehmann, R., Bartel, D. P. & Sharp, P. A. (1999) Targeted mRNA degradation by double-stranded RNA *in vitro*. *Genes & Development*, **13**, 3191–7.

Vaistij, F. E., Jones, L. & Baulcombe, D. C. (2002) Spreading of RNA targeting and DNA methylation in RNA silencing requires transcription of the target gene and a putative RNA-dependent RNA polymerase. *Plant Cell*, **14**, 857–67.

Van Blokland, R., Van der Geest, N., Mol, J. N. M. & Kooter, J. M. (1994) Transgene-mediated suppression of chalcone synthase expression in *Petunia hybrida* results from an increase in RNA turnover. *Plant Journal*, **6**, 861–77.

Van der Krol, A. R., Mur, L. A., Beld, M., Mol, J. N. M. & Stuitje, A. R. (1990) Flavonoid genes in petunia: addition of a limited number of gene copies may lead to a suppression of gene expression. *Plant Cell*, **2**, 291–9.

Van Houdt, H., Bleys, A. & Depicker, A. (2003) RNA target sequences promote spreading of RNA silencing. *Plant Physiol.*, **131**, 245–53.

Vanitharani, R., Chellappan, P. & Fauquet, C. M. (2003) Short interfering RNA-mediated interference of gene expression and viral DNA accumulation in cultured plant cells. *Proceedings of the National Academy of Sciences of the United States of America*, **100**, 9632–6.

Vaucheret, H., Nussaume, L., Palauqui, J. C., Quillere, I. & Elmayan, T. (1997) A transcriptionally active state is required for post-transcriptional silencing (cosuppression) of nitrate reductase host genes and transgenes. *Plant Cell*, **9**, 1495–504.

Vaucheret, H., Béclin, C., Elmayan, T., Feuerbach, F., Godon, C., Morel, J. -B., Mourrain, P., Palauqui, J. -C. & Vernhettes, S. (1998) Transgene-induced gene silencing in plants. *Plant Journal*, **16**, 651–9.

Vaucheret, H., Vazquez, F., Crete, P. & Bartel, D. P. (2004) The action of ARGONAUTE1 in the miRNA pathway and its regulation by the miRNA pathway are crucial for plant development. *Genes & Development*, **18**, 1187–97.

Vazquez, F., Gasciolli, V., Crete, P. & Vaucheret, H. (2004a) The nuclear dsRNA binding protein HYL1 is required for microRNA accumulation and plant development, but not posttranscriptional transgene silencing. *Current Biology*, **14**, 346–51.

Vazquez, F., Vaucheret, H., Rajagopalan, R., Lepers, C., Gasciolli, V., Mallory, A. C., Hilbert, J. L., Bartel, D. P. & Crete, P. (2004b) Endogenous *trans*-acting siRNAs regulate the accumulation of *Arabidopsis* mRNAs. *Molecular Cell*, **16**, 69–79.

Verdel, A., Jia, S., Gerber, S., Sugiyama, T., Gygi, S., Grewal, S. I. & Moazed, D. (2004) RNAi-mediated targeting of heterochromatin by the RITS complex. *Science*, **303**, 672–6.

Voinnet, O. & Baulcombe, D. C. (1997) Systemic signaling in gene silencing. *Nature*, **389**, 553.

Voinnet, O., Lederer, C. & Baulcombe, D. C. (2000) A viral movement protein prevents spread of the gene silencing signal in *Nicotiana benthamiana*. *Cell*, **103**, 157–67.

Volpe, T. A., Kidner, C., Hall, I. M., Teng, G., Grewal, S. I. S. & Martienssen, R. A. (2002) Regulation of heterochromatic silencing and histone H3 lysine-9 methylation by RNAi. *Science*, **297**, 1833–7.

Volpe, T., Schramke, V., Hamilton, G. L., White, S. A., Teng, G., Martienssen, R. A. & Allshire, R. C. (2003) RNA interference is required for normal centromere function in fission yeast. *Chromosome Research*, **11**, 137–46.

Waterhouse, P. M. & Helliwell, C. A. (2003) Exploring plant genomes by RNA-induced gene silencing. *Nature Reviews. Genetics*, **4**, 29–38.

Waterhouse, P. M., Graham, H. W. & Wang, M. B. (1998) Virus resistance and gene silencing in plants can be induced by simultaneous expression of sense and antisense RNA, *Proceedings of the National Academy of Sciences of the United States of America*, **95**, 13959–64.

Wesley, S. V., Helliwell, C. A., Smith, N. A., Wang, M., Rouse, D. T., Liu, Q., Gooding, P. S., Singh, S. P., Abbott, D., Stoutjesdijk, P. A., Robinson, S. P., Gleave, A. P., Green, A. G. & Waterhouse, P. M. (2001) Construct design for efficient, effective and high-throughput gene silencing in plants. *Plant Journal*, **27**, 581–90.

Wesley, S. V., Helliwell, C., Wang, M. B. & Waterhouse, P. (2004) Posttranscriptional gene silencing in plants. *Methods in Molecular Biology*, **265**, 117–29.

Xie, Z., Fan, B., Chen, C. & Chen, Z. (2001) An important role of an inducible RNA-dependent RNA polymerase in plant antiviral defense. *Proceedings of the National Academy of Sciences of the United States of America*, **98**, 6516–21.

Xie, Z., Kasschau, K. D. & Carrington, J. C. (2003) Negative feedback regulation of Dicer-Like1 in *Arabidopsis* by microRNA-guided mRNA degradation. *Current Biology*, **13**, 784–9.

Xie, Z., Johansen, L. K., Gustafson, A. M., Kasschau, K. D., Lellis, A. D., Zilberman, D., Jacobsen, S. E. & Carrington, J. C. (2004) Genetic and functional diversification of small RNA pathways in plants. *PLoS Biology*, **2**, E104.

Yoo, B. C., Kragler, F., Varkonyi-Gasic, E., Haywood, V., Archer-Evans, S., Lee, Y. M., Lough, T. J. & Lucas, W. J. (2004) A systemic small RNA signaling system in plants. *Plant Cell*, **16**, 1979–2000.

Yu, D., Fan, B., MacFarlane, S. A. & Z., C. (2003) Analysis of the involvement of an inducible *Arabidopsis* RNA-dependent RNA polymerase in antiviral defense. *Molecular Plant–Microbe Interaction*, **16**, 206–16.

Zhang, H., Kolb, F. A., Brondani, V., Billy, E. & Filipowicz, W. (2002) Human Dicer preferentially cleaves dsRNAs at their termini without a requirement for ATP. *EMBO Journal*, **21**, 5875–85.

Zilberman, D., Cao, X. & Jacobsen, S. E. (2003) ARGONAUTE4 control of locus-specific siRNA accumulation and DNA and histone methylation. *Science*, **299**, 716–19.

Zilberman, D., Cao, X., Johansen, L. K., Xie, Z., Carrington, J. C. & Jacobsen, S. E. (2004) Role of Arabidopsis ARGONAUTE4 in RNA-directed DNA methylation triggered by inverted repeats. *Current Biology*, **14**, 1214–20.

Zuo, J., Niu, Q. W. & Chua, N. H. (2000) Technical advance: an estrogen receptor-based transactivator XVE mediates highly inducible gene expression in transgenic plants. *Plant Journal*, **24**, 265–73.

Zuo, J., Niu, Q. W., Moller, S. G. & Chua, N. H. (2001) Chemical-regulated, site-specific DNA excision in transgenic plants. *Nature Biotechnology*, **19**, 157–61.

3   RNA-directed DNA methylation

Marjori Matzke, Tatsuo Kanno, Bruno Huettel, Estelle Jaligot, M. Florian Mette, David P. Kreil, Lucia Daxinger, Philipp Rovina, Werner Aufsatz and Antonius J. M. Matzke

3.1   Introduction

3.1.1   RNA interference

RNA interference (RNAi) has provided a new paradigm for understanding gene regulation in many eukaryotic organisms. Although plant scientists laid much of the groundwork for the discovery of RNAi (Matzke & Matzke, 2004), it was the identification of double-stranded RNA as the trigger for gene silencing in *Caenorhabditis elegans* by Fire *et al.* (1998) that provided the means to downregulate gene expression reproducibly in plants, animals and many fungi. The canonical RNAi pathway is now well established: double-stranded RNA is processed by a ribonuclease III–type enzyme called Dicer into short interfering RNAs (siRNAs) of around 21–24 nt in length. The siRNAs associate with the RNA-induced silencing complex (RISC) and guide cleavage of complementary mRNAs (Novina & Sharp, 2004). A second type of small RNA, the microRNA (miRNA), is also produced via Dicer cleavage of longer duplex RNAs. The miRNAs associate with an RISC-like complex and, depending on the degree of complementarity to the target mRNA, elicit either mRNA cleavage or translational repression. In addition to Dicer, other core proteins of the RNAi machinery include Argonaute, an RISC component that binds small RNAs and in some cases executes mRNA cleavage, and RNA-dependent RNA polymerase (RDR), which can synthesize double-stranded RNA from single-stranded RNA templates to initiate or amplify the RNAi reaction (Bartel, 2004; He & Hannon, 2004).

In addition to making substantial contributions to elucidating the classical RNAi pathway, plant scientists have been at the forefront of research on RNA-mediated epigenetic modifications. Epigenetics refers to mitotically and/or meiotically heritable changes in gene expression that do not involve a change in DNA sequence. Epigenetic modifications include DNA cytosine (C) methylation and various posttranslational modifications of histones, including acetylation and methylation. The connection between RNAi and epigenetic regulation is exemplified by the phenomenon of RNA-directed DNA methylation (RdDM), which is the subject of this chapter.

3.1.2   Discovery and characteristics of RNA-directed DNA methylation

Originally detected in viroid-infected tobacco plants, RdDM was the first RNA-guided epigenetic modification of the genome to be reported (Wassenegger *et al.*, 1994). Viroids are minute plant pathogens consisting exclusively of a non-protein-coding, circular rod-shaped RNA several hundred base pairs in length (Tabler & Tsagris, 2004). In the experiments that revealed RdDM, viroid cDNAs integrated as transgenes into tobacco chromosomes became methylated *de novo* during RNA–RNA replication of the cognate viroid. Further analysis demonstrated that Cs in all sequence contexts are modified and that methylation is largely confined to the region of RNA–DNA sequence homology (Pélissier *et al.*, 1999). DNA regions as short as 30 bp can be targeted for methylation in the RdDM pathway (Pélissier & Wassenegger, 2000).

The identification of RdDM occurred in the framework of homology-dependent gene-silencing phenomena, which were postulated at the time to be due variously to DNA–DNA, DNA–RNA or RNA–RNA interactions (Jorgensen, 1992; Matzke & Matzke, 1993). Indeed, one of the initial papers describing 'cosuppression', a process in which a resident gene is silenced in the presence of a homologous transgene, proposed that transgene RNA might interact with DNA of the resident gene to block transcription (van der Krol *et al.*, 1990). Even though cosuppression turned out to be a posttranscriptional gene-silencing phenomenon that is now considered the plant equivalent of RNAi (De Carvalho *et al.*, 1992; Matzke & Matzke, 2004; van Blokland *et al.*, 1994), this example illustrates that RNA–DNA interactions were considered as a possible trigger of silencing during the early stages of gene-silencing research. The viroid experiments supplied experimental documentation of this possibility and revealed the characteristic features of RdDM. Shortly thereafter, cosuppression (also known as post-transcriptional gene silencing (PTGS)) of a reporter transgene in tobacco was shown to be accompanied by *de novo* methylation of the corresponding genomic DNA sequences (Ingelbrecht *et al.*, 1994). Thus, a transgene subject to posttranscriptional regulation could also be targeted for DNA methylation, suggesting that cytoplasmic and nuclear events are induced by a common silencing mechanism.

A further connection between PTGS and DNA methylation was observed in transgenic pea plants infected with an RNA virus that replicates exclusively in the cytoplasm. Virus replication initiated PTGS of a transgene encoding the viral replicase gene, which was accompanied by *de novo* methylation of replicase transgene sequences integrated into nuclear DNA (Jones *et al.*, 1998). The pattern of methylation was consistent with RdDM, occurring at Cs in both symmetrical (CG and CNG, where N is A, T or C) and asymmetrical (CNN) sequence contexts within the region of RNA–DNA sequence homology. These findings suggested that a sequence-specific signal, most likely RNA, produced during virus replication and/or PTGS diffused from the cytoplasm into the nucleus to induce methylation of homologous DNA sequences. In an extension of these experiments, replicating potato virus X RNA vectors modified to contain sequences homologous to the 35S promoter were able to trigger methylation and transcriptional silencing of 35S promoter–driven nuclear transgenes (Jones *et al.*, 1999).

The experiments with viroids and viruses hinted that RdDM is initiated by a double-stranded RNA because these RNA pathogens replicate via a double-stranded RNA intermediate. An unequivocal demonstration that double-stranded RNA is required for RdDM came from studies on a non-pathogenic, transgenic system. These experiments grew from work that suggested a role for RNA in mediating methylation of transgene promoters (Mette *et al.*, 1999; Park *et al.*, 1996). To further investigate this possibility, RNAs that contained promoter sequences were tested for their ability to induce methylation and transcriptional silencing of unlinked homologous promoters in transgenic tobacco plants. Cre-*lox*-mediated recombination was used to convert a transcribed direct repeat of target promoter sequences into a transcribed inverted repeat *in planta*. Methylation and silencing of a homologous target promoter was observed only after generation of the inverted repeat and initiation of double-stranded RNA synthesis (Mette *et al.*, 2000). The double-stranded RNA that induced RdDM was shown to be processed to short RNAs 21–24 nt in length, similar to those involved in RNAi, reinforcing a link between the two silencing processes. Double-stranded RNAs containing promoter sequences have subsequently been used to silence and methylate different transgene promoters in *Arabidopsis thaliana* (Aufsatz *et al.*, 2002a, 2002b; Kanno *et al.*, 2004) and several endogenous promoters in petunia (Sijen *et al.*, 2001) and *Arabidopsis* (Melquist & Bender, 2003). Therefore, RNA-mediated transcriptional gene silencing (TGS) and promoter methylation appears to be a general process in plants.

3.2   RNAi-mediated pathways in the nucleus

Viewed as a plant-specific phenomenon for several years after its discovery, RdDM is now regarded as one of several RNAi-mediated pathways in the nucleus. In addition to RdDM, the other nuclear pathways include:

(1) RNAi-mediated heterochromatin formation that has been observed in fission yeast, plants, *Drosophila melanogaster* and vertebrates;
(2) elimination of intergenic DNA that has been packaged into heterochromatin via the RNAi pathway during nuclear differentiation in *Tetrahymena thermophila* and other ciliated protozoa;

(3) meiotic silencing of DNA that is unpaired during meiosis in *Neurospora crassa* and *C. elegans* (reviewed by Matzke & Birchler, 2005).

Here we focus on RdDM and on RNAi-mediated heterochromatin formation, particularly as it relates to DNA methylation.

3.2.1 RNAi-mediated heterochromatin formation

The involvement of the RNAi pathway in heterochromatin formation was first discovered in fission yeast (*Schizosaccharomyces pombe*). Mutants defective in the core RNAi proteins – Dicer, Argonaute and RDR – were shown to be unable to assemble functional heterochromatin at the centromeres (Volpe *et al.*, 2002, 2003). In wild-type cells, transient double-stranded RNA molecules originating from forward and reverse transcription of centromere outer repeats are processed by Dicer to siRNAs (Reinhart & Bartel, 2002). The siRNAs are thought to guide histone H3 lysine 9 (H3K9) methylation that is catalyzed by cryptic loci regulator 4 (Clr4), the *S. pombe* ortholog of the histone methyl-transferase Su(var)3–9 (from *Drosophila* suppressor of variegation 3–9). Swi6, the *S. pombe* ortholog of *Drosophila* heterochromatin protein 1 (HP1) can bind via its chromodomain to histone H3 methylated on K9, which promotes lateral spreading of heterochromatin from the siRNA-targeted nucleation site. The formation of heterochromatin leads to binding of cohesin, which fosters sister chromatid cohesion and proper chromosome segregation. Chromatin-associated RDR primes RNA synthesis from the reverse transcript of the outer centromere repeat, which is continuously transcribed, ensuring a constant supply of double-stranded RNA even when transcription of the forward strand is repressed by heterochromatin formation (reviewed by Maison & Almouzni, 2004).

   A similar process occurs at the *S. pombe* silent mating type locus, which contains a copy of the centromeric outer repeat (Hall *et al.*, 2002). In addition, synthetic hairpin RNAs can induce heterochromatin formation at the sites of genes that are normally euchromatic (Schramke & Allshire, 2003). Another natural target of RNAi-mediated heterochromatin in *S. pombe* is retrotransposon long terminal repeats (LTRs) that are transcribed bidirectionally. Spreading of heterochromatin from LTRs into adjacent genes can contribute to cellular differentiation by silencing stage-specific gene expression (Schramke & Allshire, 2003). A nuclear complex called RNA-induced initiation of transcriptional gene silencing (RITS) has been isolated, which directly links siRNAs to heterochromatin in *S. pombe*. The RITS complex contains siRNAs from known heterochromatic regions, such as centromeres, as well as Chp1, a chromodomain protein that binds centromeres, the *S. pombe* ortholog of Ago1, and Tas3, a serine-rich protein that is found only in *S. pombe* (Verdel *et al.*, 2004). The RITS complex is tethered to silenced loci that contain H3K9 methylation. This tethering faciliates siRNA production and maintenance of heterochromatin (Noma *et al.*, 2004).

The RNAi-mediated heterochromatin pathway is conserved in animals. Heterochromatin in *Drosophila*, assessed by H3K9 methylation and localization of the heterochromatin proteins HP1 and HP2, is disrupted in mutants defective in piwi and aubergine, which are both members of the PAZ/piwi domain family that contains Argonaute, and in spindle-E (homeless), a DEAD-box RNA helicase important for RNAi (Pal-Bhadra *et al.*, 2004). In vertebrate cells, formation of centromeric heterochromatin depends on functional Dicer activity. This was shown by generating a conditional loss-of-function Dicer mutant in chicken–human hybrid cells that contain human chromosome 21. Aberrant accumulation of transcripts from human centromeric α-satellite repeats and delocalization of two heterochromatin proteins, Rad21 cohesin and the mitotic checkpoint protein BubR1, were observed in the Dicer-deficient cells, which also displayed mitotic defects and frequently died in interphase (Fukagawa *et al.*, 2004).

The most comprehensive study on RNAi-mediated heterochromatin in plants has been carried out on the heterochromatic knob on chromosome 4 of *A. thaliana*. This work is considered in the context of RdDM (see Section 3.2.2).

3.2.2   *RdDM and RNAi-mediated heterochromatin assembly: one pathway or two?*

While RdDM and RNAi-mediated heterochromatin are likely to be inter-related, it is not yet clear whether they are the outcomes of a single pathway or two separate pathways. Both RdDM and RNAi-mediated heterochromatin formation are initiated by double-stranded RNAs that are substrates of Dicer cleavage, but they might diverge after this step to yield distinct primary epigenetic marks that differ in terms of chemistry, mitotic heritability and reversibility.

Historically, RdDM has been studied from the angle of C methylation, which we assume is the primary epigenetic mark associated with this process. This assumption is based on the pattern of methylation, which is largely confined to the region of RNA–DNA sequence homology and which can encompass sequences as short as 30 bp. This contrasts to histone modifications that occur in a nucleosomal context that comprises 147 bp of DNA. Often referred to as the 'fifth base' in eukaryotic DNA (Hergersberg, 1991), 5-methylcytosine represents a covalent modification of the DNA molecule itself. By contrast, the primary epigenetic mark that is evaluated for RNAi-mediated heterochromatin formation involves covalent modification of histone proteins, most typically methylation of lysine 9 of histone H3, which is a hallmark of silent heterochromatin.

C methylation is carried out by enzymes known as DNA methyltransferases. The cytosine-5-methyltransferases transfer a methyl group from S-adenosyl-L-methionine to carbon 5 of C residues (Finnegan & Kovac, 2000). When present in symmetrical CG and CNG nucleotide groups, methylation can be copied faithfully during DNA replication and passed on to daughter cells (see Figure 3.1). In addition to passive loss of methylation, which results from deficiencies in maintenance methylation, there is increasing evidence that CG methylation can be actively removed from non-replicating DNA through the activity of DNA glycosylases in plants (Choi *et al.*, 2002; Gong *et al.*, 2002; Kinoshita *et al.*, 2004) and in vertebrates (reviewed by Kress *et al.*, 2001). Thus, CG and CNG methylation are mitotically heritable but also potentially reversible epigenetic modifications. It is not yet clear how histone modifications can be inherited through rounds of DNA replication and cell divisions. Indeed, if the term 'epigenetic' implies mitotic heritability, then some authors have questioned whether histone modifications fully qualify owing to uncertainties

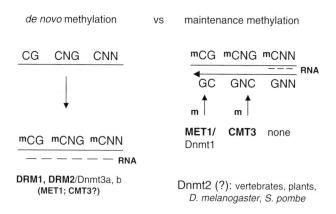

Figure 3.1 DNA methyltransferases catalyzing *de novo* and maintenance methylation. Plant enzymes are shown in bold. *De novo* methylation refers to methylation of a previously unmodified DNA sequence. One of the best-documented signals for *de novo* methylation is double-stranded RNA that can be processed by Dicer-like activity into short RNAs (dashed lines; not drawn to scale). DNA methyltransferases that are generally classified as *de novo* activities include mammalian Dnmt3a/b and the plant homologs, DRM1, DRM2. Methylation can be maintained in symmetrical CG (and in plants, CNG) nucleotide groups through subsequent rounds of DNA replication (arrow) by the action of so-called maintenance methyltransferases, which recognize the methylated C in the parental strand and catalyze methylation of the opposite C in the newly synthesized strand. The CG maintenance activity in mammals is Dnmt1; the plant homolog is MET1. Plants have a special DNA methyltransferase, CMT3, that can maintain methylation in CNG trinucleotides. Recent work on RdDM indicates that MET1 and CMT3 can also participate in *de novo* methylation in the presence of RNA signals. Methylation in asymmetric CNNs cannot be maintained and requires the continuous presence of the triggering RNA. Members of the enigmatic Dnmt2 class of DNA methyltransferases, whose function is not known, are present in vertebrates and plants and in organisms not normally thought to methylate their DNA, such as *Drosophila melanogaster* and fission yeast (*Schizosaccharomyces pombe*).

about their inheritance during mitosis (Bird, 2002). Moreover, even though arginine methylation in histones can be altered by enzymes that convert methylated arginines to citrullines (Cuthbert *et al.*, 2004; Wang *et al.*, 2004), enzymes that can remove lysine methylation in histones have not yet been identified (Zhang, 2004). Thus, despite an incomplete understanding of its mode of inheritance, H3K9 methylation is thought to be a means for ensuring relatively stable, long-term silencing.

RdDM and RNAi-mediated heterochromatin might be further distinguished by the manner in which short RNAs interact with the homologous target sequence (see Figure 3.2). The failure of RdDM to spread substantially beyond the region of RNA–DNA sequence homology hints that direct RNA–DNA base pairing provides a substrate for methylation. By contrast, models in which short RNAs base-pair to nascent RNAs transcribed from the target locus have been proposed for RNAi-mediated heterochromatin assembly (Grewal & Moazed, 2003), which might somehow facilitate spreading beyond the siRNA-targeted nucleation site.

DNA methylation can influence histone modifications and vice versa, but the order of events appears to vary depending on the system under investigation (Lund & van Lohuizen, 2004; Mutskov & Felsenfeld, 2004; Tariq & Paszkowski, 2004). For example, in *N. crassa*, all DNA methylation is catalyzed by one DNA methyltransferase, DIM-2 (Kouzminova & Selker, 2001); this DNA methylation, however, is fully dependent on DIM-5, a histone H3K9 methyltransferase (Tamaru & Selker, 2001). In *Neurospora*, therefore, histone methylation clearly precedes – and is a prerequisite for – DNA methylation. However, *Neurospora* might be exceptional. The structure of the DIM-2 DNA methyltransferase differs from the four major families of DNA

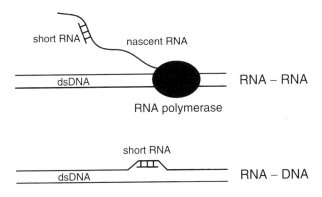

Figure 3.2 Possible modes of interaction between short RNAs and the homologous target locus. Short RNAs can base-pair to a nascent RNA transcribed from the target locus (RNA–RNA) or to one strand of DNA (RNA–DNA).

methyltransferase (see Section 3.3.2.2) (Goll & Bestor, 2005). Moreover, unlike many other eukaryotes, *Neurospora* does not seem to use the RNAi machinery to guide epigenetic modifications (Chicas *et al.*, 2004; Freitag *et al.*, 2004). Instead, signals for *de novo* DNA methylation are generated by the unusual process of repeat-induced point mutation (RIP), which produces C:G to T:A transition mutations in duplicated DNA regions by a poorly understood mechanism. 'RIPed' sequences are preferentially targeted for methylation, apparently because of their A:T richness and high density of TA dinucleotides (Tamaru & Selker, 2003).

In another example, silencing of the redundant X chromosome in female mammals appears to involve initially methylation of histone H3 at Lys9 and Lys27, followed by DNA methylation (Okamoto *et al.*, 2004). In *Arabidopsis*, there are conflicting reports about whether DNA methylation precedes (Malagnac *et al.*, 2002; Soppe *et al.*, 2002; Tariq *et al.*, 2003) or follows (Gendrel *et al.*, 2002; Jackson *et al.*, 2002; Johnson *et al.*, 2002) histone H3K9 methylation.

Despite these uncertainties and differences among species, it is clear that RNAi-mediated H3K9 methylation can be induced in the absence of detectable DNA methylation, as is the case in *S. pombe*. Thus, the mechanisms of DNA methylation and H3K9 methylation are not obligatorily coupled. It is conceivable that RdDM represents a distinct pathway that leads only secondarily to histone modifications. In this chapter, we discuss RdDM as a pathway of RNA-guided *de novo* methylation in which histone modifications are imposed in later steps to maintain and/or reinforce C methylation.

3.3 Mechanism of RNA-directed DNA methylation: RNA and protein requirements

3.3.1 Systems used for genetic analyses of RdDM and transcriptional silencing

Genetic investigations performed in our laboratory on RdDM and RNA-mediated transcriptional silencing have exploited well-defined, two-component transgene systems in *A. thaliana*. In these systems, methylation of a transgene promoter is induced experimentally by a double-stranded RNA that is encoded by a second, unlinked transgene complex (Matzke *et al.*, 2004). We have carried out genetic screens on two promoter systems. The nopaline synthase promoter (NOSpro) is a moderately active, constitutive plant promoter; the α′ promoter (α′pro) is a strong seed-specific promoter. Both these promoters are several hundred base pairs in length and have the same overall GC content (~45%) but the NOSpro contains about twice as many CG dinucleotides as the α′pro. Thus, one might anticipate *a priori* that

NOSpro would be more sensitive to CG methylation than the α'pro. Indeed, as described below, the differences in sequence composition between the two promoters are a likely explanation for the different types of mutations that have been recovered for each system.

Although transgenes provide well-defined and manipulatable systems for analysis, it is important to use the knowledge gained with them to understand RNA-mediated transcriptional regulation of endogenous genes. Parallel genetic analyses of several endogenous genes that are silenced (or likely to be silenced) by double-stranded RNA – such as *SUPERMAN* (*SUP*), *PAI2 and FWA* – are proving particularly informative with regard to RdDM. Before discussing the results of genetic analyses, we briefly describe these three endogenous gene systems.

SUP encodes a zinc finger transcription factor that is important for flower development (Ito *et al.*, 2003). For reasons that are not understood, *SUP* becomes highly methylated and silenced in *ddm1* (decrease in DNA methylation 1) and methyltransferase 1 (*met1*) mutants that otherwise display genome-wide hypomethylation (Jacobsen *et al.*, 2000). The silent loss-of-function 'epialleles' of *SUP*, termed the *clark kent* (*clk*) alleles, are hypermethylated at Cs in all sequence contexts throughout the transcribed region and at the transcription start site (Jacobsen & Meyerowitz, 1997). An approximately 350-bp region near the 5′ end is rich in asymmetric CNNs and CNGs, but contains only one CG dinucleotide (Cao & Jacobsen, 2002). To facilitate genetic analyses for the identification of loci important for methylation and silencing of *SUP*, a stable, non-reverting *clk* allele, *clk-st*, was created by introducing an additional *SUP* locus into *clk-3* plants (Lindroth *et al.*, 2001). The extra *SUP* sequences at this locus are arranged as an inverted repeat (Cao & Jacobsen, 2002), which is probably transcribed to produce a double-stranded RNA that stabilizes silencing and methylation of the endogenous *SUP* gene.

The tryptophan biosynthetic (phosphoribosylanthranilate isomerase, *PAI*) gene family in the Wassilewskija (WS) strain of *Arabidopsis* comprises four copies of the *PAI* gene that are densely methylated at CGs and non-CGs throughout their regions of sequence identity. Due to a naturally occurring duplication (perhaps generated by transposons), two copies – *PAI1*–*PAI4* – are arranged as an inverted repeat. *PAI2* and *PAI3* are unlinked singlet copies. In contrast, ecotype Columbia (Col) has three singlet, unmethylated *PAI* genes at the analogous loci (Melquist *et al.*, 1999). In WS, only *PAI1* and *PAI2* encode functional enzymes but *PAI1* is the sole expressed copy because it is transcribed from an unrelated upstream promoter that is unmethylated. The *PAI1*–*PAI4* inverted repeat induces *de novo* methylation of unmethylated *PAI* sequences, apparently by encoding a double-stranded RNA that targets the singlet copies for methylation by RdDM (Melquist & Bender, 2003, 2004). Genetic screens have been carried out for suppressors of hypermethylation and silencing of the *PAI2* gene.

The flowering Wageningen (*FWA*) gene (M. Koornneef, personal communication) encodes a homeodomain transcription factor that displays imprinted (maternal origin–specific) expression in endosperm. Normally *FWA* is kept silent during the plant vegetative phase by methylation in two transposon-derived direct repeats in the promoter region (Soppe *et al.*, 2000; Lippman *et al.*, 2004). Loss of methylation from the maternal allele in the central cell of the gametophyte through the activity of the DNA glycosylase protein DEMETER leads to maternal expression only in endosperm (Kinoshita *et al.* 2004). Short RNAs originating from the direct repeats suggest that transcriptional repression of *FWA* occurs by an RNAi-mediated pathway (Lippman *et al.*, 2004). In contrast to *SUP*, which becomes hypermethylated in *ddm1* and *met1* mutants, *FWA* becomes hypomethylated in these mutants, producing gain-of-function epialleles that condition a late flowering phenotype (Kakutani, 1997).

3.3.2   Steps in the RdDM pathway

The RdDM pathway can be divided into three steps:

(1) synthesis and processing of double-stranded RNA;
(2) *de novo* methylation of Cs in all sequence contexts in the presence of RNA signals;
(3) maintenance of CG and CNG methylation in the absence of RNA signals (or reinforcement of methylation if RNA signals remain available).

Forward and reverse genetic approaches are identifying the molecular machinery needed for each step.

3.3.2.1   Double-stranded RNA synthesis and processing

The issue of double-stranded RNA synthesis and processing is complicated in plants because of:

(1) different ways to synthesize double-stranded RNA, including through the activity of RDR, which is encoded by a multi-gene family in *Arabidopsis*;
(2) the existence of multiple Dicer-like (DCL) enzymes and Argonaute (AGO) proteins that are differentially distributed into nuclear or cytoplasmic compartments;
(3) the production of two size classes of short RNAs that appear to be functionally distinct: a longer class that is ~24 nt in length has been implicated in DNA and histone methylation and a shorter size class that is ~21 nt in length is involved in the mRNA degradation step of PTGS.

As discussed below, this distinction has generally held up but it is not yet known whether the 24-nt class is exclusively capable of eliciting epigenetic modifications in all systems.

Double-stranded RNA can be synthesized by:

(1) bidirectional transcription of double-stranded DNA, producing sense and antisense RNAs that can anneal with each other;
(2) transcription through an inverted DNA repeat, which forms an RNA that is self-complementary and can fold back on itself to form a hairpin RNA;
(3) transcription of a single-stranded RNA template, which requires the activity of an RDR (see Figure 3.3).

Only the third way of producing double-stranded RNA is compromised by mutations in genes encoding RDRs. The *A. thaliana* genome contains at least three expressed genes that encode RDRs: RDR1, RDR2 and RDR6 (Xie *et al.*, 2004). Forward genetic screens have demonstrated that RDR6 (SGS2/SDE1) is required for PTGS triggered by sense transgenes (Dalmay *et al.*, 2000; Mourrain *et al.*, 2000). Mutations in RDR6 reduce DNA methylation associated with PTGS of sense transgenes but not inverted repeat transgenes (Béclin *et al.*, 2002; Dalmay *et al.*, 2000; Mourrain *et al.*, 2000). Reverse genetics approaches have revealed that RDR2 is required for DNA methylation and H3K9 methylation of several endogenous repeats and for accumulation of siRNAs originating from these repeats (Chan *et al.*, 2004; Xie *et al.*, 2004). Mutations of RDR1 had no effect on the accumulation of these endogenous siRNAs and its function is unknown.

The *Arabidopsis* genome encodes four DCL activities (Schauer *et al.*, 2002). Three of these – DCL1, DCL2 and DCL3 – appear to be localized predominantly in the nucleus as assessed by the subcellular distribution of DCL–GFP fusion proteins (Papp *et al.*, 2003; Xie *et al.*, 2004). Forward genetic screens for mutants defective in RNA-mediated TGS of transgene promoters have not yet recovered mutants defective in any of the four DCL enzymes. This may be due to redundancy of these proteins, as has been observed for the two Dicers in *Neurospora* (Catalanotto *et al.*, 2004). Another possibility is that unprocessed double-stranded RNA can participate to some extent in RdDM. Indeed, we have not yet found a way to eliminate siRNAs originating from the NOSpro or α'pro, which would allow us to confirm they are needed for RdDM in these systems. Neither RdDM of a target NOSpro nor accumulation of NOSpro siRNAs is impaired in *dcl1* partial loss of function mutants (Papp *et al.*, 2003). Accumulation of siRNAs generated from hairpin RNAs that induce PTGS is also not reduced in *dcl1* partial loss of function mutants (Finnegan *et al.*, 2003). Instead, the primary role of DCL1 appears to be the processing of miRNA precursors (Park *et al.*, 2002; Reinhart *et al.*, 2002). Whether this is the only function of DCL1 is unknown.

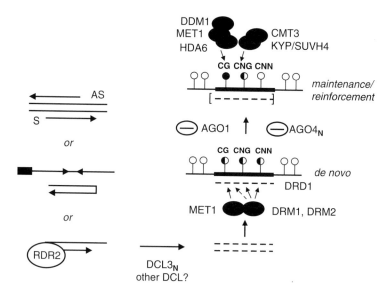

Figure 3.3 Pathway of RNA-directed DNA methylation as determined from genetic analyses in *Arabidopsis*. Left: Double-stranded RNA can be synthesized in several ways including transcription of both DNA strands, producing overlapping sense (S) and antisense (AS) transcripts; transcription of an inverted DNA repeat, producing a hairpin RNA; or through the activity of RNA-dependent RNA polymerase (RDR) on a single-stranded RNA template. Double-stranded RNAs are processed by a dicer-like (DCL) activity to generate short RNAs (dashed lines). RDR2 has been implicated together with nuclear-localized (N) DCL3 in a nuclear pathway that produces 24-nt short RNAs that elicit DNA and histone methylation. Whether this pathway applies to all cases of RdDM is not yet known. Right: In response to RNA signals, site-specific DNA methyltransferases (MET1 for CGs and DRM1, DRM2 for CGs and non-CGs) cooperate to catalyze an intermediate level of *de novo* methylation (half-filled circles). Methylation can be either maintained (−RNA) or reinforced (+RNA) by distinct DNA methyltransferases and histone-modifying enzymes, which team up to preserve specific patterns of cytosine methylation. HDA6 and MET1 maintain primarily CG methylation; CMT3 and KYP/SUVH4 maintain primarily CNG methylation. CNN methylation cannot be maintained in the absence of the trigger RNA. Two SNF2-like proteins, DRD1 and DDM1, are involved in *de novo* and maintenance methylation, respectively. AGO1 and nuclear localized (N) AGO4 are thought to bind to short RNAs and facilitate their interaction with the homologous target locus (DNA or nascent RNA; see Figure 3.2).

Reverse genetics has established a role for DCL3 in generating 24-nt small RNAs that are important for DNA and H3K9 methylation (Chan *et al.*, 2004; Xie *et al.*, 2004). Whether DCL3 is the only DCL activity involved in producing short RNAs that elicit epigenetic modifications is unclear (see Figure 3.3). DCL3 lacks double-stranded RNA-binding domains, which are present in the other three DCL proteins of *Arabidopsis* (Schauer *et al.*, 2002). DCL3 might act as a heterodimer with another DCL activity or in a complex with another double-stranded RNA-binding protein. Despite its apparent nuclear location, DCL2 has been implicated in producing siRNAs from RNA viruses that replicate in the cytoplasm (Xie *et al.*, 2004). No function has yet been assigned to DCL4, which contains a predicted nuclear localization signal (Schauer *et al.*, 2002).

Argonaute proteins are core components of RISC (Carmell *et al.*, 2002), RITS (Noma *et al.*, 2004; Verdel *et al.*, 2004), and presumably other silencing effector complexes. As members of the PAZ/piwi domain family, Argonaute proteins are able to bind short RNAs through their PAZ domains (Lingel *et al.*, 2003; Yan *et al.*, 2003a) and they are thought to 'shepherd' short RNAs to their site of action (Carmell *et al.*, 2002). At least one Argonaute protein, human Ago2, contains an RNaseH-like domain that catalyzes cleavage of target mRNAs and hence corresponds to the long-sought 'slicer' activity in the RNAi pathway (Liu *et al.*, 2004; Meister *et al.*, 2004). The *Arabidopsis* genome encodes ten AGO proteins (Morel *et al.*, 2002). Four of these proteins have been assigned a function and two – AGO1 and AGO4 – are involved in pathways leading to epigenetic modifications (see Figure 3.3).

AGO1, the founding member of the Argonaute family (Bohmert *et al.*, 1998), is required for PTGS and DNA methylation mediated by sense transgenes (Beclin *et al.*, 2002; Boutet *et al.*, 2003) as well as the miRNA pathway and plant development (Vaucheret *et al.*, 2004). AGO1 also helps to regulate silencing and epigenetic modifications of a small subset of transposons in *Arabidopsis* (Lippman *et al.*, 2003). AGO4, a nuclear protein (Xie *et al.*, 2004), is proposed to be specialized for RNA-mediated chromatin modifications. Indeed, AGO4 – identified as a suppressor of *SUP* silencing and hypermethylation – is the only classical RNAi protein recovered in a forward genetic screen of a presumed RNAi-mediated TGS system (Zilberman *et al.*, 2003). AGO4, together with RDR2 and DCL3, is also needed for maintenance of DNA methylation at several endogenous repetitive loci in *Arabidopsis* (Xie *et al.*, 2004) and for *de novo* methylation of an *FWA* transgene (Chan *et al.*, 2004). A mutation in AGO4 does not impair *de novo* methylation triggered by inverted repeats but does reduce maintenance methylation of these sequences (Zilberman *et al.*, 2004). Two other AGO proteins for which functional information is available have not yet been implicated in epigenetic regulation but are important for plant development: ZIP/AGO7 (ZIPPY) controls vegetative phase changes (Hunter *et al.*, 2003) and PNH/ZLL/AGO10 (PINHEAD/ZWILLE) is required for meristem maintenance (Lynn *et al.*, 1999; Moussian *et al.*, 1998, 2003).

It is not yet known whether both sense and antisense RNAs are needed for RdDM. Comparable amounts of each polarity are observed on Northern blots of short RNAs in the NOSpro and α'pro systems (Aufsatz *et al.*, 2002a, b; Kanno *et al.*, 2004; Mette *et al.*, 2000). Moreover, similar numbers of sense and antisense NOSpro short RNAs were cloned from preparations of size-fractioned RNAs (Papp *et al.*, 2003). miRNAs accumulate predominantly in the antisense orientation owing to stabilization by base pairing to target mRNA and/or preferred incorporation of antisense miRNAs into silencing effector complexes. The equivalent accumulation of NOSpro short RNAs of both polarities would be consistent with stabilization by base pairing to both strands of DNA, although this proposal remains to be substantiated experimentally. Sense and antisense small RNAs of the 24-nt size class accumulate from different retroelements in tobacco and *Arabidopsis* (Hamilton *et al.*, 2002). These include the *Arabidopsis* SINE element AtSN1, which appears to be a target for small RNA-induced DNA and histone methylation (Hamilton *et al.*, 2002; Lippman *et al.*, 2003; Xie *et al.*, 2004; Zilberman *et al.*, 2003, 2004). Considering the possibility that small RNAs elicit epigenetic modifications by base pairing to DNA (see Figure 3.2), a single polarity should in principle be sufficient to nucleate methylation in the homologous DNA region provided 'maintenance' methyltransferases are available (see Figure 3.4). However, RdDM might be more efficient if small RNAs of both orientations are present. On the other hand, the recent report indicating that miRNAs can trigger DNA methylation downstream of the miRNA complementary region (Bao *et al.*, 2004) might be more consistent with an RNA recognition model, in which the antisense miRNA base pairs to a nascent RNA (see Figure 3.2). This may lead to DNA methylation in the vicinity of the miRNA–RNA duplex, which does not necessarily correspond to the region RNA–DNA sequence homology. Whether this 'off-target' methylation is the result of unique capabilities of miRNAs as compared with siRNAs in inducing DNA methylation or whether it reflects the spreading of epigenetic modifications associated with the RNA recognition model (see Section 3.2.2) is not known. Determining the way(s) in which various kinds of short RNA interact with homologous target loci to elicit epigenetic modifications remains one of the most pressing questions in the field (see Section 3.5).

3.3.2.2   *DNA methyltransferases and histone-modifying enzymes*

DNA methyltransferases have been classified traditionally according to whether they catalyze *de novo* or maintenance methylation. *De novo* methylation refers to methylation of a previously unmodified DNA sequence. Maintenance methylation involves the perpetuation of methylation in symmetric CG (and in plants, CNG) nucleotide groups during successive rounds of DNA replication (see Figure 3.1). In animals, methylation is generally thought to be restricted to CG dinucleotides. In plants, however, the situation is more complex because methylation is frequently observed in Cs in all sequence contexts (Meyer *et al.*, 1994), which is the pattern observed in cases of RdDM. As genetic approaches are used to identify DNA methyltransferases required for RdDM, it is increasingly clear that the strict division of these enzymes into *de novo* and maintenance activities is untenable. Therefore, in the following discussion, we categorize DNA methyltransferases on the basis of their site specificity; in other words, whether they catalyze CG or non-CG methylation. Because methylation in asymmetric CNNs cannot be maintained in the absence of the RNA trigger (see Figure 3.1), it is considered here to be a measure of continuous *de novo* methylation.

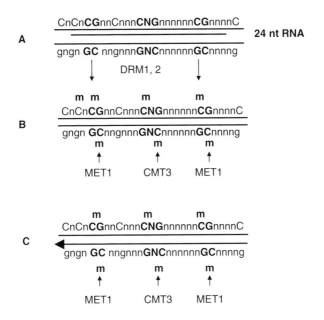

Figure 3.4 Cooperation between DRM1, DRM2 and MET1 to catalyze *de novo* methylation. (A) A short RNA (shown here as 24 nt in length) base pairs to the complementary (top) DNA strand and DMR1, DMR2 catalyze *de novo* methylation of Cs in all sequence contexts within the region of RNA–DNA sequence homology. (B) The resulting hemimethylated CGs and CNGs are then recognized by MET1 and CMT3, respectively, which catalyze methylation of the opposite Cs on the bottom DNA strand. This can occur on non-replicating DNA in the absence of the complementary RNA signal and is essentially *de novo* methylation of the bottom DNA strand. The DNA is now fully methylated at CGs and CNGs (bold), whereas CNN methylation is still limited to the top strand. (C) During DNA replication (leftward arrow), methylation at CGs and CNGs will be maintained by MET1 and CMT3, respectively, while all CNN methylation is lost in the absence of RNA. This hypothetical model provides an explanation for the *de novo* activity of MET1 in the presence of RNA signals, which is related to its conventional maintenance activity that relies on recognition of a hemimethylated substrate during DNA replication.

There are four major families of C-5 DNA methyltransferases in eukaryotes (Goll & Bestor, 2005) and plants contain representatives of them all (Finnegan & Kovac, 2000) (see Figure 3.1):

(1) The MET1 family comprises homologs of the mouse Dnmt1 methyltransferase, which is typically described as CG maintenance methyltransferase although it displays *de novo* activity *in vitro* (Goll & Bestor, 2005). There are at least four members of this gene family in *Arabidopsis*, with MET1 being the most highly expressed representative.

(2) The domains-rearranged (DRM) family, which contains two members in *Arabidopsis*, is related to the mouse Dnmt3 group of methyltransferases that are usually ascribed a *de novo* function (Cao *et al.*, 2000). DRM2 is the major

activity in *Arabidopsis* and has been reported to catalyze methylation of Cs in all sequence contexts (Cao & Jacobsen, 2002). There are noteworthy distinctions between mammalian Dnmt3 and plant DRM enzymes. First, the arrangement of ten motifs in the C-terminal catalytic domain differs. In Dnmt3, the order is I–X; in DRM, the order is VI–X, followed by I–V. Despite the rearrangement, DRM and Dnmt3 are predicted to fold into similar structures that are equally efficient in catalyzing C methylation. In addition, the N termini of the DRM class contains several ubiquitin-associated domains, which might be involved in ubiquitin binding and proteosome degradation pathways. In contrast, the N terminus of Dnmt3 comprises a PWWP domain (a module of 100–150 amino acids containing a conserved proline–tryptophan–tryptophan– proline motif) and cysteine-rich domains (Cao *et al.*, 2000). The significance of these differences between the otherwise highly related Dnmt3 and DRM DNA methyltransferases is not known.

(3) The chromomethylases (CMTs) are a plant-specific family of C methyltransferases that contain a chromodomain (chromatin organization modifier), which is a highly conserved motif found in several chromatin associated proteins. Initially found in *Drosophila* HP1 and Polycomb proteins, the ∼50 amino acid chromodomain binds to histones methylated at different sites (Brehm *et al.*, 2004). The *Arabidopsis* genome contains three CMT genes. The major activity, CMT3, catalyzes methylation primarily in CNG trinucleotides (Bartee *et al.*, 2001; Lindroth *et al.*, 2001).

(4) The Dnmt2 family is the most phylogenetically widespread, conserved and enigmatic group of DNA methyltransferases (Goll & Bestor, 2005). No function has been assigned to any Dnmt2 protein, including the single one encoded in the *Arabidopsis* genome. Dnmt2 homologs are present even in organisms that have little or no DNA methylation, such as *S. pombe* and *Drosophila*. Remarkably, the *Drosophila* Dnmt2 catalyzes a small amount of non-CG methylation early in development (Kunert *et al.*, 2003), but the purpose of this methylation is obscure. The *Dnmt2* gene in *S. pombe* is mutated and the protein is unable to catalyze C methylation, but the gene is nevertheless expressed for reasons that are not known (Wilkinson *et al.*, 1995). One suggestion is that Dnmt2 proteins do not methylate Cs but methylate an unusual structure that is only present in low amounts or under special conditions that have not been reproduced in the laboratory (Goll & Bestor, 2005). Alternatively, Dnmt2 proteins might serve as a structural component of chromatin that is independent of the ability to catalyze DNA methylation (Matzke *et al.*, 2004).

Initially, it was not clear whether RdDM requires a special DNA methyltransferase dedicated to this process. The plant-specific CMT3, for example, was suggested as a possible candidate for carrying out RdDM (Habu *et al.*, 2001; Matzke *et al.*, 2001) after the chromodomain of the *Drosophila* histone acetyltransferase MOF was shown to have RNA-binding activity (Akhtar *et al.*, 2000). However, genetic approaches in *Arabidopsis* have subsequently revealed that multiple, conventional DNA methyltransferases cooperate to carry out RdDM.

Reverse genetics was used to examine the role of DRM1 DRM2 and CMT3 in RdDM of the NOSpro. All RNA-directed *de novo* methylation was blocked in *drm1 drm2* double mutants, indicating that these DNA methyltransferases catalyze the bulk of C methylation induced by RNA (Cao *et al.*, 2003). However, DRM1 DRM2 are not the only enzymes required for full RNA-directed *de novo* methylation. Forward genetic screens of the NOSpro system recovered two mutants, *rts1* and *rts2* (RNA-mediated transcriptional silencing). The *rts2* mutant turned out to be defective in MET1. In subsequent experiments that assessed methylation in '*de novo*' and 'maintenance' set-ups, MET1 was shown to be required not only for preservation of CG methylation but also for full CG *de novo* methylation of the NOSpro (Aufsatz *et al.*, 2004). It was concluded that the DRM1 DRM2 and MET1 collaborate to establish *de novo* methylation in response to RNA signals (see Figure 3.3). This might actually occur in a manner that resembles CG maintenance methylation during DNA replication, in that hemimethylated CG dinucleotides in non-replicating DNA are recognized by MET1 (see Figure 3.4). CMT3 is usually referred to as a maintenance methyltransferase for CNG trinucleotides (Bartee *et al.*, 2001; Lindroth *et al.*, 2001), but it might also have some *de novo* activity, as has been observed in the *PAI* gene system (Malagnac *et al.*, 2002). Thus, all three major types of DNA methyltransferase in *Arabidopsis* probably contribute to varying extents to RNA-directed *de novo* methylation of different sequences, probably in a manner that depends on the abundance of CG, CNG or CNN nucleotide groups.

In the context of RdDM, maintenance methylation is defined as methylation that persists after the RNA signal is withdrawn (see Figures 3.1 and 3.3). Although there is a general consensus that asymmetrical CNN methylation is not maintained in the absence of the RNA trigger, it is still not clear how efficiently methylation in symmetrical CG and CNG nucleotide groups is preserved once the inducing RNA is removed. In one study, which used an RNA virus to induce methylation and TGS of the 35S promoter, residual CG methylation was observed in uninfected progeny plants, indicating legitimate maintenance methylation in this system (Jones *et al.*, 2001). Similarly, at least some CG methylation remains at the NOSpro after crossing out the silencing locus (Aufsatz *et al.*, 2002a). However, all methylation is lost from the α'pro in the absence of the RNA signal (A. Matzke & T. Kanno, unpublished data). Thus, for reasons that are not yet understood, promoters vary in their ability to maintain methylation induced by RNA. One possibility is that CG and/or CNG methylation is maintained most efficiently in promoters that have a high density of the respective di- or trinucleotide.

The word 'reinforcement' has been used to refer to enhancement of methylation in cases where the RNA signal remains available after the initial *de novo* step (Aufsatz *et al.*, 2002b) (see Figure 3.3). Different histone-modifying activities appear to be required to reinforce CG and CNG methylation, respectively. For example, in the NOSpro system, the *rts1* mutant was found to correspond to the histone deacetylase HDA6 (Aufsatz *et al.*, 2002b). The primary methylation defect in the *hda6* mutant was a failure to increase CG methylation above the '*de novo*' level (30–50%) initially induced by RNA. Thus, both of the *rts* mutants identified for the NOSpro system were impaired in some aspect of CG methylation. These results indicate that the NOSpro is silenced by CG methylation, which in turn is consistent with its constitutive nature and relatively high density of CG dinucleotides (Matzke *et al.*, 2004).

Other promoters appear to be silenced by CNG methylation, which is maintained (or reinforced) by a different histone modification (see Figure 3.3). Forward genetic screens of suppressors of hypermethylation and silencing of the *SUP* and *PAI2* genes identified mutants in CMT3 (Bartee *et al.*, 2001; Lindroth *et al.*, 2001) and in the histone H3K9 methyltransferase KRYPTONITE/SUVH4 (KYP) (Jackson *et al.*, 2002; Malagnac *et al.*, 2002). The fact that CMT3 and KYP/SUVH4 were both identified in two independent genetic screens indicates that they frequently team up to maintain CNG methylation. The sensitivity of *SUP* to CNG methylation is consistent with the aforementioned sequence composition of the 5′ region, which is deficient in CG dinucleotides but enriched in CNGs and asymmetric CNNs.

The DNA methyltransferases and histone-modifying enzymes described above are probably not dedicated exclusively to the RdDM pathway. For example, mutations in *MET1* (Saze *et al.*, 2003; Vongs *et al.*, 1993) and *HDA6* (Furner *et al.*, 1998; Murfett *et al.*, 2001; Probst *et al.*, 2004) have been identified in screens for mutants defective in TGS (or probable TGS) that has not yet been shown to be mediated by RNA. It is therefore likely that DNA methyltransferases in *Arabidopsis* can respond to multiple signals for *de novo* methylation, of which RNA is only one. Remarkably, HDA6, which is one of 18 histone deacetylases encoded in the *Arabidopsis* genome, is the only member of this class of proteins that has been identified in forward genetic screens for suppressors of silencing and DNA methylation. The significance of this result is not yet clear. Reverse genetic approaches have revealed requirements for HDA1 in *Arabidopsis* development (Tian & Chen, 2001; Tian *et al.*, 2003, 2004).

3.3.2.3 *SNF2-like chromatin remodeling ATPases and DNA methylation*
The stage of the cell cycle when RdDM occurs is not known. Clearly maintenance methylation needs to take place during DNA replication. However, *de novo* methylation can potentially be catalyzed on non-replicating DNA that is packaged into chromatin, which can present a barrier to proteins and other

regulatory factors. Therefore, it is perhaps not too surprising that forward genetic screens to recover mutants defective in DNA methylation, including RdDM in plants, have identified nucleosome-remodeling proteins of the sucrose non-fermenter (SNF2) family. These proteins use the energy of ATP to displace nucleosomes or modulate histone–DNA contacts, thus exposing DNA to regulatory factors (Lusser & Kadonaga, 2003). To date, members of three subfamilies of SNF2-like proteins have been implicated in DNA methylation in either plants or mammals or in both:

(1) plant DDM1 (decrease in DNA methylation) and its mammalian homolog lymphoid-specific helicase (Lsh);
(2) mammalian ATRX (alpha-thalassemia, mental retardation, X-linked);
(3) plant DRD1 (defective in RdDM). While DDM1/Lsh1 and ATRX are thought to be involved primarily in maintaining methylation at various repetitive sequences, DRD1 is likely to be a factor required for RNA-directed *de novo* methylation.

DDM1 was the first SNF2-like factor to be linked to DNA methylation. This protein was identified in a screen for mutants defective in methylation of centromeric and ribosomal DNA repeats in *Arabidopsis* (Vongs *et al.*, 1993). Genomic levels of 5-methylcytosine are reduced more than 70% in *ddm1* mutants and dramatic reductions in CG and non-CG methylation are observed in centromeric and ribosomal repeats. Although initially normal in appearance, *ddm1* mutants display developmental abnormalities after several generations of inbreeding (Kakutani *et al.*, 1996). Some of these abnormalities are due to gene disruption by reactivated transposons (Miura *et al.*, 2001) whereas others result from hypomethylation or hypermethylation of genes. Thus, *ddm1* mutants are good sources of 'epimutations', including the aforementioned hypermethylated *clk* epialleles of *SUP* (Jacobsen *et al.*, 2000) and the hypomethylated epialleles of *FWA* (Soppe *et al.*, 2000). DDM1 was eventually identified as a putative SNF2-like protein (Jeddeloh *et al.*, 1999) and has been shown recently to promote ATP-dependent nucleosome remodeling *in vitro* (Brzeski & Jermanowski, 2003).

DDM1 is usually mentioned in connection with maintaining DNA methylation of transposons (Lippman *et al.*, 2003; Singer *et al.*, 2001) as well as sequences targeted for RdDM such as the *PAI2* gene (Jeddeloh *et al.*, 1998), sense transgenes that are silenced by PTGS (Morel *et al.*, 2000) and the NOSpro (Aufsatz *et al.*, 2002a). Indeed, a role for DDM1 in *de novo* methylation seems inconsistent with the hypermethylation of *SUP* in *ddm1* mutants (Finnegan & Kovac, 2000). However, the involvement of DDM1 in H3K9 methylation targeted by siRNAs (Gendrel *et al.*, 2002) suggests that DDM1 can participate in the establishment of epigenetic modifications, at least at some loci.

The role of DDM1 in RNAi-mediated chromatin-based silencing has been studied in detail in *Arabidopsis* at the 'knob', a region of interstitial heterochromatin on chromosome 4. The chromosome 4 knob provides a paradigm for heterochromatin formation. It actually corresponds to one half of a segmental duplication that – unlike its partner – has become riddled with transposon insertions and related repeats, which supply the DNA foundation for creating a heterochromatic structure (Lippman *et al.*, 2004). These repetitive sequences are preferentially targeted for DNA and histone methylation whereas adjacent single-copy genes are free of these modifications. As demonstrated by a microarray analysis, DNA methylation and H3K9 methylation are lost from transposon sequences in the knob in *ddm1* mutants. The transposons acquiring epigenetic modifications are sources of siRNAs, which were proposed to guide DDM1 activity specifically to transposon sequences (Lippman *et al.*, 2004). Consistent with this suggestion, some of the transposon short RNAs are reduced in *ddm1* mutants, possibly because they are normally stabilized by interactions with DDM1. Short RNAs homologous to the transposon-derived tandem repeats of the *FWA* gene have also been detected. Since DDM1 contributes to the regulation of FWA expression (Soppe *et al.*, 2000), this finding links DDM1 activity to the regulation of transposons and to the regulation of genes that are adjacent to transposons via the RNAi-mediated heterochromatin pathway (Lippman *et al.*, 2004). Many quiescent transposons that are reactivated in *ddm1* mutants are also reactivated in *met1* and *hda6* mutants, suggesting that these proteins act together in a complex to maintain the transcriptionally silent state of these repetitive elements (Lippman *et al.*, 2003) (see Figure 3.3).

DDM1 has a mammalian homolog, Lsh, which is important for genome-wide CG methylation in mammals (Dennis *et al.*, 2001). In a manner similar to DDM1 in plants, Lsh controls CG methylation and heterochromatin structure at pericentromeric regions comprising satellite repeats (Yan *et al.*, 2003b). Lsh does not appear to regulate directly single-copy genes but is targeted specifically to repetitive elements, perhaps via small RNAs although there is no evidence as yet for this proposal (Huang *et al.*, 2004). Because of its role in silencing repetitive elements, Lsh has been referred to as 'a guardian of heterochromatin' (Huang *et al.*, 2004).

A second SNF2-like protein that is important for DNA methylation, at least in mammals, is ATRX. In humans, mutations in *ATRX* can cause ATRX syndrome (alpha-thalassemia mental retardation, X-linked), which is characterized by severe mental retardation, facial abnormalities, alpha-thalassemia, genital aberrations and epileptic seizures. *Arabidopsis* has an *ATRX* homolog (At1g08600), but its function has not yet been investigated. In contrast to *ddm1* mutants in *Arabidopsis*, which retain only a fraction of wild-type levels of methylation, the total amount of methylcytosine in the genome of ATRX individuals is not significantly different from normal (Gibbons *et al.*, 2000).

Effects of *ATRX* mutations on DNA methylation in humans are diverse. For example, ribosomal DNA is hypomethylated, Y-specific satellite repeats display increased methylation, and a subtelomeric/interstitial repeat shows changes that cannot be interpreted as either an increase or a decrease in methylation. Although alpha-globin expression is dysregulated in ATRX patients, there are no detectable changes in methylation of this gene, suggesting that ATRX might act indirectly to regulate alpha-globin gene expression. So far there is no evidence that ATRX is guided to sites of activity by short RNAs. However, ATRX interacts with HP1 and it is present in pericentromeric regions (McDowell *et al.*, 1999), which are targets of RNAi-mediated pathways in vertebrates and other organisms (see Section 3.2). Indeed, functional ablation of ATRX in mouse oocytes compromises centromere structure and function during meiosis (De La Fuente *et al.* 2004). It will be interesting to learn whether ATRX operates in RNAi-mediated pathways, particularly given its possible relationship to DRD1, a plant-specific SNF2-like protein required for RdDM.

DRD1 was identified in a forward genetic screen for mutants defective in RdDM and silencing of the embryo-specific α′pro in *Arabidopsis* (Kanno *et al.*, 2004). Two other mutants, *drd2* and *drd3*, were also recovered in this screen. DRD1 is a member of a previously uncharacterized, plant-specific subfamily of SNF2-like proteins (see CHR35 in proposed Clade A in the Plant Chromatin Database: http://www.chromdb.org and a modified description of the DRD1 subfamily in Kanno *et al.* (2004)). The closest non-plant homologs are members of the Rad54/ATRX subfamily of SNF2-like proteins. In *drd1* mutants, non-CG methylation of the α′pro is nearly eliminated while CG methylation is essentially unchanged. Because of the strong reduction of CNN methylation, which cannot be maintained in the absence of the RNA trigger (Aufsatz *et al.*, 2002a), DRD1 was proposed to be necessary for RNA-directed *de novo* methylation of the α′pro (Kanno *et al.*, 2004) (see Figure 3.3). Recent tests using a '*de novo*' set-up provide support for this proposal (T. Kanno, W. Aufsatz & A. Matzke, unpublished data).

Experiments are currently under way to identify the natural genomic targets of DRD1. Neither CG nor non-CG methylation at centromeric and ribosomal DNA repeats is reduced in *drd1* mutants (Kanno *et al.*, 2004) and the phenotype of these plants is relatively normal after four generations of inbreeding homozygous *drd1* plants (T. Kanno & M. Matzke, unpublished data). Thus, unlike DDM1, which acts preferentially on repetitive sequences, DRD1 does not appear to be important for determining global patterns of methylation. Moreover, analyses of several known transposons indicate that they are not strongly reactivated in *drd1* mutants (T. Kanno, B. Huettel, M. Matzke, unpublished data). This includes transcriptionally silent information (TSI), a pericentromeric repeat that is derived from *Athila* retrotransposons (Steimer *et al.*, 2000). In contrast, TSI is reactivated in many other epigenetic mutants including *met1*, *hda6*, *ddm1* and *mom1*. The MOM1 protein contains half of a SNF2 helicase domain (Amedeo *et al.*, 2000), but whether it acts in a manner similar to other SNF2-like proteins to remodel chromatin is not known. Mutants defective in MOM1 release transcriptional silencing of TSI and of a complex transgene locus without a concomitant reduction in DNA methylation. MOM1 thus appears to act at a different level than DDM1 (and possibly other SNF2-like proteins that participate in DNA methylation) to reinforce TGS (Mittelsten Scheid *et al.*, 2002). There is no evidence as yet that MOM1 operates in an RNA-mediated pathway.

DRD1 is the first SNF2-like protein to be implicated in an RNA-guided epigenetic modification of the genome. The involvement of DRD1 in RdDM suggests that chromatin remodeling is required for RNA to gain access to target DNA in a chromatin context. DRD1 appears to act locally at the RNA-targeted site to help to create a substrate for *de novo* methylation (see Figure 3.3). It might do this by displacing nucleosomes and facilitating DNA unwinding, allowing formation of an RNA–DNA hybrid that is recognized by DNA methyltransferases. Because DRD1 does not contain a recognizable domain that binds DNA or chromatin, it probably acts together with other proteins in a complex to induce RdDM. The identification of the *drd2* and *drd3* mutants will likely contribute to our understanding of DRD1 activity in the context of the RdDM pathway.

3.4 RdDM in other organisms

When considering the possible occurrence of RdDM outside the plant kingdom, we can ask whether the C methylation patterns characteristic of RdDM have been detected in other organisms and whether the necessary machinery for RdDM, as determined by genetic analyses in *Arabidopsis*, is present in other organisms.

3.4.1 Pattern of methylation

CG and non-CG methylation that is co-extensive with the region of RNA–DNA sequence homology is the hallmark of RdDM in plants. As mentioned previously, most methylation in animal genomes is typically thought to be restricted to CG dinucleotides. However, there are convincing examples of non-CG methylation in animals. Significant non-CG methylation has been detected early during development in *Drosophila* (Lyko, 2001) and in mouse embryonic stem cells (Ramsahoye *et al.*, 2000). Conceivably, this methylation is guided by RNA. The observed non-CG methylation coincides with the activity of the *de novo* DNA methyltransferases, Dnmt3a in mammals and dDnmt2, the single DNA methyltransferase in *Drosophila*. The activity of these enzymes declines as development proceeds. Consequently, non-CG methylation would be passively lost during subsequent cell divisions, providing a potential explanation for why it is not usually detected in adult cells (Matzke *et al.*, 2004). Non-CG methylation as well as hemimethylation at CG dinucleotides has also been detected in the hypermethylated promoter of a human LINE (long interspersed element) L1 retrotransposon in humans (Woodcock *et al.*, 1997). This methylation pattern could possibly be directed by RNA. Intriguingly, LINE elements have been implicated as 'way stations' or 'boosters' (Riggs, 1990) that promote the *cis*-spreading of X chromosome inactivation in female mammals (Lyon, 1998, 2003). Since *Xist* RNA is involved in the spreading phenomenon, this may reflect a general tendency of LINE elements to become methylated via RdDM.

3.4.2 RdDM machinery

Do other organisms possess the RdDM machinery as is currently defined in plants? Overall, the RdDM mechanism does not seem to require special DNA methyltransferases for catalyzing and maintaining C methylation. As explained above, mammals have homologs of the DNA methyltransferases MET1 (*viz.* Dnmt1) and DRM1 DRM2 (*viz.* Dnmt3a/b) (see Figure 3.1). Although the rearranged motifs in the catalytic domains of the DRM group as compared to mammalian Dnmt3 should be kept in mind, there is no evidence to suggest that this feature influences the catalytic activity of the DRM enzymes. The plant-specific DNA methyltransferase CMT3 contributes to RdDM of CNG trinucleotides. However, this type of methylation might be inconsequential for gene expression in mammals, which lack a CMT3 homolog and hence cannot maintain CNG methylation. The histone-modifying enzymes identified so far in forward genetic screens of RdDM systems in *Arabidopsis* – HDA6 and KYP/SUVH4 – have homologs in other organisms. The SNF2-like protein DRD1 appears to be a plant-specific component of the RdDM machinery, but it is not clear whether this protein is needed for all cases of RdDM. Furthermore, it is conceivable that ATRX, which is the closest non-plant homolog of DRD1, could substitute for DRD1 in an RdDM pathway mammals.

One potential deficiency in mammalian cells is a source of nuclear small RNAs, including those of the 24-nt size class, that can potentially induce RdDM. While *Arabidopsis* clearly has nuclear-localized DCL activities to produce short RNAs in the nucleus (see Section 3.3.2.1), the single mammalian Dicer is located in the cytoplasm (Billy *et al.*, 2001). However, this subcellular location does not necessarily preclude the presence of small RNAs in mammalian cell nuclei. Mature miRNAs have been detected in nuclear fractions of mammalian cells, suggesting a means to translocate the products of Dicer processing from the cytoplasm to the nucleus (Meister *et al.*,

2004). Even if short RNAs produced in the cytoplasm of mammalian cells are not actively transported into the nucleus, they might be able to interact with DNA when the nuclear envelope disassembles during cell divisions (Kawasaki & Taira, 2004).

Curiously, efforts to clone small RNAs from mammalian systems – including mouse embryonic stem cells and a number of mouse organs – have identified many 21-nt miRNAs but have failed to recover abundant repeat-associated siRNAs (Houbaviy et al., 2003; Lagos-Quintana et al., 2002, 2003). In contrast, repeat-associated siRNAs comprise a major fraction of the short RNAs cloned from other organisms. For example, of the 560 non-redundant short RNAs cloned from *Drosophila*, 178 were derived from repetitive sequences, including all known *Drosophila* transposons, and many of the cloned short RNAs are in the range of 24–26 nt, consistent with a possible role in chromatin modifications (Aravin et al., 2003). Of 1368 non-redundant small RNAs cloned from *Arabidopsis*, 366 represented retroelements and other transposons and many are 24 nt in length (Xie et al., 2004). All the short RNAs cloned from fission yeast are derived from the outer centromeric repeats and they range in size from 20 to 25 nt (Reinhart & Bartel, 2002). The puzzling lack of ~24 nt repeat-associated siRNAs in mammals suggests that these organisms might not regularly use the RNAi pathway for heterochromatin assembly at repetitive regions of the genome. Arguing against this suggestion, however, are recent reports that Dicer is required for centromeric heterochromatin in vertebrates (see Section 3.2.1) (Fukagawa et al., 2004). The extent to which the RNAi pathway is involved in heterochromatin formation in mammals remains an open question.

3.4.3   RNA-directed DNA methylation of promoters in human cells

Four recent papers have addressed whether RdDM occurs in mammalian cells. Two of these papers claimed positive results and two of them described negative findings. In the two positive studies, synthetic siRNAs were reported to trigger CG methylation and transcriptional silencing of several endogenous promoters and a transgene promoter in human cells. The endogenous promoters were those of genes encoding elongation factor 1 alpha (EF1A) (Morris et al., 2004) and E-cadherin (Kawasaki & Taira, 2004). The transgene construct comprised the EF1A promoter-driving expression of a green fluorescent protein reporter gene (Morris et al., 2004). In addition to CG methylation, promoter siRNAs also induced H3K9 methylation of the endogenous E-cadherin promoter (Kawasaki & Taira, 2004). Interestingly, the study of Kawasaki and Taira (2004) used synthetic siRNAs that were 24 nt in length to silence the E-cadherin promoter, presumably because this size class has been implicated in eliciting epigenetic modifications in plants (Hamilton et al.,

2002; Xie *et al.*, 2004). Maximum downregulation of the E-cadherin promoter (around four-fold) was attained with ten siRNAs – eight targeting the top DNA strand and two the bottom strand – that were distributed throughout the length of the target promoter (Kawasaki & Taira, 2004). By contrast, Morris *et al.* (2004) used a single 21-nt siRNA, which resulted in a substantial downregulation of the EF1A transgene promoter ($\sim$300-fold) but only a marginal reduction of the corresponding endogenous EF1A promoter ($\sim$two-fold) for reasons that are unknown. Both groups were concerned about whether the synthetic siRNAs were efficiently transported into nuclei. Kawasaki and Taira (2004) assumed that at least small amounts of siRNAs entered the nuclei or that the siRNAs were able to interact with DNA when the nuclear envelope broke down during cell divisions. Morris *et al.* (2004) depended either on permeabilization of the nuclear envelope by the lentivirus vector used to introduce the EF1A-*GFP* transgene or on MPG, a bipartite amphipathic peptide that facilitates transport of nucleic acids into nuclei. Neither group reported on the incidence of non-CG methylation and only Kawasaki and Taira (2004) examined spreading of methylation beyond the region of RNA–DNA sequence homology, which occurred to a modest degree.

The two attempts that yielded negative results used either long (Svoboda *et al.*, 2004) or short (Park *et al.*, 2004) double-stranded RNAs homologous to protein-coding regions. Although the long double-stranded RNA used by Svoboda *et al.* (2004) was able to initiate RNAi of a *Mos* gene in mouse oocytes (these cells do not have an interferon response), no significant change in the methylation pattern of the corresponding nuclear DNA sequence was detected by bisulfite sequencing (Svoboda *et al.*, 2004). Similarly, a short hairpin RNA (21 bp) targeted to the protein-coding region of the human huntingtin gene was unable to elicit methylation of the homologous DNA sequences in human glioblastoma cell lines (Park *et al.*, 2004). While these failed efforts might reflect differences in the ability of promoters and protein-coding regions to acquire methylation, both types of sequence can be methylated via RdDM in plants (see Section 3.1.2).

Thus, even though some promising results have been obtained in mammalian cells, it remains to be verified that classical RdDM as it is defined in plants occurs regularly in mammals. Note that promoter siRNAs might be able to trigger methylation and silencing of promoters by mechanisms that are unrelated to RdDM. For example, promoter siRNAs might bind to transcription factors, leading to methylation and silencing of target promoters by default (Cassiday & Maher, 2002). Even if the RdDM pathway does not operate in mammals, other types of non-coding RNAs (such as Xist and long non-coding RNAs involved in imprinting), whose mode of action is not yet understood, can trigger epigenetic modifications in these organisms (Morey & Avner, 2004).

3.5 How short RNAs interact with a target locus: RNA–DNA or RNA–RNA?

The way in which short RNAs interact with the homologous target locus to elicit epigenetic modifications is unknown. Short RNAs can potentially base-pair with a single strand of DNA or with a nascent RNA that is transcribed from the target locus (see Section 3.2.2) (see Figure 3.2). Definitive experiments that would distinguish between these two possibilities have not yet been conducted in any silencing system. A crucial question is whether a target locus must be transcribed for it to acquire C methylation. Ideally, this requirement would be tested by employing Cre-*lox*-mediated recombination to remove the promoter that transcribes a target sequence. If methylation of the target DNA is observed only when the transcribing promoter is present, this would be a good indication that the target locus must be transcribed to become methylated.

As we have discussed throughout this chapter, promoter sequences themselves can be targets of RdDM that are induced by transgene-encoded double-stranded RNAs (Aufsatz *et al.*, 2002a; Melquist and Bender, 2003; Sijen *et al.*, 2001). Whether this requires transcription of the target promoters in the systems described so far is unknown. Endogenous promoters are not normally thought to be transcribed, which would limit natural RdDM of these regulatory regions if this process indeed requires an RNA–RNA interaction. On the other hand, the unexpectedly high levels of transcription of non-coding and anti-sense RNAs in eukaryotic genomes (MacIntosh *et al.*, 2001; Suzuki & Haya-shizaki, 2004) suggests that there is ample opportunity for promoter sequences to be included in transcription units. In addition, a sizeable proportion of endogenous short RNAs cloned from *Arabidopsis* originate from intergenic regions (Xie *et al.*, 2004) and could conceivably target enhancers or promoters for methylation.

The most compelling argument that RdDM results from the alternate type of interaction – RNA–DNA base pairing – is provided by the pattern of DNA methylation itself, which usually does not extend too far beyond the region of RNA–DNA sequence homology (Aufsatz *et al.*, 2002a; Kanno *et al.*, 2004; Vogt *et al.*, 2004). This indicates negligible spreading from the initial site of methylation, although exceptions have been reported (Fojtova *et al.*, 2003). In contrast, RNAi-mediated heterochromatin in fission yeast can spread kilobases from the RNA-targeted nucleation site in a manner that depends on SWI6, the fission yeast ortholog of HP1, which binds to H3 methylated on lysine 9 (Hall *et al.*, 2002; Schramke & Allshire, 2003). In *Arabidopsis*, there is a single HP1 homolog, LHP1 (like HP1) that is thought to form heterochromatin-like complexes that repress the reproductive program (Gaudin *et al.*, 2001). Despite initial reports that LHP1 might provide a bridge between methylated lysine 9 on histone H3 and CMT3 (Jackson *et al.*, 2002), a second study disputed this role for LHP1 (Malagnac *et al.*, 2002). Whether LHP1 can promote spreading of histone or DNA methylation is not known. However, the restriction of H3K9 methylation and DNA methylation to transposon sequences in the chromosome 4 knob, with little or no infiltration into adjacent genes (Lippman *et al.*, 2004), may reflect an inability of LHP1 to foster the spread of heterochromatin.

3.6 Functions of RNA-directed DNA methylation: genome defense, development, others?

The function of DNA methylation in eukaryotes has been a contentious issue for years (Bird, 2003). A regulatory role in development by means of programmed methylation and demethylation of tissue-specific genes was suggested independently by Holliday and Pugh (1975) and Riggs (1975). This view was later challenged, however, because there was little convincing evidence that genes are actually regulated by methylation (Walsh & Bestor, 1999). An alternative proposal posited that the primary role of DNA methylation in plants and animals is to control transposons, with secondary roles in parental imprinting and X chromosome inactivation in mammals (Yoder *et al.*, 1997). Indeed, a number of recent reports have documented that DNA methylation is essential for curtailing transposon activity in plants (Hirochika *et al.*, 2000; Kato *et al.*, 2003; Lippman *et al.*, 2003; Miura *et al.*, 2001; Singer *et al.*, 2001; Tompa *et al.*, 2002). On the other hand, there is renewed interest in DNA methylation as a regulator of cellular differentiation as a result of studies showing that tissue-specific expression of certain genes requires active C demethylation in plants (Choi *et al.*, 2002; Kinoshita *et al.*, 2004) and in mammals (Bruniquel & Schwartz, 2003). The two views on the functions of DNA methylation can be reconciled by recognizing that the transcriptional regulatory regions of plant and animal genes often contain remnants of transposons, which render host genes targets of the genome defense function of DNA methylation (Matzke *et al.*, 1999, 2000). Thus, DNA methylation probably has a dual role in protecting the host from unchecked transposition and in regulating the expression of host genes that contain transposon-related sequences. This dual function is exemplified by the *FWA* gene, which is regulated by methylation that is targeted via siRNAs to transposon-derived direct repeats in the promoter region (Lippman *et al.*, 2004).

Given that RNA is probably only one type of signal for *de novo* methylation (Müller *et al.*, 2002), can any specific functions be assigned to RdDM? Silencing at the transcriptional level usually entails DNA and/or histone modifications and as discussed in previous sections, it is increasingly recognized that these modifications can be guided to specific regions of the genome by small RNAs. Recent results have confirmed that transposons and related repeats are sources of siRNAs that can potentially target epigenetic modifications to the cognate DNA sequences (Hamilton *et al.*, 2002; Lippman *et al.*, 2003, 2004; Xie *et al.*, 2004; Zilberman *et al.*, 2003). However, whether this occurs via the RdDM pathway described here (in which C methylation is presumed to be the primary epigenetic mark) or by the more conserved pathway of RNAi-mediated heterochromatin assembly (which can, in principle, occur independently of DNA methylation) is uncertain. To discern whether there are exclusive targets of RdDM it is necessary to determine the endogenous genomic sequences that are modified directly by this process. The natural targets can be identified by screening for genes that are reactivated in mutants that are believed to be specific for the RdDM pathway. An example is the *drd1* mutant, which has been identified only in a genetic screen of a well-defined RdDM system (Kanno *et al.*, 2004) (see Section 3.3.2.3). The observation that several types of transposons are not unleashed in a *drd1* mutant (T. Kanno, B. Huettel, & A. Matzke, unpublished data) even though they are reactivated in *met1* and *ddm1* mutants suggests that DRD1 has a primary function other than repressing transposons. In addition, even though a role for the DRD1 protein in regulating plant development cannot be excluded, *drd1* mutants do not display a strong phenotype, even after several generations of inbreeding (T. Kanno & A. Matzke, unpublished data). Thus, the full range of functions of RdDM might extend beyond transposon control and development. One possibility is that RdDM is involved in stress responses. RdDM might allow rapid changes in gene expression that would facilitate adaptation in a changing environment by altering the gene expression profile in non-dividing cells. The fact that the DRD1 subfamily of SNF2-like proteins has expanded to include at least four members, all of which are expressed, suggests that members of this family carry out important plant functions, which may involve not only RdDM but other types of chromatin modification.

3.7   Concluding remarks

The discovery of RdDM in plants provided the first indication that RNA could feed back on the genome to induce epigenetic modifications of the cognate DNA sequence. As gene-silencing research in plants progressed and eventually merged with work on RNAi in animals, it became clear that RdDM was one of several intersecting pathways of sequence-specific silencing mediated by double-stranded RNA. Genetic analyses in *Arabidopsis* are identifying the RNA and protein requirements for the RdDM machinery. In contrast to mammals, plants clearly have a nuclear pathway for processing double-stranded RNAs to short RNAs and this pathway can produce 24-nt short RNAs that might be specialized for eliciting epigenetic modifications.

Multiple conventional DNA methyltransferases act in a site-specific manner to catalyze *de novo* methylation in response to RNA signals. *De novo* methylation requires a plant-specific SNF2-like protein, DRD1, perhaps to render the target DNA accessible to RNA signals. Specific DNA methyltransferases and histone-modifying enzymes team up to maintain distinct patterns of C methylation. The sequence composition of a promoter can determine its susceptibility to methylation of Cs in different sequence contexts and possibly also to its ability to maintain CG or CNG methylation in the absence of the trigger RNA. Whether 'classical' RdDM occurs in mammalian cells is still unclear, as is the exact function of RdDM in plants. Future challenges include determining the natural genomic targets of RdDM and understanding the contribution of this RNA-mediated gene silencing pathway to plant physiology, development and adaptation to the environment.

Acknowledgments

Financial support in the Matzke laboratory is provided by the Austrian Fonds zur Förderung der wissenschaftlichen Forschung (grant number P-15611-B07) and the European Union (EC contract number HPRN-CT-2002-00025).

References

Akhtar, A., Zink, D. & Becker, P. B. (2000) Chromodomains are protein–RNA interaction modules. *Nature*, **407**, 405–409.

Amedeo, P., Habu, Y., Afsar, K., Mittelsten Scheid, O. & Paszkowski, J. (2000) Disruption of the plant gene *MOM* releases transcriptional silencing of methylated genes. *Nature*, **405**, 203–206.

Aravin, A. A., Lagos-Quintana, M., Yalcin, A., Zavolan, M., Marks, D., Snyder, B., Gaasterland, T., Meyer, J. & Tuschl, T. (2003) The small RNA profile during *Drosophila melanogaster* development. *Developmental Cell*, **5**, 337–50.

Aufsatz, W., Mette, M. F., van der Winden, J., Matzke, A. J. M. & Matzke, M. (2002a) RNA-directed DNA methylation in *Arabidopsis*. *Proceedings of the National Academy of Science of the United States of America*, **99**, 16499–506.

Aufsatz, W., Mette, M. F., van der Winden, J., Matzke, M. & Matzke, A. J. M. (2002b) HDA6, a putative histone deacetylase needed to enhance DNA methylation induced by double stranded RNA. *EMBO Journal*, **21**, 6832–41.

Aufsatz, W., Mette, M. F., Matzke, A. J. M. & Matzke, M. (2004) The role of MET1 in RNA-directed *de novo* and maintenance methylation of CG dinucleotides. *Plant Molecular Biology*, **54**, 793–804.

Bao, N., Lye, K. W. & Barton, M. K. (2004) MicroRNA binding sites in *Arabidopsis* class III HD-ZIP mRNAs are required for methylation of the template chromosome. *Developmental Cell*, **7**, 653–62.

Bartee, L., Malagnac, F. & Bender, J. (2001) *Arabidopsis* cmt3 chromomethylase mutations block non-CG methylation and silencing of an endogenous gene. *Genes & Development*, **15**, 1753–8.

Bartel, D. P. (2004) MicroRNAs: genomics, biogenesis, mechanism, and function. *Cell*, **116**, 281–97.

Béclin, C., Boutet, S., Waterhouse, P. & Vaucheret, H. (2002) A branched pathway for transgene-induced RNA silencing in plants. *Current Biology*, **12**, 684–8.

Billy, E., Brondani, V., Zhang, H., Muller, U. & Filipowicz, W. (2001) Specific interference with gene expression induced by long, double stranded RNA in mouse embryonal teratocarcinoma cell lines. *Proceedings of the National Academy of Sciences of the United States of America*, **98**, 14428–33.

Bird, A. (2002) DNA methylation patterns and epigenetic memory. *Genes & Development*, **16**, 6–21.

Bird, A. (2003) IL2 transcription unleashed by active DNA demethylation. *Nature Immunology*, **4**, 208–209.

Bohmert, K., Camus, I., Bellini, C., Bouchez, D., Caboche, M. & Benning, C. (1998) *AGO1* defines a novel locus of *Arabidopsis* controlling leaf development. *EMBO Journal*, **17**, 170–80.

Boutet, S., Vazquez, F., Liu, J., Béclin, C., Fagard, M., Gratias, A., Morel, J. B., Crété. P. Chen, X. & Vaucheret, H. (2003) *Arabidopsis* HEN1: a genetic link between endogenous miRNA controlling development and siRNA controlling transgene silencing and virus resistance. *Current Biology*, **13**, 843–8.

Brehm, A., Tufteland, K. R., Aasland, R. & Becker, P. B. (2004) The many colours of chromodomains. *BioEssays*, **26**, 133–40.

Bruniquel, D. & Schwartz, R. H. (2003) Selective, stable demethylation of the interleukin-2 gene enhances transcription by an active process. *Nature Immunology*, **4**, 235–40.

Brzeski, J. & Jermanowski, A. (2003) Deficient in DNA methylation 1 (DDM1) defines a novel family of chromatin remodeling factors. *Journal of Biological Chemistry*, **278**, 828–8.

Cao, X. & Jacobsen, S. E. (2002) Role of the *Arabidopsis DRM* methyltransferases in *de novo* DNA methylation and gene silencing. *Current Biology*, **12**, 1138–44.

Cao, X., Springer, N. M., Muszynski, M. G., Phillips, R. L., Kaeppler, S. & Jacobsen, S. E. (2000) Conserved plant genes with similarity to mammalian *de novo* DNA methyltransferases. *Proceedings of the National Academy of Sciences of the United States of America*, **97**, 4979–84.

Cao, X., Aufsatz, W., Zilberman, D., Mette, M. F., Huang, M. S., Matzke, M. & Jacobsen, S. E. (2003) Role of the DRM and CMT3 methyltransferases in RNA-directed DNA methylation. *Current Biology*, **13**, 2212–17.

Carmell, M. A., Xuan, Z., Zhang, M. Q. & Hannon, G. J. (2002) The Argonaute family: tentacles that reach into RNAi, developmental control, stem cell maintenance, and tumorigenesis. *Genes & Development*, **16**, 2733–42.

Cassiday, L. A. & Maher, L. J. (2002) Having it both ways: transcription factors that bind DNA and RNA. *Nucleic Acids Research*, **30**, 41118–26.

Catalanotto, C., Pallotta, M., ReFalo, P., Sachs, M. S., Vayssie, L., Macino, G. & Cogoni, C. (2004) Redundancy of the two Dicer genes in transgene-induced posttranscriptional gene silencing in *Neurospora crassa*. *Molecular and Cellular Biology*, **24**, 2536–45.

Chan, S. W. -L., Zilberman, D., Xie, Z., Johansen, L. K., Carrington, J. C. & Jacobsen, S. E. (2004) RNA silencing genes control *de novo* methylation. *Science*, **303**, 1336.

Chicas, A., Cogoni, C. & Macino, G. (2004) RNAi-dependent and RNAi-independent mechanisms contribute to the silencing of RIPed sequences in *Neurospora crassa*. *Nucleic Acids Research*, **32**, 4237–43.

Choi, Y., Gehring, M., Johnson, L., Hannon, M., Harada, J. J., Goldberg, R. B., Jacobsen, S. E. & Fischer, R .L. (2002) DEMETER, a DNA glycosylase domain protein, is required for endosperm gene imprinting and seed viability in *Arabidopsis*. *Cell*, **110**, 33–42.

Cuthbert, G. L., Daujat, S., Snowden, A. W., Erdjument-Bromage, H., Hagiwara, T., Yamada, M., Schenider, R., Gregory, P. D., Tempst, P., Bannister, A. J. & Kouzarides, T. (2004) Histone deimination antagonizes arginine methylation. *Cell*, **118**, 545–53.

Dalmay, T., Hamilton, A., Rudd, S., Angell, S. & Baulcombe, D. C. (2000) An RNA-dependent RNA polymerase gene in *Arabidopsis* is required for posttranscriptional gene silencing mediated by a transgene but not by a virus. *Cell*, **101**, 543–53.

De Carvalho, F., Gheysen, G., Kushnir, S., van Montagu, M., Inzé & Castresana, C. (1992) Suppression of β-1,3-glucanase transgene expression in homozygous plants. *EMBO Journal*, **11**, 2595–602.

De La Fuente, R., Viveiros, M. M., Wigglesworth, K. & Eppig, J. J. (2004) ATRX, a member of the SNF2 family of helicases/ATPases, is required for chromosome alignment and meiotic spindle organization in metaphase II stage mouse oocytes. *Developmental Biology*, **272**, 1–14.

Dennis, K., Fan, T., Geiman, T., Yan, Q. & Muegge, K. (2001) Lsh, a member of the SNF2 family, is required for genome-wide methylation. *Genes & Development*, **15**, 2940–44.

Finnegan, E. J. & Kovac, K. A. (2000) Plant DNA methyltransferases. *Plant Molecular Biology*, **43**, 189–201.

Finnegan, E. J., Margis, R. & Waterhouse, P. (2003) Posttranscriptional gene silencing is not compromised in the *Arabidopsis* CARPEL FACTORY (DICER-LIKE1) mutant, a homolog of Dicer-1 from *Drosophila*. *Current Biology*, **13**, 236–40.

Fire, A., Xu, S., Montgomery, M. K., Kostas, S. A., Driver, S. E. & Mello, C. C. (1998) Potent and specific genetic interference by double-stranded RNA in *Caenorhabditis elegans*. *Nature*, **391**, 806–11.

Fojtova, M., Van Houdt, H., Depicker, A. & Kovarik, A. (2003) Epigenetic switch from posttranscriptional to transcriptional silencing is correlated with promoter hypermethylation. *Plant Physiology*, **133**, 1240–50.

Freitag, M., Lee, D. W., Kothe, G. O., Pratt, R. J., Aramayo, R. & Selker, E. U. (2004) DNA methylation is independent of RNA interference in *Neurospora*. *Science*, **304**, 1939.

Fukagawa, T., Nogami, M., Yoshikawa, M., Ikeno, M., Okazaki, T., Takami, Y., Nakayama, T. & Oshimura, M. (2004) Dicer is essential for formation of the heterochromatin structure in vertebrate cells. *Nature Cell Biology*, **6**, 784–91.

Furner, I. J., Sheikh, M. A. & Collett, C. E. (1998) Gene silencing and homology-dependent gene silencing in *Arabidopsis*: genetic modifiers and DNA methylation. *Genetics*, **149**, 651–62.

Gaudin, V., Libault, M., Pouteau, S., Juul, T., Zhao, G., Lefebvre, D. & Grandjean, O. (2001) Mutations in *LIKE HETEROCHROMATIN PROTEIN 1* affect flowering time and plant architecture in *Arabidopsis*. *Development*, **128**, 4847–58.

Gendrel, A. -V., Lippman, Z., Yordan, C., Colot, V. & Martienssen, R. A. (2002) Dependence of heterochromatic histone H3 methylation patterns on the *Arabidopsis* gene DDM1. *Science*, **297**, 1871–73.

Gibbons, R. J., McDowell, T. L., Raman, S., O'Rourke, D. M., Garrick, D., Ayyub, H. & Higgs, D. R. (2000) Mutations in *ATRX*, encoding a SWI/SNF-like protein, cause diverse changes in the pattern of DNA methylation. *Nature Genetics*, **24**, 368–71.

Goll, M. G. & Bestor, T. H. (2005) Eukaryotic cytosine methyltransferases. *Annual Review of Biochemistry*, (in press).

Gong, Z., Morales-Ruiz, T., Ariza, R. R., Roldán-Arjona, T., David, L., & Zhu, J. K. (2002) ROS1, a repressor of transcriptional gene silencing in *Arabidopsis*, encodes a DNA glycosylase/lyase. *Cell*, **111**, 803–14.

Grewal, S. I. S. & Moazed, D. (2003) Heterochromatin and epigenetic control of gene expression. *Science*, **301**, 798–802.

Habu, Y., Kakutani, T. & Paszkowski, J. (2001) Epigenetic developmental mechanisms in plants: molecules and targets of plant epigenetic regulation. *Current Opinion in Genetics and Development*, **11**, 215–20.

Hall, I. M., Shankararayana, G. D., Noma, K., Ayoub, N., Cohen, A. & Grewal, S. I. S. (2002) Establishment and maintenance of a heterochromatic domain. *Science*, **297**, 2232–7.

Hamilton, A., Voinnet, O., Chappell, L. & Baulcombe, D. (2002) Two classes of short interfering RNA in RNA silencing. *EMBO Journal*, **21**, 4671–9.

He, L. & Hannon, G. J. (2004) MicroRNAs: small RNAs with a big role in gene regulation. *Nature Reviews. Genetics*, **5**, 522–31.

Hergersberg, M. (1991) Biological aspects of cytosine methylation in eukaryotic cells. *Experientia*, **47**, 1171–85.

Hirochika, H., Okamoto, H. & Kakutani, T. (2000) Silencing of retrotransposons in *Arabidopsis* and reactivation by the *ddm1* mutation. *Plant Cell*, **12**, 357–68.

Holliday, R. & Pugh, J. E. (1975) DNA modification mechanisms and gene activity during development. *Science*, **187**, 226–32.

Houbaviy, H. B., Murray, M. F. & Sharp, P. A. (2003) Embryonic stem cell-specific microRNAs. *Developmental Cell*, **5**, 351–8.

Huang, J., Fan, T., Yan, Q., Zhu, H., Fox, H., Issaq, H. J., Best, L., Gangi, L., Munroe, D. & Muegge, K. (2004) Lsh, an epigenetic guardian of repetitive elements. *Nucleic Acids Research*, **32**, 5019–28.

Hunter, C., Sun, H. & Poethig, R. S. (2003) The *Arabidopsis* heterochronic gene ZIPPY is an ARGONAUTE family member. *Current Biology*, **13**, 1734–9.

Ingelbrecht, I., Van Houdt, H., Van Montagu, M. & Depicker, A. (1994) Posttranscriptional silencing of reporter transgenes in tobacco correlates with DNA methylation. *Proceedings of the National Academy of Sciences of the United States of America*, **91**, 10502–506.

Ito, T., Sakai, H. & Meyerowitz, E. M. (2003) Whorl-specific expression of the SUPERMAN gene of *Arabidopsis* is mediated by *cis* elements in the transcribed region. *Current Biology*, **13**, 1524–30.

Jackson, J. P., Lindroth, A. M., Cao, X. & Jacobsen, S. E. (2002) Control of CpNpG DNA methylation by the KRYPTONITE histone H3 methyltransferase. *Nature*, **416**, 556–60.

Jacobsen, S. E. & Meyerowitz, E. M. (1997) Hypermethylated SUPERMAN epigenetic alleles in *Arabidopsis*. *Science*, **277**, 1100–103.

Jacobsen, S. E., Sakai, H., Finnegan, E. J., Cao, X. & Meyerowitz, E. M. (2000) Ectopic hypermethylation of flower-specific genes in *Arabidopsis*. *Current Biology*, **10**, 179–86.

Jeddeloh, J. A., Bender, J. & Richards, E. J. (1998) The DNA methylation locus DDM1 is required for maintenance of gene silencing in *Arabidopsis*. *Genes & Development*, **12**, 1714–25.

Jeddeloh, J. A., Stokes, T. L. & Richards, E. J. (1999) Maintenance of genomic methylation requires a SWI2/SNF2-like protein. *Nature Genetics*, **22**, 94–7.

Johnson, L., Cao, X. & Jacobsen, S. E. (2002) Interplay between two epigenetic marks. DNA methylation and histone H3 lysine 9 methylation. *Current Biology*, **12**, 1360–67.

Jones, A. L., Thomas, C. L. & Maule, A. J. (1998) *De novo* methylation and co-suppression induced by a cytoplasmically replicating plant RNA virus. *EMBO Journal*, **17**, 6385–93.

Jones, A. L., Hamilton, A. J., Voinnet, O., Thomas, C. L., Maule A. J. & Baulcombe, D. C. (1999) RNA–DNA interactions and DNA methylation in post-transcriptional gene silencing. *Plant Cell*, **11**, 2291–301.

Jones, A. L., Ratcliff, F. & Baulcombe, D. C. (2001) RNA-directed transcriptional gene silencing in plants can be inherited independently of the RNA trigger and requires Met1 for maintenance. *Current Biology*, **11**, 747–57.

Jorgensen, R. A. (1992) Silencing of plant genes by homologous transgenes. *Agricultural Biotechnology News Information*, **4**, 265N–73N.

Kakutani, T. (1997) Genetic characterization of late-flowering traits induced by DNA hypomethylation mutation in *Arabidopsis thaliana*. *Plant Journal*, **12**, 1447–51.

Kakutani, T. (2002) Epi-alleles in plants: inheritance of epigenetic information over generations. *Plant Cell Physiology*, **43**, 1106–11.

Kakutani, T., Jeddeloh, J. A., Flowers, S. K., Munakata, K. & Richards, E. J. (1996) Developmental abnormalities and epimutations associated with DNA hypomethylation mutations. *Proceedings of the National Academy of Sciences of the United States of America*, **93**, 12406–11.

Kanno, T., Mette, M. F., Kreil, D. P., Aufsatz, W., Matzke, M. & Matzke, A. J. M. (2004) Involvement of putative SNF2 chromatin remodeling protein DRD1 in RNA-directed DNA methylation. *Current Biology*, **14**, 801–805.

Kato, M., Miura, A., Bender, J., Jacobsen, S. E. & Kakutani, T. (2003) Role of CG and non-CG methylation in immobilization of transposons in *Arabidopsis*. *Current Biology*, **3**, 421–6.

Kawasaki, H. & Taira, K. (2004) Induction of DNA methylation and gene silencing by short interfering RNAs in human cells. *Nature*, **431**, 211–17.

Kinoshita, T., Miura, A., Choi, Y., Kinoshita, Y., Cao, X., Jacobsen, S. E., Fischer, R. L. & Kakutani, T. (2004) One-way control of *FWA* imprinting in *Arabidopsis* endosperm by DNA methylation. *Science*, **303**, 521–3.

Kouzminova, E. A. & Selker, E. U. (2001) *Dim-2* encodes a DNA methyltransferase responsible for all known cytosine methylation in *Neurospora*. *EMBO Journal*, **20**, 4309–323.

Kress, C., Thomassin, H. & Grange, T. (2001) Local DNA demethylation in vertebrates: how could it be performed and targeted? *FEBS Letters*, **494**, 135–40.

Kunert, N., Marhold, J., Stanke, J., Stach, D. & Lyko, F. (2003) A Dnmt2-like protein mediates DNA methylation in *Drosophila*. *Development*, **130**, 5083–90.

Lagos-Quintana, M., Rauhut, R., Yalcin, A., Meyer, J., Lendeckel, W. & Tuschl, T. (2002) Identification of tissue-specific microRNAs from mouse. *Current Biology*, **12**, 735–9.

Lagos-Quintana, M., Rauhut, R., Meyer, J., Borkhardt, A. & Tuschl, T. (2003) New microRNAs from mouse and human. *RNA*, **9**, 175–9.

Lindroth, A. M., Cao, X., Jackson, J. P., Zilberman, D., McCallum, C., Henikoff, S. & Jacobsen, S. E. (2001) Requirement of CHROMOMETHYLASE3 for maintenance of CpXpG methylation. *Science*, **292**, 2077–2080.

Lingel, A., Simon, B., Izaurralde, E. & Sattler, M. (2003) Structure and nucleic-acid binding of the *Drosophila* Argonaute 2 PAZ domain. *Nature*, **426**, 465–9.

Lippman, Z., May, B., Yordan, C., Singer, T. & Martienssen, R. (2003) Distinct mechanisms determine transposon inheritance and methylation via small interfering RNA and histone modification. *PLoS Biology*, **1**, 420–28.

Lippman, Z., Gendrel, A. -M., Black, M., Vaughn, M. W., Dedhia, N., McCombie, W. R., Lavine, K., Mittal, V., May, B., Kasschau, K. D., Carrington, J. C., Doerge, R. W., Colot, V. & Martienssen, R. (2004) Role of transposable elements in heterochromatin and epigenetic control. *Nature*, **430**, 471–6.

Liu, J., Carmell, M. A., Rivas, F. V., Marsden, C. G., Thomson, J. M., Song, J. -J., Hammond, S. M., Joshua-Tor, L. & Hannon, G. J. (2004) Argonaute2 is the catalytic engine of mammalian RNAi. *Science*, **305**, 1437–41.

Lund, A. H. & van Lohuizen, M. (2004) Epigenetics and cancer. *Genes & Development*, **18**, 2315–35.

Lusser, A. & Kadonaga, J. T. (2003) Chromatin remodeling by ATP-dependent molecular machines. *BioEssays*, **25**, 1192–200.

Lyko, F. (2001) DNA methylation learns to fly. *Trends in Genetics*, **17**, 169–72.

Lynn, K., Fernanadez, A., Aida, M., Sedbook, J., Tasaka, M., Masson, P. & Barton, M. K. (1999) The PINHEAD/ZWILLE gene acts pleiotropically in *Arabidopsis* development and has overlapping functions with the ARGONAUTE1 gene. *Development*, **126**, 469–81.

Lyon, M. F. (1998) X-chromosome inactivation: a repeat hypothesis. *Cytogenetics and Cell Genetics*, **80**, 133–7.

Lyon, M. F. (2003) The Lyon and the LINE hypothesis. *Seminars in Cell and Developmental Biology*, **14**, 313–18.

MacIntosh, G. C., Wilkerson, C. & Green, P. J. (2001) Identification and analysis of *Arabidopsis* expressed sequence tags characteristic of non-coding RNAs. *Plant Physiology*, **127**, 765–76.

Maison, C. & Almouzni, G. (2004) HP1 and the dynamics of heterochromatin. *Nature Reviews. Molecular Cell Biology*, **5**, 296–304.

Malagnac, F., Bartee, L. & Bender, J. (2002) An *Arabidopsis* SET domain protein required for maintenance but not establishment of DNA methylation. *EMBO Journal*, **21**, 6842–52.

Matzke, M. A. & Birchler, J. A. (2005) RNAi-mediated pathways in the nucleus. *Nature Reviews. Genetics*, **6**, 24–35.

Matzke, M. A. & Matzke, A. J. M. (1993) Genomic imprinting in plants: parental effects and *trans*-inactivation phenomena. *Annual Review of Plant Physiology and Plant Molecular Biology*, **44**, 53–76.

Matzke, M. A. & Matzke, A. J. M. (2004) Planting the seeds of a new paradigm. *PLoS Biology*, **2**, 582–6.

Matzke, M. A., Mette, M. F., Aufsatz, W., Jakowitsch, J. & Matzke, A. J. M. (1999) Host defenses to parasitic sequences and the evolution of epigenetic control mechanisms. *Genetica*, **107**, 271–87.

Matzke, M. A., Mette, M. F. & Matzke, A. J. M. (2000) Transgene silencing by the host genome defense: implications for the evolution of epigenetic control mechanisms in plants and vertebrates. *Plant Molecular Biology*, **43**, 401–15.

Matzke, M. A., Matzke, A. J. M. & Kooter, J. (2001) RNA: guiding gene silencing. *Science*, **293**, 1080–1083.

Matzke, M. A., Aufsatz, W., Kanno, T., Daxinger, L., Papp, I., Mette, M. F. & Matzke, A. J. M. (2004) Genetic analysis of RNA-mediated transcriptional gene silencing. *Biochimica et Biophysica Acta*, **1677**, 129–41.

McDowell, T. L., Gibbons, R. J., Sutherland, H., O'Rourke, D. M., Bickmore, W. A., Pombo, A., Turley, H., Gatter, K., Picketts, D. J., Buckle, V. J., Chapman, L., Rhodes, D. & Higgs, D. R. (1999) Localization of a putative transcriptional regulator (ATRX) at pericentromeric heterochromatin and the short arms of acrocentric chromosomes. *Proceedings of the National Academy of Sciences of the United States of America*, **95**, 13983–8.

Meister, G., Landthaler, M., Patkaniowska, A., Dorsett, Y., Teng, G. & Tuschl, T. (2004) Human Argonaute2 mediates RNA cleavage targeted by miRNAs and siRNAs. *Molecular Cell*, **15**, 185–97.

Melquist, S. & Bender, J. (2003) Transcription from an upstream promoter controls methylation signaling from an inverted repeat of endogenous genes in *Arabidopsis*. *Genes & Development*, **17**, 2036–47.

Melquist, S. & Bender, J. (2004) An internal rearrangement in an *Arabidopsis* inverted repeat locus impairs DNA methylation triggered by the locus. *Genetics*, **166**, 437–48.

Melquist, S., Luff, B. & Bender, J. (1999) *Arabidopsis PAI* gene arrangements, cytosine methylation and expression. *Genetics*, **153**, 401–13.

Mette, M. F., van der Winden, J., Matzke, M. A. & Matzke, A. J. M. (1999) Production of aberrant promoter transcripts contributes to methylation and silencing of unlinked homologous promoters *in trans*. *EMBO Journal*, **18**, 241–8.

Mette, M. F., Aufsatz, W., van der Winden, J., Matzke, M. A. & Matzke, A. J. M. (2000) Transcriptional silencing and promoter methylation triggered by double stranded RNA. *EMBO Journal*, **19**, 5194–201.

Meyer, P., Neidenhof, I. & ten Lohuis, M. (1994) Evidence for cytosine methylation of non-symmetrical sequences in transgenic *Petunia hybrida*. *EMBO Journal*, **13**, 2088–2984.

Mittelsten Scheid, O., Probst, A. V., Afsar, K. & Paszkowski, J. (2002) Two regulatory levels of transcriptional gene silencing in *Arabidopsis*. *Proceedings of the National Academy of Sciences of the United States of America*, **99**, 1359–62.

Miura, A., Yonebayashi, S., Watanabe, K., Toyama, T., Shimada, H. & Kakutani, T. (2001) Mobilization of transposons by a mutation abolishing full DNA methylation in *Arabidopsis*. *Nature*, **411**, 212–14.

Morel, J. B., Mourrain, P., Béclin, C. & Vaucheret, H. (2000) DNA methylation and chromatin structure affect transcriptional and posttranscriptional transgene silencing in *Arabidopsis*. *Current Biology*, **10**, 1591–4.

Morel, J. B., Godon, C., Mourrain, P., Béclin, C., Boutet, S., Feuerbach, F., Proux, F. & Vaucheret, H. (2002) Fertile hypomorphic ARGONAUTE (ago1) mutants impaired in post-transcriptional gene silencing and virus resistance. *Plant Cell*, **14**, 629–39.

Morey, C. & Avner, P. (2004) Employment opportunities for non-coding RNAs. *FEBS Letters*, **567**, 27–43.

Morris, K. V., Chan, S. W. L., Jacobsen, S. E. & Looney, D. J. (2004) Small interfering RNA-induced transcriptional gene silencing in human cells. *Science*, **305**, 1289–92.

Mourrain, P., Béclin, C., Elmayan, T., Feuerbach, F., Godon, C., Morel, J. -B., Jouette, D., Lacombe, A. -M., Nikic, S., Picault, N., Rémoué, K., Sanial, M., Vo, T. -A. & Vaucheret, H. (2000) *Arabidopsis* SGS2 and SGS3 genes are required for posttranscriptional gene silencing and natural virus resistance. *Cell*, **101**, 533–42.

Moussian, B., Schoof, H., Haecker, A., Jurgens, G. & Laux, T. (1998) Role of the ZWILLE gene in the regulation of central shoot meristem cell fate during *Arabidopsis* embryogenesis. *EMBO Journal*, **17**, 1799–809.

Moussian, B., Haecker, A. & Laux, T. (2003) ZWILLE buffers meristem stability in *Arabidopsis thaliana*. *Development Genes and Evolution*, **213**, 534–40.

Murfett, J., Wang, X. J., Hagen, G. & Guilfoyle, T. (2001) Identification of *Arabidopsis* histone deacetylase HDA6 mutants that affect transgene expression. *Plant Cell*, **13**, 1047–61.

Müller, A., Marins, M., Kamisugi, Y. & Meyer, P. (2002) Analysis of hypermethylation in the RPS element suggests a signal function for short inverted repeats in *de novo* methylation. *Plant Molecular Biology*, **48**, 383–99.

Mutskov, V. & Felsenfeld, G. (2004) Silencing of transgene transcription precedes methylation of promoter DNA and histone H3 lysine 9. *EMBO Journal*, **23**, 138–49.

Noma, K. -I., Sugiyama, T., Cam, H., Verdel, A., Zofall, M., Jia, S., Moazed, D. & Grewal, S. I. S. (2004) RITS acts in *cis* to promote RNA interference-mediated transcriptional and post-transcriptional silencing. *Nature Genetics*, **36**, 1174–80.

Novina, C. D. & Sharp, P. A. (2004) The RNAi revolution. *Nature*, **430**, 161–4.

Okamoto, I., Otte, A. P., Allis, C. D., Reinberg, D. & Heard, E. (2004) Epigenetic dynamics of imprinted X inactivation during early mouse development. *Science*, **303**, 644–9.

Pal-Bhadra, M., Leibovitch, B. A., Gandhi, S. G., Rao, M., Bhadra, U., Birchler, J. A. & Elgin, S. C. R. (2004) Heterochromatic silencing and HP1 localization in *Drosophila* are dependent on the RNAi machinery. *Science*, **303**, 669–72.

Papp, I., Mette, M. F., Aufsatz, W., Daxinger, L., Schauer, S. E., Ray, A., van der Winden, J., Matzke, M. & Matzke, A. J. M. (2003) Evidence for nuclear processing of plant microRNA and short interfering RNA precursors. *Plant Physiology*, **132**, 1382–90.

Park, C. W., Chen, Z., Kren, B. T. & Steer, C. J. (2004) Double-stranded siRNA targeted to the huntingtin gene does not induce DNA methylation. *Biochemical and Biophysical Research Communications*, **323**, 275–80.

Park, W., Li, J., Song, R., Messing, J. & Chen, X. (2002) CARPEL FACTORY, a Dicer homolog, and HEN1, a novel protein, act in microRNA metabolism in *Arabidopsis thaliana*. *Current Biology*, **12**, 1484–95.

Park, Y. D., Papp, I., Moscone, E. A., Iglesias, V. A., Vaucheret, H., Matzke, A. J. M. & Matzke, M. A. (1996) Gene silencing mediated by promoter homology occurs at the level of transcription and results in meiotically heritable alterations in methylation and gene activity. *Plant Journal*, **9**, 183–94.

Pélissier, T. & Wassenegger, M. (2000) A DNA target of 30 bp is sufficient for RNA-directed DNA methylation. *RNA*, **6**, 55–65.

Pélissier, T., Thalmeir, S., Kempe, D., Sänger, H. -L. & Wassenegger, M. (1999) Heavy *de novo* methylation at symmetrical and non-symmetrical sites is a hallmark of RNA-directed DNA methylation. *Nucleic Acids Research*, **27**, 1625–34.

Probst, A. V., Fagard, M., Proux, F., Mourrain, P., Boutet, S., Earley, K., Lawrence, R. J., Pikaard, C. S., Murfett, J., Furner, I., Vaucheret, H. & Mittelsten Scheid, O. (2004) *Arabidopsis* histone deacetylase HDA6 is required for maintenance of transcriptional gene silencing and determines nuclear organization of rDNA repeats. *Plant Cell*, **16**, 1021–34.

Ramsahoye, B. H., Biniszkiewicz, D., Lyko, F., Clark, V., Bird, A. P. & Jaenisch, R. (2000) Non-CG methylation is prevalent in embryonic stem cells and may be mediated by DNA methyltransferase 3a. *Proceedings of the National Academy of Sciences of the United States of America*, **97**, 5237–42.

Reinhart, B. J. & Bartel, D. P. (2002) Small RNAs correspond to centromere heterochromatic repeats. *Science*, **297**, 1831.

Reinhart, B. J., Weinstein, E., Rhoades, M., Bartel, B. & Bartel, D. (2002) MicroRNAs in plants. *Genes & Development*, **16**, 1616–26.

Riggs, A. D. (1975) X inactivation, differentiation, and DNA methylation. *Cytogenetics and Cell Genetics*, **14**, 9–25.

Riggs, A. D. (1990) Marsupials and mechanisms of X-chromosome inactivation. *Australian Journal of Zoology*, **37**, 419–41.

Saze, H., Mittelsten Scheid, O. & Paszkowski, J. (2003) Maintenance of CpG methylation is essential for epigenetic inheritance during plant gametogenesis. *Nature Genetics*, **34**, 65–9.

Schauer, S., Jacobsen, S. E., Meinke, D. & Ray, A. (2002) DICER-LIKE1: blind men and elephants in *Arabidopsis* development. *Trends in Plant Science*, **7**, 487–91.

Schramke, V. & Allshire, R. (2003) Hairpin RNAs and retrotransposon LTRs effect RNAi and chromatin-based silencing. *Science*, **301**, 1069–74.

Sijen, T., Vijn, I., Rebocho, A., van Blokland, R., Roelofs, D., Mol, J. N. & Kooter, J. M. (2001) Transcriptional and posttranscriptional gene silencing are mechanistically related. *Current Biology*, **11**, 436–40.

Singer, T., Yordan, C. & Martienssen, R. A. (2001) Robertson's mutator transposons in *A. thaliana* are regulated by the chromatin remodeling gene decrease in DNA methylation (DDM1). *Genes & Development*, **15**, 591–602.

Soppe, W. J. J., Jacobsen, S. E., Alonso-Blanco, C., Jackson, J. P., Kakutani, T., Koornneef, M. & Peeters, A. J. M. (2000) The late flowering phenotype of *fwa* mutants is caused by gain-of-function epigenetic alleles of a homeodomain gene. *Cell*, **6**, 791–802.

Soppe, W. J. J., Jasencakova, Z., Houben, A., Kakutani, T., Meister, A., Huang, M. S., Jacobsen, S. E., Schubert, I. & Fransz, P. F. (2002) DNA methylation controls histone H3 lysine 9 methylation and heterochromatin assembly in *Arabidopsis. EMBO Journal*, **21**, 6549–59.

Steimer, A., Amedeo, P., Afsar, K., Fransz, P., Mittelsten Scheid, O. & Paszkowski, J. (2000) Endogenous targets of transcriptional gene silencing in *Arabidopsis. Plant Cell*, **12**, 1165–78.

Suzuki, M. & Hayashizaki, Y. (2004) Mouse-centric comparative transcriptomics of protein coding and non-coding RNAs. *BioEssays*, **26**, 833–43.

Svoboda, P., Stein, P., Filipowicz, W. & Schultz, R. M. (2004) Lack of homologous sequence-specific DNA methylation in response to stable dsRNA expression in mouse oocytes. *Nucleic Acids Research*, **32**, 3601–606.

Tabler, M. & Tsagris, M. (2004) Viroids: petite RNA pathogens with distinguished talents. *Trends in Plant Science*, **9**, 339–48.

Tamaru, H. & Selker, E. U. (2001) A histone H3 methyltransferase controls DNA methylation in *Neurospora crassa. Nature*, **414**, 277–83.

Tamaru, H. & Selker, E. U. (2003) Synthesis of signals for *de novo* DNA methylation in *Neurospora crassa. Molecular and Cellular Biology*, **23**, 2379–94.

Tariq, M. & Paszkowski, J. (2004) DNA and histone methylation in plants. *Trends in Genetics*, **20**, 244–51.

Tariq, M., Saze, H., Probst, A. V., Lichota, J., Habu, Y. & Paszkowski, J. (2003) Erasure of CpG methylation in *Arabidopsis* alters patterns of histone H3 methylation in heterochromatin. *Proceedings of the National Academy of Sciences of the United States of America*, **100**, 8823–7.

Tian, L. & Chen, Z. J. (2001) Blocking histone deacetylation in *Arabidopsis* induces pleiotropic effects on plant gene regulation and development. *Proceedings of the National Academy of Sciences of the United States of America*, **98**, 200–205.

Tian, L., Wang, J., Fong, M. P., Chen, M., Cao, H., Gelvin, S. B. & Chen, Z. J. (2003) Genetic control of developmental changes induced by disruption of *Arabidopsis* histone deacetylase 1 (AtHD1) expression. *Genetics*, **165**, 399–409.

Tian, L., Fong, M. P., Wang, J. J., Wei, N. E., Jiang, H., Doerge, R. W. & Chen, Z. J. (2004) Reversible histone acetylation and deacetylation mediate genome-wide, promoter-dependent, and locus-specific changes in gene expression during plant development. *Genetics*, **169**, 337–45 (Epub ahead of print).

Tompa, R., McCallum, C. M., Delrow, J., Henikoff, J., van Steensel, B. & Henikoff, S. (2002) Genome-wide profiling of DNA methylation reveals transposon targets of CHROMOMETHYLASE3. *Current Biology*, **12**, 65–8.

van Blokland, R., van der Geest, N., Mol, J. & Kooter, J. (1994) Transgene-mediated suppression of chalcone synthase expression in *Petunia hybrida* results from an increase in RNA turnover. *Plant Journal*, **6**, 861–77.

van der Krol, A. R., Mur, L. A., Beld, M., Mol, J. N. M. & Stuitje, A. R. (1990) Flavonoid genes in petunia: addition of a limited number of gene copies may lead to a suppression of gene expression. *Plant Cell*, **2**, 291–9.

Vaucheret, H., Vazquez, F., Crété, P. & Bartel, D. P. (2004) The action of ARGONAUTE1 in the miRNA pathway and its regulation by the miRNA pathway are crucial for plant development. *Genes & Development*, **18**, 1187–97.

Verdel, A., Jia, S., Gerber, S., Sugiyama, T., Gygi, S., Grewal, S. I. S. & Moazed, D. (2004) RNAi-mediated targeting of heterochromatin by the RITS complex. *Science*, **303**, 672–6.

Vogt, U., Pélissier, T., Putz, A., Razvi, F., Fischer, R. & Wassenegger, M. (2004) Viroid-induced RNA silencing of GFP-viroid fusion transgenes does not induce extensive spreading of methylation or transitive silencing. *Plant Journal*, **38**, 107–18.

Volpe, T. A., Kidner, C., Hall, I. M., Teng, G., Grewal, S. I. & Martienssen, R. A. (2002) Regulation of heterochromatic silencing and histone H3 lysine-9 methylation by RNAi. *Science*, **297**, 1833–7.

Volpe, T., Schramke, V., Hamilton, G. L., White, S. A., Teng, G., Martienssen, R. A. & Allshire, R. C. (2003) RNA interference is required for normal centromere function in fission yeast. *Chromosome Research*, **11**, 137–46.

Vongs, A., Kakutani, T., Martienssen, R. A. & Richards, E. J. (1993) *Arabidopsis thaliana* DNA methylation mutants. *Science*, **260**, 1926–8.

Walsh, C. P. & Bestor, T. H. (1999) Cytosine methylation and mammalian development. *Genes & Development*, **13**, 26–34.

Wang, Y., Wysocka, J., Sayegh, J., Lee, Y. H., Perlin, J. R., Leonelli, L., Sonbuchner, L. S., McDonald, C. H., Cook, R. G., Dou, Y., Roeder, R. G., Clarke, S., Stallcup, M. R., Allis, C. D. & Coonrad, S. A. (2004) Human PAD4 regulates histone arginine methylation levels via demethylimination. *Science*, **306**, 279–83.

Wassenegger, M., Heimes, S., Riedel, L. & Sänger, H. L (1994) RNA-directed *de novo* methylation of genomic sequences in plants. *Cell*, **76**, 567–76.

Wilkinson, C. R., Bartlett, R., Nurse, P. & Bird, A. P. (1995) The fission yeast gene pmt1+ encodes a DNA methyltransferase homologue. *Nucleic Acids Research*, **23**, 203–210.

Woodcock, D. M., Lawler, C. B., Linsenmeyer, M. E., Doherty, J. P. & Warren, W. D. (1997) Asymmetric methylation in the hypermethylated CpG promoter region of the human L1 retrotransposon. *Journal of Biological Chemistry*, **272**, 7810–16.

Xie, Z., Johansen, L. K., Gustafson, A. M., Kasschau, K. D., Lellis, A. D., Zilberman, D., Jacobsen, S. E. & Carrington, J. C. (2004) Genetic and functional diversification of small RNA pathways in plants. *PLoS Biology*, **2**, 642–52.

Yan, K. S., Yan, S., Farooq, A., Han, A., Zeng, L. & Zhou, M. M. (2003a) Structure and conserved RNA binding of the PAZ domain. *Nature*, **426**, 468–74.

Yan, Q., Cho, E., Lockett, S. & Muegge, K. (2003b) Association of Lsh, a regulator of DNA methylation, with pericentromeric heterochromatin is dependent on intact heterochromatin. *Molecular and Cellular Biology*, **23**, 8416–28.

Yoder, J. A., Walsh, C. P. & Bestor, T. H. (1997) Cytosine methylation and the ecology of intragenomic parasites. *Trends in Genetics*, **13**, 335–40.

Zhang, Y. (2004) No exception to reversibility. *Nature*, **431**, 637–8.

Zilberman, D., Cao, X., & Jacobsen, S. E. (2003) ARGONAUTE4 control of locus-specific siRNA accumulation and DNA and histone methylation. *Science*, **299**, 716–19.

Zilberman, D., Cao, X., Johansen, L. K., Xie, Z., Carrington, J. C. & Jacobsen, S. E. (2004) Role of *Arabidopsis* ARGONAUTE4 in RNA-directed DNA methylation triggered by inverted repeats. *Current Biology*, **14**, 1214–20.

4 Heterochromatin and the control of gene silencing in plants

G. Reuter, A. Fischer and I. Hofmann

4.1 Introduction

Euchromatin and heterochromatin form distinct chromosomal domains in eukaryotic organisms. Euchromatin contains actively transcribed genes and becomes decondensed during interphase. Heterochromatin, in contrast, remains condensed and densely stained throughout the cell cycle. Heterochromatic regions are regularly found at centromeric (pericentromeric heterochromatin), around nucleolus organiser regions (NORs) and at telomeric regions. Heterochromatin contains only few active genes and is composed mainly of long stretches of repetitive sequences packed into a condensed and less accessible form with regularly spaced nucleosomes. Heterochromatisation of euchromatic sequences causes gene silencing, and facultative heterochromatisation represents an epigenetic mechanism for permanent gene inactivation during development. Heterochromatic gene silencing affects transposons, endogenous repeated coding sequences, transgene repeats or is found in position effect variegation (PEV). Molecular mechanisms establishing heterochromatic domains are highly conserved. Covalent modification of histone amino-terminal tails function as epigenetic marks to index heterochromatic domains. In general, heterochromatin is characterised by hypoacetylation of lysine residues in histone H3 and H4 and by methylation of histone H3 at lysine 9 and 27, and lysine 20 at histone H4. Mono-, di- and trimethylation at these lysine residues increase complexity of indexing processes in chromatin. The histone methylation marks of heterochromatin show both conserved and species-specific characteristics. In most organisms, DNA in heterochromatin is hypermethylated at cytosine residues, which is important for epigenetic maintenance of silenced heterochromatic structures. Histone-modifying enzymes (acetyl-transferases, deacetylases and methyltransferases) and DNA methyltransferases are evolutionary conserved proteins found in major eukaryotic model organisms. Histone and DNA methylation are interdependent processes and, by interaction with other factors, cause establishment and stable maintenance of densely packed and genetically repressed heterochromatic structures. Heterochromatin is late replicating and plays a role in chromosome stability and control of recombination frequencies.

4.2 Cytological, molecular and genetic characteristics of heterochromatin in plants

4.2.1 Discovery of heterochromatin and defining its cytological characteristics

In his cytological analysis of chromosome behaviour during mitosis in the liver moss *Pellia epiphylla* Heitz (1928) observed that parts of certain chromosomes do not become invisible in interphase. These parts of chromosomes, which he found frequently attached to the nucleolus, are regularly observed in nuclei of differentiated cells. In order to distinguish between heteropycnotic chromosomes (heterochromosomes, Montgomery, 1904) and heteropycnotic chromosomal regions, Heitz introduced for the latter the term heterochromatin. Those regions that become invisible in interphase he called euchromatin. Heitz (1929) furthermore stressed the direct correlation between heterochromatic chromosome regions and chromocenters of interphase nuclei and was the first to differentiate between different types of nuclei according to heterochromatin distribution (Heitz, 1932).

Recent classification differentiates between diffuse type, gradient type and simple or complex chromocenter type of nuclei (cf. Fukui & Nakayama, 1996). Plate 4.1 shows several typical examples for differential distribution of heterochromatin in interphase nuclei. The nuclei were stained with DAPI and distribution of dimethylated histone H3K9, the specific histone modification mark of heterochromatin, in plants was studied by immunocytology. Diffuse distribution of heterochromatin is found in *Hordeum vulgare* and *Allium cepa*. In *Scilla mischtschenkoana* large masses of heterochromatin, which are frequently concentrated in one half of the nucleus, are found (gradient type of nuclei) whereas chromocenter type nuclei are characteristic for many plants as shown for the liver moss *Marchantia polymorpha*, the lily *Ornithogalum longebracteanum* and *Arabidopsis thaliana*. In species with a gradient type distribution of heterochromatin, a typical Rabl orientation (Rabl, 1885) of chromosomes is found with centromeres and telomeres located at opposite nuclear territories (Abranches *et al.*, 1998; Jasencakova *et al.*, 2001). In *Arabidopsis* the number of chromocenters corresponds to the number of chromosomes. FISH analysis with probes containing centromeric and pericentromeric repeats proved that chromocenters are the nuclear domains of pericentromeric heterochromatin whereas euchromatic regions form loops around chromocenters. Frequently, an association of homologous chromocenters is found (Fransz *et al.*, 2002).

Differential staining of heterochromatic regions by base-specific fluorochromes or by the C-banding technique allowed comparable analysis of heterochromatin distribution in many plant species. Guerra (2000) compared

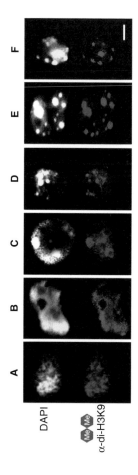

Plate 4.1 Differential heterochromatin distribution in various plant species. Diffuse heterochromatin distribution in (A) *Hordeum vulgare* and (B) *Allium cepa*. Gradient type of heterochromatin distribution in (C) *Scilla mischtschenkoana*. Chromocenter type nuclei are found in many plant species. (D) Liver moss *Marchantia polymorpha*, (E) lily *Ornithogalum longebracteanum* and (F) wall cress *Arabidopsis thaliana* (F). Upper row DAPI staining and lower row staining with α-di-H3K9. Bar represents 10 μm. See also colour plate section for colour version of this figure.

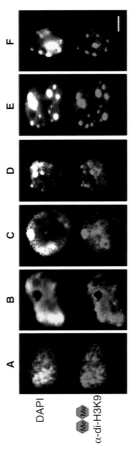

Plate 4.1 Differential heterochromatin distribution in various plant species. Diffuse heterochromatin distribution in (A) *Hordeum vulgare* and (B) *Allium cepa*. Gradient type of heterochromatin distribution in (C) *Scilla mischtschenkoana*. Chromocenter type nuclei are found in many plant species. (D) Liver moss *Marchantia polymorpha*, (E) lily *Ornithogalum longebracteanum* and (F) wall cress *Arabidopsis thaliana* (F). Upper row DAPI staining and lower row staining with α-di-H3K9. Bar represents 10 µm.

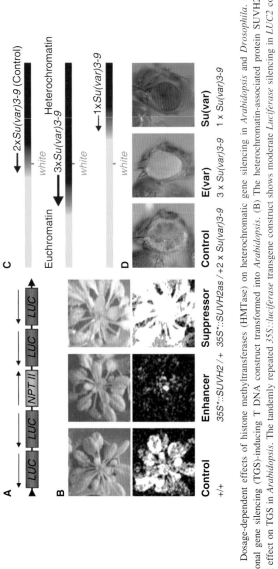

Plate 4.2 Dosage-dependent effects of histone methyltransferases (HMTase) on heterochromatic gene silencing in *Arabidopsis* and *Drosophila*. (A) Structure of a transcriptional gene silencing (TGS)-inducing T DNA construct transformed into *Arabidopsis*. (B) The heterochromatin-associated protein SUVH2 shows a dosage-dependent effect on TGS in *Arabidopsis*. The tandemly repeated *35S::luciferase* transgene construct shows moderate *Luciferase* silencing in *LUC2* control plants. After overexpression of *SUVH2* a strong enhancer effect is found whereas TGS at the *Luciferase* transgene is significantly suppressed in *SUVH2*-null plants containing an antisense *35S*::SUVH2as* construct. SUVH2 in *Arabidopsis* is a HMTase that controls mono- and dimethylation of H3K9 and H3K27 and monomethylation of H4K20 (Naumann *et al.*, 2005). (C) Variable silencing of the *white* gene in the PEV rearrangment w^{m4} depends on *Su(var)3–9* dosage. (D) An additional gene copy of *Su(var)3–9* causes strong enhancement whereas a deletion of one *Su(var)3–9* copy results in suppression of *white* gene silencing in position-effect variegation. SU(VAR)3–9 is central for dimethylation of H3K9 in *Drosophila* heterochromatin (Schotta *et al.*, 2002; Ebert *et al.*, 2004).

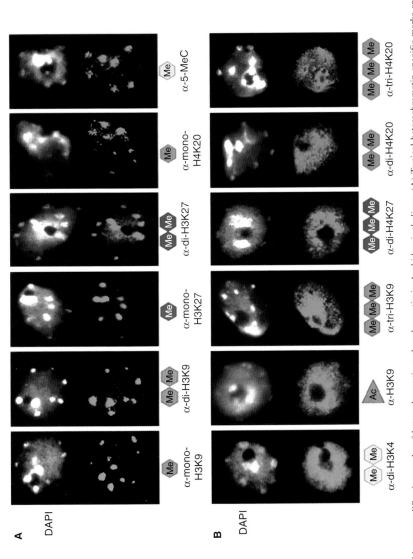

Plate 4.3 The histone modification marks of heterochromatin and euchromatin in *Arabidopsis thaliana*. (A) Typical heterochromatin specific marks are monomethyl H3K9, dimethyl H3K9, monomethyl H3K27, dimethyl H3K27, monomethyl H4K20 and 5-methylcytosine methylation. (B) Euchromatic marks are dimethylated H3K4, acetylated H3K9, trimethyl H3K9, trimethyl H3K27, dimethyl H4K20 and trimethyl H4K20.

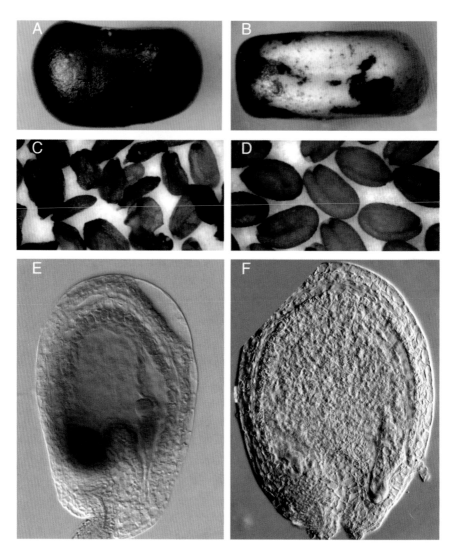

Plate 6.1 Examples of imprinted genes in plants. (A) and (B) show kernels with a maternally and a paternally inherited *R* allele of maize, respectively. The *R* allele was exposed to *R-st* in the previous generation, which strengthens the effect. (C) and (D) show *Arabidopsis* seeds with a maternally and a paternally inherited mutant *medea* allele, respectively. Maternal inheritance of a mutant *medea* allele causes seed abortion irrespective of the paternal contribution. (E) and (F) show a maternally and a paternally inherited *MEDEA-promoter::GUS* fusion construct, respectively. Only the maternally inherited allele is active and produces β-galactosidase (GUS), which is detected as a blue stain in a cytochemical reaction. Some diffusion of the blue product into the surrounding sporophytic tissue is visible.

C-band distribution patterns of 58 dicotyledonous and 32 monocotyledonous species. These studies showed that heterochromatin is preferentially found at proximal chromosomal regions (pericentromeric heterochromatin) and at NORs. Interstitial heterochromatic regions are more variable in their chromosomal positions. Interestingly, no difference in chromosomal distribution of heterochromatin is found between species with holocentric and monocentric chromosomes.

A distinguishing feature of heterochromatin is its replication in the late S phase. This was first shown by Lima-de-Faria and Jaworska (1986) and until now has been established for a wide range of organisms (Gilbert, 2002). However, heterochromatic regions in all organisms might not fall into this paradigm as demonstrated for heterochromatic centromeres and silent mating type cassettes in the fission yeast *Schizosaccharomyces pombe*, which in contrast to the heterochromatic telomeres replicate in early S phase (Kim *et al.*, 2003). Comparable analysis of DNA replication origins between transcriptional active and silenced X chromosomes in mouse revealed that heterochromatin does not prevent replication origin activity but rather delays assembly or progression of the replication machinery (Gomez & Brockdorff, 2004). In *Drosophila* the heterochromatin-associated protein suppressor of underreplication (SUUR) appears to control late replication in heterochromatin by interaction with proteins directly involved in heterochromatin formation (Zhimulev *et al.*, 2003).

Differences in packaging between euchromatin and heterochromatin at the nucleosomal level have been demonstrated in *Drosophila* by probing with nucleases (Sun *et al.*, 2001). In contrast to euchromatin with irregular nucleosomal arrays, heterochromatic regions are characterised by long-range regular spacing of nucleosomes. In animals different chromatin remodelling complexes have been identified (Tsukiyama, 2002). The ACF1–ISWI remodelling complex localises during replication with pericentromeric heterochromatin and is essential for its replication (Collins *et al.*, 2002). ACF1 depletion experiments showed that its requirement for replication progression of heterochromatic sequences correlates with DNA hypermethylation. In *Arabidopsis* the DDM1 protein was shown to act as chromatin-remodelling factor (Brzeski & Jerzmanowski, 2003). Recombinant DDM1 binds nucleosomes and promotes chromatin remodelling in an ATP-dependent manner. Enzymatic activity of DDM1 was not affected by DNA methylation *in vitro* although *in vivo* DDM1 is involved in maintenance of DNA methylation. *Ddm1* mutations cause immediate hypomethylation of repeated pericentromeric sequences but only gradual and delayed loss of methylation in single-copy sequences (Kakutani *et al.*, 1996; Vongs *et al.*, 1993). Decondensation of chromocenters in *ddm1* mutations (Probst *et al.*, 2003) indicates preferential association of DDM1 with heterochromatin.

*4.2.2   Sequence content, chromosomal and genomic organisation
          of heterochromatin*

In plants high-resolution cytogenetic mapping and DNA sequence analysis was integrated first in *A. thaliana*. In *Arabidopsis* with its small genome (about 120 Mb per haploid genome) and detailed genetic and physical maps, structural organisation of heterochromatic regions could be resolved (Copenhaver *et al.*, 1999; Fransz *et al.*, 2000). In a study of chromosome arm 4S of *A. thaliana* Fransz *et al.* (2000) showed that the majority of repeated elements are restricted to pericentromeric heterochromatin and NORs. Detailed mapping of heterochromatic regions in *Arabidopsis* also developed new tools for analysis of epigenetic processes connected with heterochromatin. Pachytene chromosome analysis allowed cytogenetic differentiation between centromeric and pericentromeric regions. The centromere region of chromosome 4 was found to be largely composed of tandemly arranged 180-bp satellite sequences and several other dispersed repeats whereas pericentromeric heterochromatin is rich in transposon sequences, e.g. Athila. In rice the centromere of chromosome 8 (*Cen8*) was molecularly analysed (Nagaki *et al.*, 2004). The region binds the rice centromere-specific histone H3, CENH3, and is localised within a larger region enriched in dimethylated histone H3K9 (pericentromeric heterochromatin). The 750-kb rice *Cen8* is comparable in size to the 420-kb centromere of a *D. melanogaster* mini-chromosome (Sun *et al.*, 1997), a maize B chromosome centromere of about 500 kb (Kaszas & Birchler, 1998) and human neocentromeres with a length of 330 and 460 kb, respectively (Lo *et al.*, 2001a, b). Sequence analysis of *Arabidopsis CEN2* and *CEN4* regions revealed, besides the 180-bp repeat sequences, retrotransposons, middle repetitive elements and telomeric repeat sequences but in comparison to other organisms (*Drosophila* and *Neurospora*) only a low amount of low-complexity DNA. A more uniform structure with only two different types of repeats was described for *Arabidopsis CEN1* (Haupt *et al.*, 2001). In rice *Cen8* several active genes were detected, a situation that can be compared with *Drosophila* where active genes in pericentromeric heterochromatin are also found at low density (Lohe & Hilliker, 1995). Gene density in *Arabidopsis* heterochromatin is also low (Mayer *et al.*, 1999). More recent work shows that all five *Arabidopsis* centromeres are located within a 180-bp satellite array of about 3 Mb but in contrast to rice *Cen8*, they do not contain active genes (Hosouchi *et al.*, 2002; Nagaki *et al.*, 2003). Identification and mapping of the full complement of *Arabidopsis* long terminal repeat (LTR) retroelements (about 4.4% of total genome) onto the genome sequence showed variable degree of clustering at pericentromeric heterochromatin (Peterson-Burch *et al.*, 2004). A strict association with pericentromeric heterochromatin is found for *Tat* and *Athila* transposons.

An analysis of DNA methylation along the euchromatic–heterochromatic transition regions revealed a gradual increase in cytosine methylation at pericentromeric heterochromatin (Mathieu *et al.*, 2002). At chromosome 5 two 5S rDNA clusters flank the about 2-Mb long pericentromeric region. The 5S rRNA genes are highly methylated although transcriptionally active. Besides high CpG methylation of satellite repeats in plants additional enrichment of CpNpG and non-symmetrical methylation is found in heterochromatic transposon sequences (Bartee *et al.*, 2001; Lindroth *et al.*, 2001).

Of special interest are heterochromatic knobs first reported in maize (McClintock, 1929). In *Arabidopsis*, chromosome arm 4S contains in several ecotypes a heterochromatic knob (Fransz *et al.*, 1998). An origin of this heterochromatic knob by inversion of pericentromeric repeated DNA to a more distal euchromatic region has been suggested (Fransz *et al.*, 2000). Heterochromatic knobs affect recombination and can even cause segregation distortion (Rhoades, 1978). Sequence analysis of heterochromatic knobs in maize and *Arabidopsis* revealed a heterochromatin-like combined organisation with tandem repeat satellite sequences and dispersally inserted retrotransposons (Ananiev *et al.*, 1998; McCombie *et al.*, 2000). Its uniform association with dimethylated histone H3K9 (Gendrel *et al.*, 2002) also supports heterochromatic structure of the knob on chromosome 4S of *Arabidopsis*.

Because in most plants, chromosome painting has been unsuccessful (Fuchs *et al.*, 1996; Lysak *et al.*, 2001), a wide distribution of repetitive heterochromatic elements throughout the genome was suggested. Analyses of sequence composition and organisation in the maize genome revealed that repeat sequences occupy 58% (Messing *et al.*, 2004). Multi-colour FISH analysis with tandemly repeated DNA sequences generated distinctive banding pattern for each of the ten maize chromosomes (Kato *et al.*, 2004a) and revealed significant variation in presence and size of repeated sequences among different inbred lines, indicating a highly dynamic behaviour of interstitial heterochromatic regions.

Accessory or B chromosomes, which are largely heterochromatic, are found sporadically in many plant species (Jones, 1995). B chromosomes do not contain major genes with specific phenotypic effects. In certain B chromosomes rRNA genes were detected, which are all inactivated (Donald *et al.*, 1997). Weak immunostaining for H4Ac5 and H4Ac8, late replication and occurrence of hypermethylated, highly repeated sequences suggest a heterochromatic nature of B chromosomes (for review cf. Puertas, 2002). These chromosomes significantly increase the total amount of heterochromatic material in the genome but no detailed data are available about their origin or possible effects on fitness under natural conditions.

Heteromorphic sex chromosomes have only been described in a few dioecious plants (Charlesworth, 2002; Vyskot & Hobza, 2004). *Silene latifolia* and *Rumex acetosa* are the best-studied species with heteromorphic sex chromosomes. The Y chromosomes of *R. acetosa* represent constitutive heterochromatin and are visible in interphase nuclei as 'sex-bodies' (Lengerova &

Vyskot, 2001). In female nuclei of the white campion *S. latifolia* higher overall DNA methylation was detected in one of the two X chromosomes (Siroky *et al.*, 1998), suggesting X chromosome inactivation by heterochromatisation although clear molecular evidence is still missing. Induced hypomethylation causes sex reversal and formation of bisexual flowers in XY plants of *S. latifolia*, indicating that heterochromatic gene silencing might be involved in control of sex-determining genes (Janousek *et al.*, 1996). Compared with mammalian sex chromosomes *S. latifolia* sex chromosomes are evolutionarily about ten times younger. Sequence analysis shows limited degeneration of Y chromosomal alleles (Atanassov *et al.*, 2001; Guttman & Charlesworth, 1998), a situation which might be comparable to the neoY chromosome in *Drosophila miranda* (Steinemann & Steinemann, 2001).

4.2.3   Heterochromatin and genetic recombination

Heterochromatin is characterised by suppression of crossing-over (Baker, 1958; Szauter, 1984). Stack (1984) showed in two angiospermous plants, *Plantago ovata* and *Lycopersicon esculentum*, significant structural differences in synaptonemal complexes between euchromatic and heterochromatic regions and suggested that the highly condensed structure of synaptonemal complexes in heterochromatin prevents recombination. In maize comparison of crossing-over frequencies between gene-rich regions and regions containing retrotransposons revealed significant crossing-over suppression in transposon-rich regions (Fu *et al.*, 2002; Yao *et al.*, 2002). Tetrad analysis allowed mapping of pericentromeric regions in all *Arabidopsis* chromosomes (Copenhaver *et al.*, 1999). The total size of all pericentromeric heterochromatin (including the centromere cores) comprises about 21 Mb. In pericentromeric regions of *Arabidopsis* recombination rates 10–30 times below the genomic average were found. Lethal mutations linked to the *CEN1* pericentromeric region were used for measuring recombination frequencies. Within the centromere core crossover is almost completely suppressed. Significant differences of crossover suppression between left and right pericentromeric regions are found. In *Arabidopsis* suppression of recombination occurs abruptly at euchromatic–heterochromatic transition regions. This is in contrast to what is found in *Drosophila* where suppression of crossover by pericentromeric heterochromatin occurs along all proximal euchromatin. Several of the dominant mutations suppressing heterochromatic gene silencing in *Drosophila* display significant recombinogenic effects within pericentromeric heterochromatin and at proximal euchromatin (Westphal & Reuter, 2002). If different non-allelic mutations are combined into one genotype, seven to eight times increase in crossover between markers flanking pericentromeric heterochromatin is observed. Recently mutations of the *Arabidopsis BRU1* gene were identified as suppressors of transcriptional gene silencing (TGS) (Takeda *et al.*,

2004). Significantly increased homologous recombination rates in *bru1* mutants might indicate an involvement of BRU1 in controlling gene silencing and recombination within heterochromatic sequences.

4.2.4   *Heterochromatin and gene silencing in position effect variegation*

In PEV, euchromatic genes juxtaposed to heterochromatin become silenced (variegated phenotype) by heterochromatisation (facultative heterochromatin). This phenomenon was discovered by Muller (1930) in X-ray-induced chromosomal rearrangements of *Drosophila*. Heterochromatic gene silencing in PEV, therefore, reflects the repressive effect of heterochromatin on active genes. Although PEV has been studied intensively in *Drosophila* and was demonstrated in mammals (Russel & Bangham, 1961), in plants only in *Oenothera blandina* an unequivocal case of PEV has been described (Catcheside, 1939, 1947). The gene affected is the *P* locus located in chromosome arm 3 of the 3.4 chromosome. Catcheside demonstrated variegated expression of the dominant *P* alleles in an X-ray-induced translocation between chromosomes 3.4 and 11.12. For the *P* locus a series of alleles has been identified. One of these allele, P^r, which produces uniform red sepals, was inserted into the translocation via crossing-over. Inactivation of the P^r in translocation becomes visible in heterozygotes with a P^s allele by variegation for green sectors. Transfer of P^r back to a normal 3.4 chromosome resulted in normal P^r expression. The *S* locus on chromosome arm 3, which is about 8 crossover units distal to *P* was also found to variegate, demonstrating polar spreading of gene silencing over a remarkably long genetic distance. Spreading of inactivation over an euchromatic region is a characteristic feature of gene silencing by heterochromatisation in PEV. No other documented case of PEV in plants has appeared to date. However, several other related phenomena of gene silencing that appear to have a similar molecular basis as PEV have been detected in plants (Matzke & Matzke, 1993, 1995; Meyer, 1995; Meyer & Saedler, 1996).

4.2.5   *Transcriptional gene silencing by heterochromatisation*

Phenomena comparable to PEV include homology-dependent gene silencing in transgenic plants and paramutation. Generally, in homology-dependent gene silencing, an increase in the copy number of particular sequences is positively correlated with a reduction in gene expression of both the endogenous and the *trans* copies. This can be due to cosuppression (Jorgensen, 1990) at a posttranscriptional level or *trans*-inactivation at the transcriptional level (Matzke & Matzke, 1990). It is generally believed that *trans*-inactivation is accompanied by a change in higher-order chromatin structure (heterochromatisation) of the affected gene. Different models have been proposed for the mechanism of *trans*-inactivation. Both the DNA–DNA and the RNA–DNA pairing model posit an exchange of epigenetic states (methylation and/or chromatin structure).

RNA is implicated in the induction of DNA methylation in tobacco transformed with cDNA of the potato spindle tuber viroid (Wassenegger *et al.*, 1994). A potential dependence of heterochromatic silencing on short interfering RNA transcripts in plants has recently been intensively studied (cf. Baulcombe, 2004; Lippman & Martienssen, 2004). Double-stranded RNA can induce *de novo* methylation in a homologous target DNA sequence (Mette *et al.*, 2000; Pelissier *et al.*, 1999). This RNA-directed DNA methylation depends on the DRM and MET1 methyltransferases and also requires DDM1 and the histone deacetylase *At*HDA6 (Aufsatz *et al.*, 2002; Jones *et al.*, 2001; Morel *et al.*, 2000). Because only a small subset of transposon silencing depends on the RNAi machinery (Lippman *et al.*, 2003), it appears more likely that different molecular mechanisms of epigenetic regulation are involved in the induction of heterochromatic gene silencing.

In most cases of *trans*-inactivation an obvious correlation between gene inactivation and DNA methylation of CpG dinucleotides and CpNpG trinucleotides has been shown (Matzke *et al.*, 1994; Meyer *et al.*, 1993). If promoters are affected, they must be duplicated (Matzke *et al.*, 1993). If the duplications occur only in the coding regions, the promoter region is not affected (Goring *et al.*, 1991), suggesting that duplication per se triggers methylation. This led to the suggestion that pairing of homologous sequences causes *trans*-inactivation (Matzke *et al.*, 1994).

Different repeat-dependent TGS or repeat-induced gene silencing (RIGS) systems were established for genetic dissection of heterochromatic gene silencing in *Arabidopsis*. One of the first systems was based on a single insertion of a transgenic construct containing the *hygromycin phosphotransferase* (*HPT*) gene flanked by two *neomycin phosphotransferase* (*NPT*) genes with different non-overlapping deletions (Assaad & Signer, 1992). Recombination events between the directly repeated *NPT* genes generated an allelic series of single-copy as well as multi-copy inserts. Because all recombinants derived from one primary single insertion, position effects could be ruled out. It could be shown that gene silencing depends strictly on the presence of repeated sequences and is correlated with a lack of run-on transcripts (Ye & Signer, 1996). The first silencing mutants could be isolated with a TGS system based on a transcriptional inactivated *HPT* line containing a multi-copy insert (Mittelsten Scheid *et al.*, 1991, 1998). At least five of the isolated *somniferous* (*som*) mutants were *ddm1* alleles. In a second screen using T DNA mutagenesis, *morpheus molecule1* (*mom1*) was isolated.

With another TGS system based on multiple inserts of a T DNA containing the *HPT*, *NPT* and *chalconsynthase* (*CHS*) genes (Davies *et al.*, 1997) the mutants *sil1*, *sil2* (modifiers of silencing) and homology-dependent silencing (*hog1*) were isolated (Furner *et al.*, 1998). The *sil1* mutation is allelic to *axe1* and both identify the gene encoding the *At*HDA6 histone deacetylase (Probst *et al.*, 2004).

Other TGS systems depend on silencing of endogenous loci. Epigenetically silenced alleles of the *SUP* locus represent the *clark kent* (*clk*) alleles. The *clk* epi-alleles like *SUP* mutants are characterised by an increased number of stamen. Based on the observation that *clk* alleles spontaneously revert to the non-silenced state, a transgenic allele called *clk-st* was used for isolation of silencing mutations. The transgenic *clk-st* line contains a single inverted repeat of the *SUP* locus, which causes silencing of the transgenic and the endogenous *SUP* locus (Cao & Jacobsen, 2002a). With the *clk-st* system a new set of silencing mutants with specific effects on DNA methylation could be isolated. The isolated mutants are *cmt3* (Lindroth *et al.*, 2001), *kyp* (Jackson *et al.*, 2002) and *ago4* (Zilberman *et al.*, 2003).

Another well-established non-transgenic silencing system in *Arabidopsis* is based on a natively silenced endogenous locus encoding the tryptophan pathway enzyme *phosphoribosylanthranilate isomerase* (*PAI*). In the *Wassilewskija* (*WS*) ecotype the *PAI2* locus is silenced due to an unlinked inverted repeat *PAI1–PAI4* gene rearrangement (Luff *et al.*, 1999). The only source of PAI enzyme activity is the *PAI1* gene because the *PAI2* locus is silenced and hypermethylated. The *WS* strain (*pai1C251Y*) contains a mutated *PAI1* gene and accumulates tryptophan pathway intermediates resulting in yellowish green leaves and other morphological abnormalities (Bartee & Bender, 2001). Second-site mutations that relieve silencing of the inactivated *PAI2* locus suppress the *PAI*-deficient phenotypes. Screens for suppressors of *PAI2* silencing yielded altogether 11 mutant alleles of the *CMT3* DNA methyltransferase and 7 alleles of the histone H3K9 methyltransferase *SUVH4* (Bartee *et al.*, 2001; Malagnac *et al.*, 2002).

Recently, another TGS system that does not depend on reactivation of genes mediating antibiotic resistance was established (Hofmann, I., Fischer, A., Fiedler, C., Thümmler, A., Scheel, D., Tschiersch, B., Reuter, G., unpublished data; Naumann *et al.*, 2005). This TGS system is based on a transgene construct containing four tandemly arranged 35S::luciferase repeats and allows direct visualisation of gene silencing by monitoring luciferase activity. This luciferase transgene repeat system can be used for analysis of both suppressors and enhancers of TGS in *Arabidopsis* (Plate 4.2).

Most of the isolated TGS suppressor mutants (*met1*, *ddm1*, *cmt3*, *kyp*, *drm1* and *drm2* alleles) affect DNA or histone methylation within chromocenter heterochromatin and at silenced genes. Their concomitant effect on pericentromeric heterochromatin and gene silencing proves that in TGS, genes are silenced by heterochromatisation. The silencing defective *ros1* mutations also cause activation of a silenced transgene. These mutations identify a putative DNA glycosylase/lyase (Gong *et al.*, 2002). Recombinant ROS1 induces

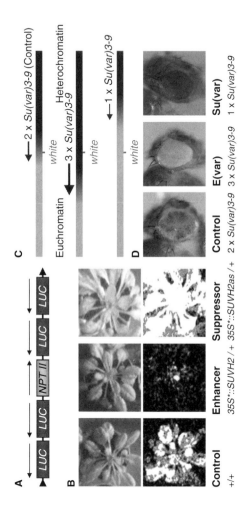

Plate 4.2 Dosage-dependent effects of histone methyltransferases (HMTase) on heterochromatic gene silencing in *Arabidopsis* and *Drosophila*. (A) Structure of a transcriptional gene silencing (TGS)-inducing T DNA construct transformed into *Arabidopsis*. (B) The heterochromatin-associated protein SUVH2 shows a dosage-dependent effect on TGS in *Arabidopsis*. The tandemly repeated *35S::luciferase* transgene construct shows moderate *Luciferase* silencing in *LUC2* control plants. After overexpression of *SUVH2* a strong enhancer effect is found whereas TGS at the *Luciferase* transgene is significantly suppressed in *SUVH2*-null plants containing an antisense *35S*::SUVH2as* construct. SUVH2 in *Arabidopsis* is a HMTase that controls mono- and dimethylation of H3K9 and H3K27 and monomethylation of H4K20 (Naumann *et al.*, 2005). (C) Variable silencing of the *white* gene in the PEV rearrangment w^{m4} depends on *Su(var)3–9* dosage. (D) An additional gene copy of *Su(var)3–9* causes strong enhancement whereas a deletion of one *Su(var)3–9* copy results in suppression of *white* gene silencing in position-effect variegation. SU(VAR)3–9 is central for dimethylation of H3K9 in *Drosophila* heterochromatin (Schotta *et al.*, 2002; Ebert *et al.*, 2004). See also colour plate section for colour version of this figure.

in vitro strand breaks in DNA containing 5-methylcytosine, indicating involvement of ROS1 in DNA demethylation through a base excision repair mechanism. Another putative DNA glycosylase is encoded by the *DEMETER* gene, which was identified by mutants interfering with imprinting of MEA (Choi *et al.*, 2002). Only *mom1* (Amedeo *et al.*, 2000) and *bru1* (Takeda *et al.*, 2004) mutations suppress gene silencing without changing DNA or histone modification and therefore might affect higher-order levels of chromatin regulation.

4.3 DNA and histone modification in plant heterochromatin

Epigenetic processes are controlled by complex DNA and histone modification systems. Distinct differences in DNA methylation and histone modification marks are found between heterochromatic and euchromatic chromosomal domains (Jenuwein & Allis, 2001; Stahl & Allis, 2000). Epigenetic stable transmission of the condensed and transcriptionally inert state of heterochromatin is directly correlated with these modifications.

In plants, involvement of histone and DNA methylation in heterochromatin formation and heterochromatic gene silencing was documented in studies of SU(VAR)3–9 homologous SUVH proteins and DNA methylation-defective mutations of the *MET1*, *CMT3*, *DRM* and *DDM1* genes (Cao & Jacobsen, 2002a; Gendrel *et al.*, 2002; Jackson *et al.*, 2002; Naumann *et al.*, 2005; Tariq *et al.*, 2003). In contrast to animals, development in plants is rather plastic and considerably more affected by environmental effects. Therefore plants might require more subtle changes in chromatin structure for fine-tuning of gene regulation. Accordingly, plants contain multi-gene families for DNA and histone modification systems. In *Arabidopsis*, three different classes of DNA methyltransferases (Martienssen & Colot, 2001), 12 putative methyl-cytosine-binding proteins (Zemach & Grafi, 2003), 37 SET domain proteins (Baumbusch *et al.*, 2001), 18 putative histone deacetylases and 12 putative histone acetyltransferases (*Arabidopsis* Genome Initiative, 2000; Pandey *et al.*, 2002) were identified.

4.3.1 *SUVH proteins and the control of heterochromatic chromatin domains*

SU(VAR)3–9-like proteins have a fundamental role in controlling heterochromatin formation in animals, yeast and plants. The *Su(var)3–9* gene was identified in *Drosophila* as a dominant suppressor of heterochromatic gene silencing in PEV (Tschiersch *et al.*, 1994). The SU(VAR)3–9 SET domain protein is evolutionary conserved (Aagaard *et al.*, 1999; Ivanova *et al.*, 1998;

Tschiersch et al., 1994) and functions as a histone H3K9 methyltransferase (Rea et al., 2000). In contrast to animals, plants contain a large number of SU(VAR)3–9 homologues (Baumbusch et al., 2001). In Arabidopsis ten different SU(VAR)3–9 homologous SUVH proteins are found and the function of the three proteins SUVH1, SUVH2 and SUVH4 (KYP) has been studied. Heterochromatin association was shown for SUVH1 and SUVH2 (Naumann et al., 2005). However, nuclear distribution of the other SUVH proteins might differ significantly as indicated by the presence of AT-hook sequences in some of the proteins (Baumbusch et al., 2001). Immunocytological analysis with antibodies specifically recognising different histone methylation marks (Peters et al., 2003) revealed accumulation of mono- and dimethyl H3K9, mono- and dimethyl H3K27 and monomethyl H4K20 in Arabidopsis heterochromatin (Lindroth et al., 2004; Naumann et al., 2005; Soppe et al., 2002). Besides methylated H3K4 and acetylated H3K9, a preferential association with euchromatic regions in interphase nuclei of Arabidopsis is also detected for trimethyl H3K9, trimethyl H3K27 and di- and trimethyl H4K20 (Plate 4.3).

In Drosophila heterochromatin-specific histone methylation marks and their molecular control have been studied in detail (Ebert et al., 2004; Schotta et al., 2002, 2004). In Drosophila heterochromatin mono-, di- and trimethylated H3K9 is found. Dimethyl H3K9 indexes the bulk of chromocenter hetero-chromatin whereas trimethyl H3K9 is preferentially associated with the chromocenter core likely representing centromeric heterochromatin. Mono-, di- and trimethylation of H3K27 at both euchromatic and heterochromatic regions is controlled by the Enhancer of Zeste [E(Z)] histone methyltransfer-ase (HMTase). Heterochromatic trimethyl H4K20 is catalysed by SUV4–20 in mammals and in Drosophila (Schotta et al., 2004). In contrast, di- and trimethyl H4K20 are preferentially associated with euchromatin in Arabidop-sis. HMTases controlling mono- and dimethylation of H4K20 in Drosophila heterochromatin are still unknown. HMTase activity of SU(VAR)3–9 is restricted to di- and trimethylation in heterochromatin. Although present, the protein does not control H3K9 dimethylation at telomeres, euchromatic sites and in the fourth chromosome where another yet unknown enzyme is sug-gested to function (Schotta et al., 2002). This finding indicates that also in other systems the presence of the enzyme might not in all cases correlate with each function as an HMTase.

Comparison between the different genetic systems (Drosophila, mammals and Arabidopsis) reveals conserved as well as species-specific elements of the histone code of epigenetic programming. In contrast to Arabidopsis and Drosophila, in mammals only strong enrichment of trimethylated H3K9 and H4K20 is found in heterochromatin (Peters et al., 2003; Schotta et al., 2004). In Drosophila all H3K9, H3K27 and H4K20 methylation states are found in heterochromatin whereas in Arabidopsis heterochromatin dimethyl H4K20 as well as trimethylated H3K9, H3K27 and H4K20 are preferentially found in

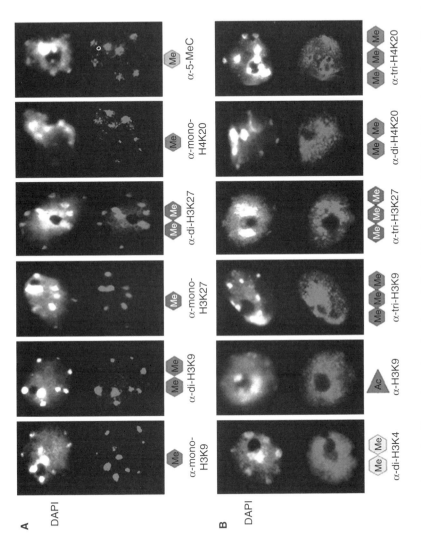

Plate 4.3 The histone modification marks of heterochromatin and euchromatin in *Arabidopsis thaliana*. (A) Typical heterochromatin specific marks are monomethyl H3K9, dimethyl H3K9, monomethyl H3K27, dimethyl H3K27, monomethyl H4K20 and 5-methylcytosine methylation. (B) Euchromatic marks are dimethylated H3K4, acetylated H3K9, trimethyl H3K9, trimethyl H3K27, dimethyl H4K20 and trimethyl H4K20. See also colour plate section for colour version of this figure.

euchromatin. These pronounced differences in nuclear distribution of histone methylation marks not only indicate a different enzymatic control but also might reflect differences in their functional role.

In *Arabidopsis* the effect of the three SUVH proteins SUVH1, SUVH2 and SUVH4 on histone methylation has been studied. For SUVH6 *in vitro* H3K9 HMTase activity was found (Jackson *et al.*, 2004). A pivotal role in the control of heterochromatic histone methylation marks was shown for the SUVH2 protein (Naumann *et al.*, 2005). *In vitro* SUVH2 shows nucleosome-dependent HMTase activity for histone H3 and histone H4. In *SUVH2*-null mutants or antisense lines, significant reductions of mono- and dimethyl H3K9, mono- and dimethyl H3K27 and monomethyl H4K20 in chromocenter heterochromatin are detected by immunocytological analysis. Western analysis of bulk histones shows identical results. In contrast, null mutations of *SUVH1* and *SUVH4* show only a significant reduction in H3K9 dimethylation whereas all other heterochromatin-specific histone methylation marks are not significantly affected. Partially redundant effects of different SUVH proteins on histone H3K9 methylation are indicated by the observation that in any of the SUVH1, SUVH2 or SUVH4-null mutants H3K9 methylation in chromocenter heterochromatin is not completely lost (Jackson *et al.*, 2004; Naumann *et al.*, 2005).

The *Drosophila Su(var)3–9* gene shows a distinct dosage-dependent effect on heterochromatic gene silencing (Ebert *et al.*, 2004; Schotta *et al.*, 2002). *Su(var)3–9* belongs to a group of genes that are characterised by a haplo-suppressor and triplo-dependent enhancer effect (Schotta *et al.*, 2003). The dosage-dependent effect together with differential silencing defects found in 19 point mutations demonstrates that the silencing potential of SU(VAR)3–9 correlates with its cellular concentration and enzymatic activity (Ebert *et al.*, 2004). Dosage-dependent effects on TGS could also be demonstrated for SUVH2 in *Arabidopsis* lines containing the *LUC2* repeated luciferase transgene (Plate 4.2). The *LUC2* transgene shows moderate luciferase silencing, which is significantly enhanced after *SUVH2* overexpression. Loss-of-function mutations or antisense lines of *SUVH2* cause strong suppression of *LUC2* transgene silencing. SUVH2 shows dosage-dependent effects on *LUC2* transgene silencing, demonstrating its central role for heterochromatic gene silencing in *Arabidopsis*. The effect of *Arabidopsis* SUVH2 on *LUC2* transgene silencing parallels the effect of *Drosophila* SU(VAR)3–9 on PEV (Plate 4.2).

4.3.2 DNA methylation and the epigenetic control of heterochromatic domains

The genome-wide level of DNA methylation in plant and animal genomes is with 60–90%, which is considerably high (Gruenbaum *et al.*, 1981). In plants besides cytosine methylation of symmetrical CpG and CpNpG sequences, a high frequency of non-symmetrical methylation at CpT, CpA and CpC dinucleotides is found (Meyer *et al.*, 1994). In *Arabidopsis* at least ten genes might encode DNA methyltransferases (*Arabidopsis* Genome Initiative, 2000). The *MET* class of genes is related to mammalian *Dnmt1* (Finnegan & Kovac, 2000). *MET1* has been intensively studied. Antisense suppression revealed a complex role of *MET1* in plant development (Finnegan *et al.*, 1996; Ronemus *et al.*, 1996). Mutant *met1* alleles were isolated either as DNA hypomethylation mutants in Southern blot screens with centromeric repeats after digestion with the methylation-sensitive endonuclease *HpaII* (Kankel *et al.*, 2003) or as suppressors of TGS (Saze *et al.*, 2003). Loss-of-function *met1* mutations result in complete loss of CpG methylation and cause strong reduction of viability in homozygotes. Three other *MET1*-related genes (*MET2a*, *MET2b* and *MET3*) have not yet been studied. The chromomethyltransferase (CMT) class of DNA methyltransferases is plant-specific and characterised by a chromodomain (Henikoff & Comai, 1998). Mutations for *CMT3* were isolated as suppressors of the hypermethylated *clk* alleles of the floral development gene *SUPERMAN* (*SUP*). The mutations decrease CpNpG methylation of *SUP* sequences as well as other sequences throughout the genome (Lindroth *et al.*, 2001). By genome-wide methylated DNA profiling it could be shown that CMT3 is preferentially targeted to transposons (Tompa *et al.*, 2002). A third class of DNA methyltransferases are the domain-rearranged methyltransferases (DRMs) characterised by a rearranged structure of conserved motifs within the methyltransferase catalytic domains (Cao *et al.*, 2000). T DNA mutant analysis indicated that the enzymes interfere with *de novo* CpG, CpNpG and non-symmetrical methylation but do not affect pre-existing methylation (Cao & Jacobsen, 2002b). The importance of non-symmetrical methylation for gene silencing was first demonstrated with a transgene from which all symmetric CpG and CpNpG sites had been removed (Dieguez *et al.*, 1998). Independent of this sequence modification the transgene becomes heavily methylated and silenced, suggesting that non-symmetrical methylation is important for gene silencing too. Furthermore, the analysis of *drm1 drm2* double and *drm1 drm2 cmt3* triple mutants showed that non-symmetrical methylation is largely controlled by the DRM methyltransferases. The mutant analyses further revealed partially redundant and locus-specific effects of the DRMs and CMT3 (Cao & Jacobsen, 2002a).

Mutations of the *DDM1* gene (decrease in DNA methylation1) cause a global reduction of DNA methylation, which is independent of the sequence context and which leads to suppression of transposon and transgene silencing (Jeddeloh *et al.*, 1998, 1999; Mittelsten Scheid *et al.*, 1998; Vongs *et al.*, 1993). The *DDM1* gene encodes a SNF2/SWI2-related chromatin remodelling protein (Brzeski & Jerzmanowski 2003; Jeddeloh *et al.*, 1999). Hirochika *et al.* (2000) introduced the tobacco retrotansposon *Tto1* into *Arabidopsis*. After an increase of copy number *Tto1* became silenced in wild-type backgrounds. In *ddm1* plants *Tto1* was transcriptionally reactivated. Similarly, silencing of the endogenous *CACTA* transposons is released in *ddm1* mutants (Kato *et al.*, 2004b). Interestingly, after its silencing was released, the transposon remained mobile in a wild-type *DDM1* background. In general, remethylation of sequences hypomethylated by the *ddm1* mutation is extremely slow or non-existing in the wild-type *DDM1* background (Kakutani *et al.*, 1999). This suggests that *ddm1* mutations induce heritable changes in epigenetic marks. In contrast, centromeric repeats hypomethylated in *met1–1* become partially remethylated when introduced into a wild-type background (Kankel *et al.*, 2003).

Several methylation mutants cause global changes in chromatin organisation within interphase nuclei. DDM1 predominantly influences DNA and histone H3K9 methylation in pericentromeric heterochromatin (Johnson *et al.*, 2002). Using cytological methods Probst *et al.* (2003) visualised complex structural alterations of chromocenter heterochromatin and at a repeated transgene in the *ddm1* mutant background. The *ddm1* mutation causes significant decondensation of chromocenter heterochromatin whereas telomeres remained unaffected. The transgenic locus showed release of silencing and significant decondensation. In the *ddm1* mutant background histone H3K9 methylation was dispersed and significantly reduced at chromocenter heterochromatin and the repeated transgenic locus. Complete erasure of CpG methylation in a specific *met1* mutant caused clear loss of histone H3K9 methylation in chromocenter heterochromatin (Tariq *et al.*, 2003). However, in contrast to *ddm1* mutations no significant relaxation of heterochromatin was found.

Strand-biased DNA methylation was found at *Arabidopsis* centromeric regions (Luo & Preuss, 2003). Centromeric sequences show characteristic patterns of nearly complete modification of one strand but limited modification of the complementary strand. Strand-biased DNA methylation was not found in other heterochromatic regions (rDNA islands, telomeres and knobs). Its abundance in centromeric heterochromatin suggests a role for strand-biased methylation in epigenetic control of centromere function.

4.3.3   Interdependence of heterochromatic DNA and histone methylation

The analysis of silencing mutants demonstrated that histone H3K9 and DNA methylation marks at heterochromatin and silenced genes are interrelated (Martienssen & Colot, 2001; Selker, 2002). Recent models suggest either that DNA methylation is triggered by histone H3K9 methylation (Jackson *et al.*, 2002; Lindroth *et al.*, 2004; Malagnac *et al.*, 2002; Tamaru & Selker, 2001) or alternatively that methylation of histone H3K9 depends on DNA methylation (Johnson *et al.*, 2002; Soppe *et al.*, 2002; Tariq *et al.*, 2003).

Genetic dissection of *SUVH2* clearly favours the latter hypothesis (Naumann *et al.*, 2005). Overexpression of *SUVH2* causes formation of ectopic heterochromatin and massive gene silencing resulting in a strong 'mini-plant' growth defect and curled cotyledon phenotypes. Mutational dissection of these phenotypes led to the isolation of a series of transgene mutations, which represent point mutations within the different domains of the SUVH2 protein. *SUVH2* overexpression results in silencing of Athila transposons accompanied by DNA hypermethylation. Although silencing of Athila is completely released by overexpression of HMTase-inactive SUVH2 the Athila sequences remain strongly hypermethylated. Analysis of a series of *SUVH2* transgene mutants revealed the hierarchic sequence of molecular events involved in SUVH2-dependent gene silencing. The N-terminal regions of SUVH2 appear to mediate target sequence recognition. Mutations in the N-terminus of SUVH2 abolish completely ectopic protein distribution after overexpression. The YDG domain and a region immediately adjacent to the preSET region appear to be important for recruitment of DNA methylation to target sequences. The latter mutations do not affect ectopic distribution and HMTase activity of SUVH2 but cause release of Athila hypermethylation. Therefore, it is suggested that SUVH2, via its YDG domain, firstly mediates DNA methylation at target sequences, which appears to be a prerequisite for consecutive histone H3 and H4 methylation by its own HMTase activity. Preference for CpNpG symmetric methylation is found for SUVH4-dependent silencing (Jackson *et al.*, 2002; Lindroth *et al.*, 2004). In contrast, gene silencing induced by SUVH2 depends on both symmetric and non-symmetric DNA methylation. In agreement with these results are the findings that SUVH4-induced silencing depends on CMT3 whereas SUVH2 is largely independent of CMT3 (Naumann *et al.*, 2005). The epistatic effect of a *met1* mutation on SUVH2-induced silencing and suppression of ectopic SUVH2 distribution in *ddm1* mutant plants shows that SUVH2 depends on MET1 and DDM1 but not CMT3.

In animals and fungi, the HP1 protein is of central importance for stable heterochromatin association of silencing complexes. HP1 binds specifically to di- and trimethylated H3K9 (Bannister *et al.*, 2001) and by protein–protein interaction restricts association of the HMTases SU(VAR)3–9 and SUVH4–20 to heterochromatin (Schotta *et al.*, 2002, 2004). Proteins recognising specific methylation marks *in vivo* are still unknown in plants. *In vitro*, a high affinity of CMT3 chromodomain homodimers to lysine 9 and lysine 27 double-methylated histone H3 tails was demonstrated (Lindroth *et al.*, 2004), indicating that proteins binding specifically histone methylation marks also exist in plants. Complex interactions between such proteins, various SUVH proteins, other SET domain proteins and DNA methyltransferases could result in a complex network of regulatory interactions.

In *Arabidopsis* a large number of genes encoding MBD proteins is found (Zemach & Grafi, 2003). These proteins could be involved in differential transition of 5-methylcytosine methylation pattern to histone methylation by recruitment of specific SUVH complexes. Furthermore, different SUVH silencing complexes might be functionally interconnected with different histone deacetylases. Deacetylation of lysine 9 in histone H3 is a perquisite for consecutive methylation of H3K9 by SU(VAR)3–9 enzymes (Czermin *et al.*, 2001; Nakayama *et al.*, 2001). Similar interconnection of histone H3K9 deacetelyation and methylation in plants is indicated by the finding that mutations of certain *At*HDA genes are suppressors of TGS (Furner *et al.*, 1998; Murfett *et al.*, 2001; Probst *et al.*, 2004; Hofmann, I., Fischer, A., Fiedler, C., Thümmler, A., Scheel, D., Tschiersch, B., Reuter, G., unpublished data).

4.4   Epigenetic inheritance in plants and heterochromatin

Possibly endogenous gene pairing–dependent inactivation in plants is found in paramutation. The phenomenon was first discovered by Brotherton (1923) in *Pisum* and Renner (1937) in *Oenothera* and subsequently studied in detail in other plant species (Brink, 1956; Hagemann, 1958; Harrison & Carpenter, 1973). Although paramutation was originally described as a genetic change, it is now generally considered to represent a directed epigenetic change (Matzke & Matzke, 1995; Mittelsten Scheid *et al.*, 2003). In heterozygotes, a sensitive paramutable allele is epigenetically silenced after its interaction with a para-mutagenic allele, and consequently acquires (in most cases) the paramutagenic status even after being separated from the original paramutagenic homologue. Mittelsten Scheid *et al.* (2003) showed formation of *HPT* epialleles in tetraploid *A. thaliana* lines, which remained stable in diploids. Partial reactivation in a *ddm1* background suggests that silencing is due to a heterochromatisation-like process. In a series of crosses it could be shown that inactivated *HPT* alleles are able to induce a heritable stable inactivation of an active *HPT* allele.

In certain genetic hybrids silencing of one of the parent's ribosomal RNA genes is found (nucleolar dominance). Derepression of the silenced rRNA genes was found after treatment with inhibitors of DNA methylation or histone deacetylation (Chen & Pikaard, 1997). These data indicate that silencing of NORs in nucleolar dominance depends on processes involved in establishment of heterochromatic structures.

The SET-domain protein encoding *MEDEA* (*MEA*) gene is regulated by imprinting with only maternally inherited alleles being active. Paternally inherited alleles are not transcribed in the young embryo and endosperm (Grossniklaus *et al.*, 1998). Because of silencing of the paternal *MEA* locus, embryos inheriting an *mea* mutant allele from the mother become aborted. The maternal effect of an *mea* mutation is rescued by *ddm1* mutations through reactivation of the paternally inherited *MEA* allele (Vielle-Calzada *et al.*, 1999). It is suggested that DDM1 is required for maintenance but not for induction of silencing at the paternal *MEA* allele. Although other factors involved in establishing heterochromatin have not yet been tested it is intriguing that silencing at the *MEA* locus is caused by heterochromatisation.

Stokes *et al.* (2002) described an interesting effect of *ddml* mutations on expression of an *Arabidopsis* pathogen resistance gene cluster. In *ddml* mutant plants the heritable dwarfing variant *bal* was generated, which resembles the phenotype of mutants that constitutively express certain pathogen defence genes. Finally, the analysis revealed that the *bal* phenotype generated in *ddml* plants was caused by overexpression of a resistance gene. The plants showed significantly reduced bacterial growth after infection with *Pseudomonas syringiae*. These findings indicate that heterochromatic gene silencing might also be involved in epigenetic control of plant resistance genes.

Vernalisation in plants depends on downregulation of the *FLC* genes, which encodes a MADS-box transcriptional regulator repressing a set of genes required for transition of apical meristem to a reproductive fate (Sheldon *et al.*, 1999). Expression of *FLC* becomes downregulated after prolonged exposure to cold. Repression of *FLC* remains epigenetically stable during subsequent development and is correlated with increased histone H3K9 and H3K27 dimethylation in discrete domains of the *FCL* locus (Bastow *et al.*, 2004).

The listed examples of epigenetic processes clearly indicate that heterochromatisation has an important impact on a wide range of gene-silencing processes during plant development. These processes might also play a fundamental role in response to a wide spectrum of environmental cues. How these might result in acquired epigenetically heritable changes in gene expression still remains to be studied in more detail. Genome-wide chromatin profiling of histone methylation marks in different tissues during plant development and under differing environmental conditions might also help to resolve many of the still unknown functional aspects of heterochromatin in plants.

References

Aagaard, L., Laible, G., Selenko, P., Schmid, M., Dorn, R., Schotta, G., Kuhfittig, S., Wolf, A., Lebersorger, A., Singh, P. B., Reuter, G. & Jenuwein, T. (1999) Functional mammalian homologues of the *Drosophila* PEV-modifier *Su(var)3–9* encode centromere-associated proteins which complex with the heterochromatin component M31. *EMBO Journal*, **18**, 1923–38.

Abranches, R., Beven, A. F., Aragon-Alcaide, L. & Shaw, P. J. (1998) Transcriptional sites are not correlated with chromosomal territories in wheat nuclei. *Journal of Cell Biology*, **143**, 5–12.

Amedeo, P., Habu, Y., Afsar, K., Mittelsten Scheid, O. & Paszkowski, J. (2000) Disruption of the plant gene MOM releases transcriptional silencing of methylated genes. *Nature*, **405**, 203–206.

Ananiev, E. V., Phillips, R. L. & Rines, H. W. (1998) Complex structure of knob DNA on maize chromosome 9: retrotransposon invasion into heterochromatin. *Genetics*, **149**, 2025–37.

Arabidopsis Genome Initiative (2000) Analysis of the genome sequence of the flowering plant *Arabidopsis thaliana*. *Nature*, **408**, 796–815.

Assaad, F. F. & Signer, E. R. (1992) Somatic and germinal recombination of a direct repeat in *Arabidopsis*. *Genetics*, **132**, 553–66.

Atanassov, I., Delichere, C., Filatov, D. A., Charlesworth, D., Negrutiu, I. & Moneger, F. (2001) Analysis and evolution of two functional Y-linked loci in a plant sex chromosome system. *Molecular Biology and Evolution*, **18**, 2162–8.

Aufsatz, W., Mette, M. F., van der Winden, J., Matzke, M. & Matzke, A. J. M. (2002) HDA6, a putative histone deacetylase needed to enhance DNA methylation induced by double stranded RNA. *EMBO Journal*, **21**, 6832–41.

Baker, W. K. (1958) Crossing over in heterochromatin. *American Naturalist*, **92**, 59–60.

Bannister, A. J., Zegermann, P., Patridge, J. F., Miska, E. A., Thomas, J. O., Allshire, T. C. & Kouzarides, T. (2001) Selective recognition of methylated lysine 9 on histone H3 by the HP1 chromo domain. *Nature*, **410**, 120–24.

Bartee, L. & Bender, J. (2001) Two *Arabidopsis* methylation-deficiency mutations confer only partial effects on a methylated endogenous gene family. *Nucleic Acids Research*, **29**, 2127–34.

Bartee, L., Malagnac, F. & Bender, J. (2001) *Arabidopsis cmt3* chromomethylase mutations block non-CG methylation and silencing of an endogenous gene. *Genes and Development*, **15**, 1753–8.

Bastow, R., Mylne, J. S., Lister, C., Lippman, Z., Mariennssen, R. A. & Dean, C. (2004) Vernalization requires epigenetic silencing of *FLC* by histone methylation. *Nature*, **427**, 164–7.

Baulcombe, D. (2004) RNA silencing in plants. *Nature*, **431**, 356–63.

Baumbusch, L. O., Thorstensen, T., Krauss, V., Fischer, A., Naumann, K., Assalkhou, R., Schulz, I., Reuter, G. & Aalen, R. (2001) The *Arabidopsis thaliana* genome contains at least 29 active genes encoding SET domain proteins that can be assigned to four evolutionary conserved classes. *Nucleic Acids Research*, **29**, 4319–33.

Brink, R. A. (1956) A genetic change associated with the R locus in maize which is directed and potentially reversible. *Genetics*, **41**, 872–89.

Brotherton, W. (1923) Further studies on the inheritance of 'rogues' in peas. *Journal of Agricultural Research*, **28**, 1247–52.

Brzeski, I. & Jerzmanowski, A. (2003) Deficient in DNA methylation 1 (DDM1) defines a novel family of chromatin-remodeling factors. *Journal of Biological Chemistry*, **278**, 823–8.

Cao, X. & Jacobsen, S. E. (2002a) Role of the *Arabidopsis* DRM methyltransferases in *de novo* DNA methylation and gene silencing. *Current Biology*, **12**, 1138–44.

Cao, X. & Jacobsen, S. E. (2002b) Locus-specific control of asymmetric and CpNpG methylation by the *DRM* and *CMT3* methyltransferase genes. *Proceedings of the National Academy of Sciences of the United States of America*, **99**, 16491–8.

Cao, X. C., Springer, N. M., Muszynski, M. G., Phillips, R. L., Kaeppler, S. & Jacobsen, S. E. (2000) Conserved plant genes with similarity to mammalian *de novo* DNA methyltransferases. *Proceedings of the National Academy of Sciences of the United States of America*, **97**, 4979–84.

Catcheside, D. G. (1939) A position effect in *Oenothera*. *Journal of Genetics*, **38**, 345–52.

Catcheside, D. G. (1947) The P-locus position effect in *Oenothera*. *Journal of Genetics*, **48**, 31–42.

Charlesworth, D. (2002) Plant sex determination and sex chromosomes. *Heredity*, **88**, 94–101.

Chen, Z. J. & Pikaard, C. S. (1997) Epigenetic silencing of RNA polymerase I transcription: a role for DNA methylation and histone modification in nucleolar dominance. *Genes & Development*, **11**, 2124–36.

Choi, Y., Gehring, M., Johnson, L., Hannon, M., Harada, J. J., Goldberg, R. I., Jacobsen, S. E. & Fischer, R. L. (2002) DEMETER, a DNA glycosylase domain protein, is required for endosperm gene imprinting and seed viability in *Arabidopsis*. *Cell*, **110**, 33–42.

Collins, N., Poot, R. A., Kukimoto, I., Garcia-Jimenez, C., Dellaire, G. & Varga-Weisz, P. D. (2002) An ACF1-ISWI chromatin-remodeling complex is required for DNA replication through heterochromatin. *Nature Genetics*, **32**, 627–32.

Copenhaver, G. P., Nickel, K., Kuromori, T., Benito, M. -I., Kaul, S., Lin, X., Bevan, M., Murphy, G., Harris, B., Parnell, L. D., McCombie, W. R., Martienssen, R. A., Marra, M. & Preuss, D. (1999) Genetic definition and sequence analysis of *Arabidopsis* centromeres. *Science*, **286**, 2468–74.

Czermin, B., Schotta, G., Hulsmann, B. B., Brehm, A., Becker, P. B., Reuter, G. & Imhof, A. 2001. Physical and functional association of SU(VAR)3–9 and HDAC1 in *Drosophila*. *EMBO Reports*, **2**, 915–9.

Davies, G. J., Sheikh, M. A., Ratcliffe, O. J., Coupland, G. & Furner, I. J. (1997) Genetics of homology-dependent gene silencing in *Arabidopsis*: a role for methylation. *Plant Journal*, **12**, 791–804.

Dieguez, M. J., Vaucheret, H., Paszkowski, J. & Mittelsten Scheid, O. (1998) Cytosine methylation at CG and CNG sites is not a prerequisite for the initiation of transcriptional gene silencing in plants, but it is required for its maintenance. *Molecular Genetics and Genomics*, **259**, 207–15.

Donald, T. M., Houben, A., Leach, C. R. & Timmis J. N. (1997) Ribosomal RNA genes specific to the B chromosomes in *Brachycome dichromosomatica* are not transcribed in leaf tissue. *Genome*, **40**, 674–81.

Ebert, A., Schotta, G., Lein, S., Kubicek, S., Krauss, V., Jenuwein, T. & Reuter, G. (2004) *Su(var)* genes regulate the balance between euchromatin and heterochromatin in *Drosophila*. *Genes & Development*, **18**, 2973–83.

Finnegan, E. J. & Kovac, K. A. (2000) Plant DNA methyltransferases. *Plant Molecular Biology*, **43**, 189–201.

Finnegan, E. J., Peacock, W. J. & Dennis, E. S. (1996) Reduced DNA methylation in *Arabidopsis thaliana* results in abnormal plant development. *Proceedings of the National Academy of Sciences of the United States of America*, **93**, 8449–54.

Fransz, P. F., Armstrong, S., Alonso-Blanco, C., Fischer, T. C., Torres-Ruiz, R. A. & Jones, G. H. (1998) Cytogenetics for the model system *Arabidopsis thaliana*. *Plant Journal*, **13**, 867–76.

Fransz, P. F., Armstrong, S., de Jong, J. H., Parnell, L. D., van Drunen, C., Dean, C., Zabel, P., Bisseling, T. & Jones, G. H. (2000) Integrated cytogenetic map of chromosome arm 4S of *A. thaliana*: structural organization of heterochromatic knob and centromere regions. *Cell*, **100**, 367–76.

Fransz, P. F., de Jong, J. H., Lysak, M., Ruffini Castiglione, M. & Schubert, I. (2002) Interphase chromosomes in *Arabidopsis* are organized as well defined chromocenters from which euchromatic loops emanate. *Proceedings of the National Academy of Sciences of the United States of America*, **99**, 14584–9.

Fu, H., Zheng, Z & Dooner, H. K. (2002) Recombination rates between adjacent genic and retrotransposon regions in maize vary by 2 orders of magnitude. *Proceedings of the National Academy of Sciences of the United States of America*, **99**, 1082–7.

Fuchs, J., Houben, A., Brandes, A. & Schubert, I. (1996) Chromosome 'painting' in plants – a feasible technique? *Chromosoma*, **104**, 315–20.

Fukui, K. & Nakayama, S. (1996) *Plant chromosomes: laboratory methods*. CRC Press, Boca Raton, Florida.

Furner, I. J., Sheikh, M. A. & Collett, C. E. (1998) Gene silencing and homology-dependent gene silencing in *Arabidopsis*: genetic modifiers and DNA methylation. *Genetics*, **149**, 651–62.

Gendrel, A. -V., Lippman, Z., Yordan, C., Colot, V. & Martienssen, R. A. (2002) Dependence of heterochromatic histone H3 methylation patterns on the *Arabidopsis* gene DDM1. *Science*, **297**, 1871–3.

Gilbert, D. M. (2002) Replication timing and transcriptional control: beyond cause and effect. *Current Opinion in Cell Biology*, **14**, 377–83.

Gomez, M. & Brockdorff, N. (2004) Heterochromatin on the inactive X chromosome delays replication timing without affecting origin usage. *Proceedings of the National Academy of Sciences of the United States of America*, **101**, 6923–8.

Gong, Z., Morales-Ruiz, T., Ariza, R. R., Roldan-Arjona, T., David, L. & Zhu, J. K. (2002) *ROS1*, a repressor of transcriptional gene silencing in *Arabidopsis*, encodes a DNA glycosylase/lyase. *Cell*, **111**, 803–14.

Goring, D. R., Thomson, L. & Rothstein, S. J. (1991) Transformation of a partial nopaline synthetase gene into tobacco suppresses the expression of a resident wild-type gene. *Proceedings of the National Academy of Sciences of the United States of America*, **88**, 1770–74.

Grossniklaus, U., Vielle-Calzada, J. -P., Hoeppner, M. A. & Gagliano, W. B. (1998) Maternal control of embryogenesis by *MEDEA*, a *Polycomb* group gene in *Arabidopsis*. *Science*, **280**, 446–50.

Gruenbaum, Y., Stein, R., Cedar, H. & Razin, A. (1981) Methylation of CpG sequences in eukaryotic DNA. *FEBS Letters*, **124**, 67–71.

Guerra, M. (2000) Patterns of heterochromatin distribution in plant chromosomes. *Genetics and Molecular Biology*, **23**, 1029–41.

Guttman, D. S. & Charlesworth, D. (1998) An X-linked gene with a degenerated Y-linked homologue in a dioecious plant. *Nature*, **393**, 263–6.

Hagemann, R. (1958) Somatische Konversion bei *Lycopersicon esculentum. Zeitschrift für Vererbungslehre*, **89**, 587–613.

Harrison, B. J. & Carpenter, R. (1973) A comparison of the instabilities at the *nivea* and *pallida* loci in *Antirrhinum majus. Heredity*, **31**, 309–23.

Haupt, W., Fischer, T. C., Winderl, S., Fransz, P. & Torres-Ruiz, R. A. (2001) The CENTROMERE (CEN1) region of *Arabidopsis thaliana*: architecture and functional impact of chromatin. *Plant Journal*, **27**, 285–96.

Heitz, E. (1928) Das Heterochromatin der Moose. *Jahrbuch für wissenschaftliche. Botanik*, **69**, 762–818.

Heitz, E. (1929) Heterochromatin, Chromocentren, Chromomeren. *Berichte der Deutschen Botanischen Gesellschaft*, **47**, 274–84.

Heitz, E. (1932) Die Herkunft der Chromocentren. Dritter Beitrag zur Kenntnis der Beziehung zwischen Kernstruktur und qualitativer Verschiedenheit der Chromosomen in ihrer Längsrichtung. *Planta*, **18**, 571–636.

Henikoff, S. & Comai, L. (1998) A DNA methyltransferase homolog with a chromodomain exists in multiple polymorphic forms in *Arabidopsis. Genetics*, **149**, 307–18.

Hirochika, H., Okamoto, H. & Kakutani, T. (2000) Silencing of retrotransposons in *Arabidopsis* and reactivation by the *ddm1* mutations. *Plant Cell*, **12**, 357–68.

Hosouchi, T., Kumekawa, N., Tsuruoka, H. & Kotani, H. (2002) Physical map-based size of the centromeric regions of *Arabidopsis thaliana* chromosomes 1, 2 and 3. *DNA Research*, **9**, 117–21.

Ivanova, A. V., Bonaduce, M. J., Ivanov, S. V. & Klar, A. J. S. (1998) The chromo and SET domains of the Clr4 protein are essential for silencing in fission yeast. *Nature Genetics*, **19**, 192–5.

Jackson, J. P., Lindroth, A. M., Cao, X. & Jacobsen, S. E. (2002) Control of CpNpG DNA methylation by the *KRYPONITE* histone H3 methyltransferase. *Nature*, **416**, 556–60.

Jackson, J. P., Johnson, L., Jasencakova, Z., Zhang, X., PerezBurgos, L., Singh, P. B., Cheng, X., Schubert, I., Jenuwein, T. & Jacobsen, S. E. (2004) Dimethylation of histone H3K9 is a critical mark for DNA methylation and gene silencing in *Arabidopsis thaliana. Chromosoma*, **112**, 308–15.

Janousek, B., Siroky, J. & Vyskot, B. (1996) Epigenetic control of sexual phenotype in a dioecious plant, *Melandrium album. Molecular Genetics and Genomics*, **250**, 483–90.

Jasencakova, Z., Meister, A. & Schubert, I. (2001) Chromatin organization and its relation to replication and histone acetylation during the cell cycle in barley. *Chromosoma*, **110**, 83–92.

Jeddeloh, J. A., Bender, J. & Richards, E. J. (1998) The DNA methylation locus *DDM1* is required for maintenance of gene silencing in *Arabidopsis. Genes & Development*, **12**, 1714–25.

Jeddeloh, J. A., Stoke, T. L. & Richards, E. J. (1999) Maintenance of genomic methylation requires a SWI2/SNF2-like protein. *Nature Genetics*, **22**, 94–7.

Jenuwein, T. & Allis, C. D. (2001) Translating the histone code. *Science*, **293**, 1074–80.

Johnson, L. M., Cao, X. & Jacobsen, S. E. (2002) Interplay between two epigenetic marks: DNA methylation and histone H3 lysine 9 methylation. *Current Biology*, **12**, 1360–67.

Jones, R. N. (1995) B chromosomes in plants. *New Phytologist*, **131**, 411–34.

Jones, L., Ratcliff, F. & Baulcombe D. C. (2001) RNA-directed gene silencing in plants can be inherited independently of the RNA trigger and requires Met1 for maintenance. *Current Biology*, **11**, 747–57.

Jorgensen, R. (1990) Altered gene expression in plants due to *trans* interactions between homologous genes. *Trends in Biotechnology*, **8**, 340–44.

Kakutani, T., Jeddeloh, J., Flowers, S., Munakata, K. & Richards, E. (1996) Developmental abnormalities and epimutations associated with DNA hypomethylation mutations. *Proceedings of the National Academy of Sciences of the United States of America*, **93**, 12406–11.

Kakutani, T., Munakata, K., Richards, E. J. & Hirochika, H. (1999) Meiotically and mitotically stable inheritance of DNA hypomethylation induced by *ddm1* mutations of *Arabidopsis thaliana. Genetics*, **151**, 831–8.

Kankel, W. M., Ramsey, E. D., Stokes, T. L., Flowers, S. K., Haag, J. R., Jeddeloh, J. A., Riddle, N. C., Verbsky, M. L. & Richards, E. J. (2003) *Arabidopsis* MET1 cytosine methyltransferase mutants. *Genetics*, **163**, 1109–22.

Kaszas, E. & Birchler, J. A. (1998) Meiotic transmission rates correlate with physical features of rearranged centromeres in maize. *Genetics*, **150**, 1683–92.

Kato, A., Lamb, J. C. & Birchler, J. A. (2004a) Chromosome painting using repetitive DNA sequences as probes for somatic chromosome identification in maize. *Proceedings of the National Academy of Sciences of the United States of America*, **101**, 13554–9.

Kato, M., Takashima, K. & Kakutani, T. (2004b) Epigenetic control of CACTA transposon mobility in *Arabidopsis thaliana*. *Genetics*, **168**, 961–9.

Kim, S. -M., Dubey, D. D. & Huberman, J. A. (2003) Early-replicating heterochromatin. *Genes & Development*, **17**, 330–35.

Lengerova, M. & Vyskot, B. (2001) Sex chromatin and nucleolar analysis in *Rumex acetosa*. *Protoplasma*, **217**, 147–53.

Lima-de-Faria, A. & Jaworska, H. (1986) Late DNA synthesis in heterochromatin. *Nature*, **217**, 138–42.

Lindroth, A. M., Cao, X. & Jackson, J. P. (2001) Requirement of chromothylase3 for maintenance of CpXpG methylation. *Science*, **293**, 1070–1073.

Lindroth, M. A., Shultis, D., Jasencakova, Z., Fuchs, J., Johnson, L., Schubert, D., Patnaik, D., Pradhan, S., Goodrich, J., Schubert, I., Jenuwein, T., Khorasanizadeh, S. & Jacobson, S. E. (2004) Dual histone H3 methylation marks at lysine 9 and 27 required for interaction with *CHROMOMETHYLASE3*. *EMBO Journal*, **23**, 4146–55.

Lippman, Z. & Martienssen, R. (2004) The role of RNA interference in heterochromatic silencing. *Nature*, **431**, 364–70.

Lippman, Z., May, B., Yordan, C., Singer, T. & Martienssen, R. (2003) Distinct mechanisms determine transposon inheritance and methylation via small interfering RNA and histone modification. *PloS Biology*, **1**, 420–28.

Lo, A. W. I., Craig, J. M., Saffery, R., Kalitsis, P., Irvine, D. V., Earle, E., Magliano, D. J. & Choo, K. H. A. (2001a) A 330kb CENP-A binding domain and altered replication timing at a human neocentromere. *EMBO Journal*, **20**, 2087–96.

Lo, A. W. I., Magliano, D. J., Sibson, M. C., Kalitsis, P., Craig, J. M. & Choo, K. H. A. (2001b) A novel chromatin immunoprecipitation and array (CIA) analysis identifies a 460-kb CENP-A binding neocentromere DNA. *Genome Research*, **11**, 448–57.

Lohe, A. R. & Hilliker, A. J. (1995) Return of the H-word (heterochromatin). *Current Opinion in Genetics and Development*, **5**, 746–55.

Luff, B., Pawlowski, L. & Bender, J. (1999) An inverted repeat triggers cytosine methylation of identical sequences in *Arabidopsis*. *Molecular Cell*, **3**, 505–11.

Luo, S. & Preuss, D. (2003) Strand-biased DNA methylation associated with centromeric regions in *Arabidopsis*. *Proceedings of the National Academy of Sciences of the United States of America*, **19**, 11133–8.

Lysak, M. A., Fransz, P. F., Ali, H. B. M. & Schubert, I. (2001) Chromosome painting in *Arabidopsis thaliana*. *Plant Journal*, **28**, 689–97.

Malagnac, F., Bartee, L. & Bender, J. (2002) An *Arabidopsis* SET domain protein required for maintenance but not establishment of DNA methylation. *EMBO Journal*, **21**, 6842–52.

Martienssen, R. A. & Colot, V. (2001) DNA methylation and epigenetic inheritance in plants and filamentous fungi. *Science*, **293**, 1070–1074.

Mathieu, O., Picard, G. & Tourmente, S. (2002) Methylation of a euchromatin–heterochromatin transition region in *Arabidopsis thaliana* chromosome 5 left arm. *Chromosome Research*, **10**, 455–66.

Matzke, A. J. M., Neuhuber, F., Park, Y. D, Ambros, P. F & Matzke, M. A. (1994) Homology-dependent gene silencing in transgenic plants: epistatic silencing loci contain multiple copies of methylated transgenes. *Molecular Genetics and Genomics.*, **244**, 219–29.

Matzke, M. A. & Matzke, A. J. M. (1990) Gene interactions and epigenetic variations in transgenic plants. *Developmental Genetics*, **11**, 214–23.

Matzke, M. A. & Matzke, A. J. M. (1993) Genomic imprinting in plants: parental effects and *trans*-inactivation phenomena. *Annual Review of Plant Physiology and Plant Molecular Biology*, **44**, 53–76.

Matzke, M. A. & Matzke, A. J. M. (1995) Homology-dependent gene silencing in transgenic plants: what does it really tell us? *Trends in Genetics*, **11**, 1–3.

Matzke, M. A., Neuhuber, F. & Matzke, A. J. M. (1993) A variety of epistatic interactions can occur between partially homologous transgene loci brought together by sexual crossing. *Molecular Genetics and Genomics*, **236**, 379–86.

Mayer, K., Schüller, C., Warmbutt, R., Murphy, G., Volckaert, G., Pohl, T., Dusterhoft, A., Stiekema, W., Entian, K. D., Terryn, N. *et al.* (1999) Sequence analysis of chromosome 4 of the plant *Arabidopsis thaliana. Nature*, **402**, 769–77.

McClintock, B. (1929) Chromosome morphology in *Zea mays. Science*, **69**, 629.

McCombie, W. R., de la Bastide, M., Habermann, K., Parnell, L., Dedhia, N., Gnoj, L., Schutz, K., Huang, E., Spiegel, L., Yordan, C. *et al.* (2000) The complete sequence of a heterochromatic island from a higher eukaryote. *Cell*, **100**, 377–86.

Messing, J., Bharti, A. K., Karlowski, W. M., Gundlach, H., Kim, H. R., Yu, Y., Wie, F., Fuks, G., Soderlund, C. A., Mayer, K. F. X. & Wing, R. A. (2004) Sequence composition and genome organization of maize. *Proceedings of the National Academy of Sciences of the United States of America*, **101**, 14349–54.

Mette, M. F., Aufsatz, W., van der Winden, J., Matzke, M. A. & Matzke, A. J. M. (2000) Transcriptional silencing and promotor methylation triggered by double-stranded RNA. *EMBO Journal*, **19**, 5194–201.

Meyer, P. (1995) DNA methylation and transgene silencing in *Petunia hybrida*. In Meyer, P. (ed.) *Gene silencing in higher plants and related phenomena in other eukaryotes*. Springer Verlag, Berlin, pp. 15–28.

Meyer, P. & Saedler, H. (1996) Homology-dependent gene silencing in plants. *Annual Reviews of Plant Physiology and Plant Molecular Biology*, **47**, 23–48.

Meyer, P., Heidmann, I. & Niedenhof, I. (1993) Differences in DNA-methylation are associated with a paramutation phenomenon in transgenic *petunia. Plant Journal*, **4**, 86–100.

Meyer, P., Niedenhof, I. & ten Lohuis, M. (1994) Evidence for cytosine methylation of non-symmetrical sequences in transgenic *Petunia hybrida. EMBO Journal*, **13**, 2084–8.

Mittelsten Scheid, O., Paszkowski, J. & Potrykus, I. (1991) Reversible inactivation of a transgene in *Arabidopsis thaliana. Molecular Genetics and Genomics*, **228**, 104–112.

Mittelsten Scheid, O., Afsar, K. & Paszkowski, J. (1998) Release of epigenetic gene silencing by *trans*-acting mutations in *Arabidopsis. Proceedings of the National Academy of Sciences of the United States of America*, **95**, 632–7.

Mittelsten Scheid, O., Afsar, K. & Paszkowski, J. (2003) Formation of stable epialleles and their paramutation-like interaction in tetraploid *Arabidopsis thaliana. Nature Genetics*, **4**, 450–54.

Montgomery, T. H. (1904) Some observations and considerations upon the maturation phenomena of the germ cells. *Biological Bulletin*, **6**, 137–57.

Morel, J. B., Mourrain, P., Beclin, C. & Vaucheret, H. (2000) DNA methylation and chromatin structure affect transcriptional and post-transcriptional transgene silencing in *Arabidopsis. Current Biology*, **10**, 1591–4.

Muller, H. J. (1930) Types of visible variations induced by X-rays in *Drosophila. Journal of Genetics*, **22**, 299–334.

Murfett, J., Wang, X. Y., Hagen, G. & Guilfoyle, T. (2001) Identification of *Arabidopsis* histone deacetylase *hda6* mutants that affect transgene expression. *Plant Cell*, **13**, 1047–61.

Nagaki, K., Talbert, P. B., Zhong, C. X., Henikoff, S. & Jiang, J. (2003) Chromatin immunoprecipitation reveals that the 180-bp satellite repeat is the key functional DNA element of *Arabidopsis thaliana* centromeres. *Genetics*, **163**, 1221–5.

Nagaki, K., Cheng, Z., Ouyang, S., Talbert, P. B., Kim, M., Jones, K. M., Henikoff, S., Buell, C. R. & Jiang, J. (2004) Sequencing of a rice centromere uncovers active genes. *Nature Genetics*, **36**, 138–45.

Nakayama, J., Rice, J. C., Strahl, B. D., Allis, C. D. & Grewal, S. I. (2001) Role of histone H3 lysine 9 methylation in epigenetic control of heterochromatin assembly. *Science*, **292**, 110–13.

Naumann, K., Fischer, A., Hofmann, I., Krauss, V., Phalke, S., Irmler, K., Hause, G., Aurich, A. C., Dorn, R., Jenuwein, T. & Reuter, G. (2005) Pivotal role of *At*SUVH2 in control of heterochromatic histone methylation and gene silencing in *Arabidopsis. EMBO Journal*, **24**, 1418–29.

Pandey, R., Muller, A., Napoli, C. A., Selinger, D. A., Pikaard, C. S., Richards, E. J., Bender, J., Mount, D. W. & Jorgensen, R. A. (2002) Analysis of histone acetyltransferase and histone deacetylase families of *Arabidopsis thaliana* suggests functional diversification of chromatin modification among multicellular eukaryotes. *Nucleic Acids Research*, **30**, 5036–5055.

Pelissier, T., Thalmeir, S., Kempe, D., Sänger, H. -L. & Wassenegger, M. (1999) Heavy *de novo* methylation at symmetrical and non-symmetrical sites is a hallmark of RNA-directed DNA methylation. *Nucleic Acids Research*, **27**, 1625–34.

Peters, A. H. F. M., Kubicek, S., Mechtler, K., O'Sullivan, J., Derijck, A. A. H. A., Perez-Burgos, L., Kohlmaier, A., Opravil, S., Tachibana, M., Shinkai, Y., Martens, J. H. A. & Jenuwein, T. (2003) Partitioning and plasticity of repressive histone methylation states in mammalian chromatin. *Molecular Cell*, **12**, 1577–89.

Peterson-Burch, B. D., Nettleton, D. & Voytas, D. F. (2004) Genomic neighborhoods for *Arabidopsis* retrotransposons: a role for targeted integration in the distribution of the Metaviridae. *Genome Biology*, **5**, R78.

Probst, A. V., Fransz, P. F., Paszkowski, J. & Mittelsten Scheid, O. (2003) Two means of transcriptional reactivation within heterochromatin. *Plant Journal*, **33**, 743–9.

Probst, A. V., Fagard, M., Proux, F., Mourrain, P., Boutet, S., Earley, K., Lawrence, R. J., Pikaard, C. S., Murfett, J., Furner, I., Vaucheret, H. & Mittelsten Scheid, O. (2004) *Arabidopsis* histone deacetylase *HDA6* is required for maintenance of transcriptional gene silencing and determines nuclear organization of rDNA repeats. *Plant Cell*, 16, 1021–34.

Puertas, M. J. (2002) Nature and evolution of B chromosomes in plants: a non-coding but information-rich part of plant genomes. *Cytogenetics and Genome Research*, **96**, 198–205.

Rabl, C. (1885) Über Zelltheilung. *Morphologisches Jahrbuch*, **10**, 14–330.

Rea, S., Eisenhaber, F., O'Carroll, D., Strahl, B. D., Sun, Z. -W., Schmid, M., Opravil, S., Mechtler, K., Ponting, C. P., Allis, C. D. & Jenuwein, T. (2000) Regulation of chromatin structure by site-specific histone H3 methyltransferases. *Nature*, **406**, 593–9.

Renner, O. (1937) Über *Oenothera atrovirens* Sh. et Bartl. und über somatische Konversion im Erbgang des *cruciata*-Merkmals der *Oenotheren*. *Zeitschrift für Induktive Abstammungs- and Vererbungslehre*, **74**, 91–124.

Rhoades, M. M. (1978) Genetic effects of heterochromatin in maize. In B. D. Walden (ed.) *Maize Breeding and Genetics*. Wiley, New York, pp. 641–672.

Ronemus, M. J., Galbiati, M., Ticknor, C., Chen, J. & Dellaporta, S. L. (1996) Demethylation-induced developmental pleiotropy in *Arabidopsis*. *Science*, **273**, 654–7.

Russel, L. B. & Bangham, J. W. (1961) Variegated type position effects in the mouse. *Genetics*, **46**, 509–525.

Saze, H., Mittelsten Scheid, O. & Paszkowski, J. (2003) Maintenance of CpG methylation is essential for epigenetic inheritance during plant gametogenesis. *Nature Genetics*, 34, 65–9.

Schotta, G., Ebert, A., Krauss, V., Fischer, A., Hoffmann, J., Rea, S., Jenuwein, T. & Reuter, G. (2002) Central role of *Drosophila* SU(VAR)3–9 in histone H3-K9 methylation and heterochromatic gene silencing. *EMBO Journal*, **21**, 1121–31.

Schotta, G., Ebert, A. & Reuter, G. (2003) Position-effect variegation and the genetic dissection of chromatin regulation in *Drosophila*. *Seminars in Cell and Developmental Biology*, **14**, 67–75.

Schotta, G., Lachner, M., Sarma, K., Ebert, A., Sengupta, R., Reuter, G., Reinberg, D. & Jenuwein, T. (2004) A silencing pathway to induce H3-K9 and H4-K20 tri-methylation at constitutive heterochromatin. *Genes & Development*, **18**, 1251–62.

Selker, E. U. (2002) Repeat-induced gene silencing in fungi. *Advances in Genetics*, **46**, 439–50.

Sheldon, C. C., Burn, J. E., Perez, P. P., Metzger, J., Peacock, W. J. & Dennis, E. S. (1999) The *FLF* MADS box gene: a repressor of flowering in *Arabidopsis* regulated by vernalization and methylation. *Plant Cell*, **11**, 445–58.

Siroky, J., Ruffini Castiglione, M. & Vyskot, B. (1998) DNA methylation patterns of *Melandrium album* chromosomes. *Chromosome Research*, **6**, 441–6.

Soppe, W. J., Jasencakova, Z., Houben, A., Kakutani, T., Meister, A., Huang, M. S., Jacobsen, S. E., Schubert, I. & Fransz, P. (2002) DNA methylation controls histone H3 lysine 9 methylation and heterochromatin assembly in *Arabidopsis*. *EMBO Journal*, **21**, 6549–59.

Stack, S. M. (1984) Heterochromatin, the synaptonemal complex and crossing over. *Journal of Cell Science*, **71**, 159–76.

Stahl, B. D. & Allis, C. D. (2000) The language of covalent histone modifications. *Nature*, **403**, 41–5.

Steinemann, S. & Steinemann, M. (2001) Biased distribution of repetitive elements: landmarks for neo-Y chromosome evolution in *Drosophila miranda*. *Cytogenetics and Cell Genetics*, **93**, 228–33.

Stokes, T. L., Kunkel, B. N. & Richards, E. J. (2002) Epigenetic variation in *Arabidopsis* disease resistance. *Genes & Development*, **16**, 171–82.

Sun, F. L. Cuaycong, M. H. & Elgin, S. C. (2001) Long-range nucleosome ordering is associated with gene silencing in *Drosophila melanogaster* pericentric heterochromatin. *Molecular and Cellular Biology*, **21**, 2867–79.

Sun, X., Wahlstrom, J. & Karpen, G. (1997) Molecular structure of a functional *Drosophila* centromere. *Cell*, **91**, 1007–19.

Szauter, P. (1984) An analysis of regional constraints on exchange in *Drosophila melanogaster* using recombination-defective meiotic mutants. *Genetics*, **106**, 45–71.

Takeda, S., Tadele, Z., Hofmann, I., Probst, A. V., Angelis, K. J., Kaya, H., Araki, T., Mengiste, T., Mittelsten Scheid, O., Shibahara, K., Scheel, D. & Paszkowski, J. (2004) BRU1, a novel link between response to DNA damage and epigenetic gene silencing in *Arabidopsis*. *Genes & Development*, **18**, 782–93.

Tamaru, H. & Selker, E. U. (2001) A histone H3 methyltransferase controls DNA methylation in *Neurospora crassa*. *Nature*, **414**, 277–83.

Tariq, M., Saze, H., Probst, A., Lichota, J., Habu, Y. & Paszkowski, J. (2003) Erasure of CpG methylation in *Arabidopsis* alters patterns of histone H3 methylation in heterochromatin. *Proceedings of the National Academy of Sciences of the United States of America*, **100**, 8823–7.

Tompa, R., McCallum, C. M., Deltrow, J., Henikoff, J. G., van Steensel, B. & Henikoff, S. (2002) Genome-wide profiling of DNA methylation reveals transposon targets of CHROMOMETHYLASE3. *Current Biology*, **12**, 65–8.

Tschiersch, B., Hofmann, A., Krauss, V., Dorn, R., Korge, G. & Reuter, G. (1994) The protein encoded by the *Drosophila* position-effect variegation suppressor gene *Su(var)3–9* combines domains of antagonistic regulators of homeotic gene complexes. *EMBO Journal*, **13**, 3822–31.

Tsukiyama, T. (2002) The *in vivo* functions of ATP-dependent chromatin-remodelling factors. *Nature Reviews. Molecular Cell Biology*, **3**, 422–9.

Vielle-Calzada, J. -P., Thomas, J., Spillane, C., Coluccio, A., Hoeppner, M. A. & Grossniklaus, U. (1999) Maintenance of genomic imprinting at the *Arabidopsis medea* locus requires zygotic *DDM1* activity. *Genes & Development*, **13**, 2971–82.

Vongs, A., Kakutani, T., Matiensson, R. A. & Richards, E. J. (1993) *Arabidopsis thaliana* DNA methylation mutants. *Science*, **260**, 1926–8.

Vyskot, B. & Hobza, R. (2004) Gender in plants: sex chromosomes are emerging from the fog. *Trends in Genetics*, **20**, 432–8.

Wassenegger, M., Heimes, S., Riedel, L. & Sänger, H. L. (1994) RNA-directed *de novo* methylation of genomic sequences in plants. *Cell*, **76**, 567–76.

Westphal, T. & Reuter, G. (2002) Recombinogenic effects of suppressors of position-effect variegation in *Drosophila*. *Genetics*, **160**, 609–21.

Yao, H., Zhou, Q., Li, J., Smith, H., Yandeau, M., Nikolau, B. J. & Schnable, P. S. Molecular characterization of meiotic recombination across the 140-kb multigenic *a1-sh2* interval. (2002) *Proceedings of the National Academy of Sciences of the United States of America*, **99**, 6157–62.

Ye, F. & Signer E. R. (1996) RIGS (repeat-induced gene silencing) in *Arabidopsis* is transcriptional and alters chromatin configuration. *Proceedings of the National Academy of Sciences of the United States of America*, **93**, 10881–6.

Zemach, A. & Grafi, G. (2003) Characterization of *Arabidopsis thaliana* methyl-CpG-binding domain (MBD) proteins. *Plant Journal*, **34**, 565–72.

Zhimulev, I. F., Belayaeva, E. S., Semeshin, V. F., Shloma, V. V., Makunin, I. V. & Volkova, E. I. (2003) Overexpression of the SUUR gene induces reversible modification at pericentric, telomeric and intercalary heterochromatin of *Drosophila melanogaster* polytene chromosomes. *Journal of Cell Science*, **116**, 169–76.

Zilberman, D., Cao, X. & Jacobson, S. E. (2003) ARGONAUTE4 control of locus-specific siRNA accumulation and DNA and histone methylation. *Science*, **299**, 716–19.

5   When alleles meet: paramutation

Marieke Louwers, Max Haring and Maike Stam

5.1   Introduction

More than 50 years ago, genetic experiments led to unexpected results (Bateson & Pellew, 1915; Hagemann, 1958; Lilienfeld, 1929; Renner, 1959). Mutant phenotypes appeared at a frequency much higher than observed for classical mutations, and they were not as stable; in some cases, reversions to the original state could be observed. In addition, crosses between mutant and wild-type plants gave rise to progeny with mutant phenotypes, indicating that the wild-type allele was changed into a mutant allele. These observations were in contradiction to Mendel's first law, which states that alleles segregate unchanged from each other during meiosis. Brink (1958) termed this phenomenon 'paramutation', as it shares similarities with, but is also distinct from, genetic mutations. Today, paramutation has been reported not only for plants, but also for animals and fungi (reviewed in Brink, 1973; Chandler & Stam, 2004). Definitions concerning paramutation are listed in Box 5.1.

All examples of paramutation involve a *trans*-interaction between alleles that results in a heritable change in gene expression of one of the alleles (Figure 5.1A). This *trans*-interaction does not cause a change in DNA sequence, but rather a change in DNA methylation (see Luff *et al.*, 1999; Meyer *et al.*, 1993; Rassoulzadegan *et al.*, 2002; Sidorenko & Peterson, 2001; Walker & Panavas, 2001) and/or chromatin structure (Chandler *et al.*, 2000; Stam *et al.*, 2002a; van Blokland *et al.*, 1997), with transcriptional silencing as the usual consequence (Hollick *et al.*, 2000; Meyer *et al.*, 1993; Mittelsten Scheid *et al.*, 2003; Patterson *et al.*, 1993; van West *et al.*, 1999), and it is therefore a classic example of an epigenetic phenomenon. The alleles involved in paramutation are called paramutagenic and paramutable. These are, in general, two different epigenetic states of the same allele and therefore not true alleles, but epialleles. The paramutagenic allele changes the epigenetic state of the paramutable allele, which becomes a paramutated allele. In genetic nomenclature the paramutated allele is often marked with a prime (e.g. *B'*: *B-prime*) to indicate that the allele is derived from its paramutable counterpart. In several cases, once the paramutable allele has been changed by the paramutagenic allele, it displays secondary paramutation. The paramutable allele has become paramutagenic itself: it can alter naive paramutable alleles. In a number of

Paramutation:
A *trans*-interaction between alleles resulting in a heritable change in gene expression of one of the alleles

Paramutable allele:
An allele susceptible to a change in epigenetic state induced by a paramutagenic allele

Paramutagenic allele:
An allele able to induce a change in the epigenetic state of a paramutable allele

Paramutated allele:
The epigenetic state of a paramutable allele after *trans*-inactivation by a paramutagenic allele

Paramutation alleles:
All alleles (paramutable, paramutated and paramutagenic) participating in paramutation

Secondary paramutation:
Upon paramutation, the paramutable allele becomes paramutagenic itself

Spontaneous paramutation:
A paramutable allele spontaneously changes into a paramutated allele

Neutral alleles:
Alleles not participating in paramutation

Paramutability:
The capacity of a paramutable allele to become paramutated, when combined in one nucleus with a paramutagenic allele

Paramutagenicity:
The capacity of a paramutagenic allele to paramutate a paramutable allele, when combined in one nucleus

Paramutation sequences:
Sequences required for paramutability and/or paramutagenicity

Box 5.1

cases, the epigenetic state of the paramutable allele can also spontaneously change into the paramutagenic state (Figure 5.1B; Bateson & Pellew, 1915; Coe, 1959; English & Jones, 1998; Hollick *et al.*, 1995; Meyer *et al.*, 1993). Paramutagenic and paramutable alleles are rare. In fact, most alleles of a gene are neither paramutagenic nor paramutable, and are therefore called neutral alleles. The change in epigenetic state generally involves a change in gene expression. In some cases, however, paramutation involves other processes: transposition (Harrison & Carpenter, 1973; van Houwelingen *et al.*, 1999), recombination (Rassoulzadegan *et al.*, 2002), genomic imprinting (Duvillie *et al.*, 1998; Forne *et al.*, 1997) or the susceptibility to diabetes (Bennett *et al.*, 1997). In this review, for simplicity, the term 'gene expression' is used to indicate all processes that can be affected by paramutation.

Paramutation often involves genetically identical paramutable and paramutagenic alleles (Meyer *et al.*, 1993; Stam *et al.*, 2002a), but can also involve different, but homologous alleles (Qin & von Arnim, 2002; Rassoulzadegan *et al.*, 2002; Walker & Panavas, 2001). More recently the term paramutation has been broadened to include *trans*-interactions between non-allelic homologous sequences. Especially transgenic approaches led to discoveries of non-allelic

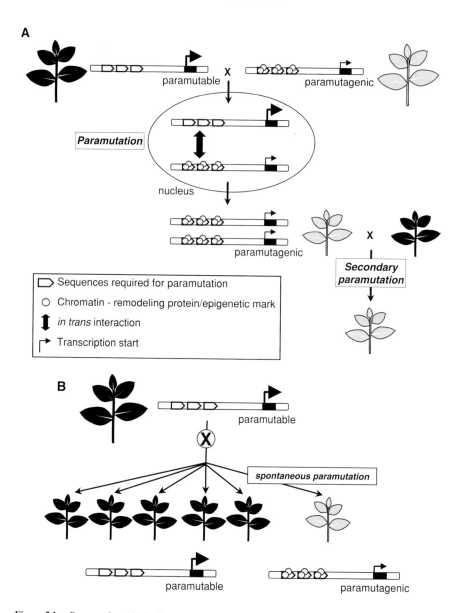

Figure 5.1 Paramutation. (A) A plant carrying a paramutable allele (giving rise to a dark phenotype) is crossed to a plant containing a paramutagenic allele (giving rise to a light phenotype). When combined in one nucleus, the paramutagenic allele *trans*-interacts with the paramutable allele, heritably downregulating the expression of the paramutable allele. In several cases, once the paramutable allele is downregulated, it displays secondary paramutation; it has become paramutagenic itself. As a result, crosses between the F1 and a plant carrying naive paramutable alleles only yield light phenotypes. (B) Spontaneous paramutation. The paramutable state of an allele can spontaneously change into the paramutagenic state with a certain frequency. As a result, plants carrying a paramutable allele, when self-fertilized, give rise to a certain percentage of progeny plants containing a paramutagenic allele.

paramutation-like phenomena (Qin & von Arnim, 2002; Sidorenko & Peterson, 2001; van West *et al.*, 1999). For the sake of simplicity, throughout this review the term ''alleles'' is used for the interacting loci.

Most reported paramutation phenomena deal with visible phenotypes, such as changes in pigment or drug resistance, which facilitated their discovery. It is to be expected that the recent genome-wide approaches will reveal that para-mutation is a far more widespread phenomenon than originally anticipated, also affecting loci not influencing a visible phenotype.

This chapter describes in detail several paramutation phenomena in plants, animals and fungi, and points out features they share and those they differ in. RNA- and pairing-based models are discussed, and *trans*-acting mutations affecting paramutation are described. Finally, possible roles and the evolu-tionary significance of paramutation are discussed.

5.2   Paramutation across kingdoms

5.2.1   *Paramutation in plants*

Paramutation is most extensively studied in plants. Below we discuss several examples.

5.2.1.1   *Paramutation at the b1 locus in maize*

In maize (*Zea mays*), the loci known to undergo paramutation encode tran-scription factors involved in the activation of the anthocyanin biosynthesis pathway (Chandler *et al.*, 2000). Anthocyanins are red and purple plant pigments that are not essential for plant survival. Paramutation at the *booster1* (*b1*), *pericarp color1* (*p1*), *purple plant1* (*pl1*) and *red1* (*r1*) loci is readily visualized by changes in pigmentation of specific plant tissues (Brink, 1956; Hollick *et al.*, 1995; Patterson *et al.*, 1993; Sidorenko & Peterson, 2001). These phenomena are therefore ideal model systems for studying allelic *trans*-interactions. One of the best examined cases of paramutation occurs at the *b1* locus. This locus codes for a basic helix–loop–helix transcription factor that activates anthocyanin biosynthesis in the epidermal cell layer of most vegetative parts of the plant (reviewed in Chandler *et al.*, 2000).

Many *b1* alleles have been identified (Selinger & Chandler, 1999), but most of these are neutral to paramutation. The alleles involved in paramutation are the paramutable *B-I* and paramutagenic *B'* allele (Coe, 1966; Patterson *et al.*, 1993). *B-I* confers dark, intense anthocyanin pigmentation. The *B-I* epigenetic state is unstable, 1–10% of the progeny of homozygous *B-I* plants show a light pigmented phenotype. In the light-colored plants the paramutable *B-I* state is changed into the paramutagenic *B'* state, which is transcribed at a level 20 times lower than the *B-I* state (Patterson *et al.*, 1993). When the light-colored

B' plants are crossed with dark-colored B-I plants, *trans*-interactions between the B' and the B-I allele always lead to a heritable change of the B-I into the B' state, resulting in a light-colored offspring. The B' state is very stable, reversions to a B-I state have never been seen (~100,000 plants looked at; Coe, 1966; Patterson & Chandler, 1995). Although in the literature B-I and B' are referred to as being different alleles, they actually represent two different epigenetic states of the same allele (Stam *et al.*, 2002a).

The sequences required for *b1* paramutation are located in a 6-kb region 100 kb upstream of the transcribed sequences (Stam *et al.*, 2002b). This region contains seven directly repeated copies of a 853-bp sequence otherwise unique in the maize genome (Stam *et al.*, 2002a). Neutral *b1* alleles have only one copy of this sequence. The 853-bp sequence is AT-rich (60%) and does not show significant similarity to a known gene. The repeats are required for both paramutation and high *b1* expression. An allele with five repeats can become fully paramutagenic, an allele with three repeats shows a decreased paramutagenicity, and alleles with one copy do not participate in paramutation, they are neutral. Furthermore, only alleles with multiple repeats can drive high *b1* expression.

Generally, differences in expression level correlate with differences in DNA methylation level. No difference in DNA methylation level could however be detected between B' and B-I around and within the transcribed region (Patterson *et al.*, 1993). They do however differ in chromatin structure at this region, i.e. B-I is more nuclease-sensitive near the transcription start site than B' (Chandler *et al.*, 2000). B-I and B' show differences in DNA methylation at the repeated region (M. Stam & R. Bader, unpublished data; Stam *et al.*, 2002a), and the B-I repeats are more nuclease-sensitive than the B' repeats. The differences in DNA methylation and chromatin structure are specific for the repeated sequences; the regions flanking the repeats show no differences between B-I and B' (Stam *et al.*, 2002a).

5.2.1.2 *Paramutation at the pl1 locus in maize*

Another well-studied case of paramutation in maize occurs at the *pl1* locus (Hollick *et al.*, 1995). *Pl1* codes for a myb-related transcription factor regulating the anthocyanin biosynthesis pathway (Cone *et al.*, 1993). The paramutable *pl1* allele is called *Pl-Rhoades* (*Pl-Rh*) and results in dark pigmented vegetative tissues, anthers, seeds and young seedlings. *Pl-Rh* can spontaneously change to the low-expressed paramutagenic *pl1* state called *Pl'* (Hollick *et al.*, 1995). In *Pl'*/*Pl-Rh* heterozygous plants, *Pl'* always heritably alters the *Pl-Rh* state into a *Pl'* state. The reduction in *pl1* expression is visible in all tissues where *pl1* is expressed, but is most obvious in the anthers. The degree of reduction can vary, resulting in a series of phenotypes, ranging from complete loss to only a slight reduction of anther pigmentation (Hollick *et al.*, 1995). The pigment levels reflect *pl1* RNA levels (Hollick *et al.*,

2000). The amount of anther pigmentation varies considerably among individuals, but is uniform within a given plant. The expression levels of *Pl'* alleles correlate with the level of *Pl'* paramutagenicity; alleles directing a low expression level are more paramutagenic than alleles directing an intermediate level of expression. The lower expression states are relatively stable while alleles with intermediate levels of expression are less stable and can show either increased or decreased levels of expression relative to their parents.

The sequences of *Pl-Rh*, *Pl'* and the neutral *Pl-blotched* allele are 99.8% identical over a 5.5-kb region spanning the coding region, suggesting that, similar to the situation at the *b1* locus, the *cis*-acting sequences required for *pl1* paramutation are located either further upstream or downstream of the coding region (Hollick *et al.*, 2000). *Pl1* paramutation is associated with an 18.5-fold reduction in *pl1* RNA levels, while the transcription is only reduced threefold. This suggests *pl1* paramutation involves both transcriptional as well as post-transcriptional components. As observed for *b1* paramutation, extensive restriction analyses could not detect any differences in cytosine methylation levels between *Pl-Rh* and *Pl'* within a region of about 15 kb encompassing the *pl1*-coding region.

5.2.1.3 Paramutation at the sulfurea locus in tomato

A classic example of paramutation occurs at the *sulfurea* (*sulf*) locus in tomato (Hagemann, 1969, 1993; Wisman *et al.*, 1983). The *sulf* locus is involved in the chlorophyll content of the leaves. The paramutable *sulf*⁺ allele gives rise to green leaves and does not show spontaneous paramutation. It can however be heritably changed by paramutagenic *sulf* alleles. The paramutagenic *sulf* alleles (*sulf^{vag}* (*variegata*), *sulf^{pura}* and SC148) give rise to different degrees of chlorophyll-deficient tomato plants, and were isolated upon X-ray treatment (Hagemann, 1969) or regeneration after tissue culture (Wisman *et al.*, 1983). Plants homozygous for the *sulf^{vag}* allele have green cotyledons and variegated leaves, whereas plants homozygous for the *sulf^{pura}* allele carry completely yellow cotyledons and leaves. Although this has not been demonstrated, it is likely that the *sulf^{vag}* and *sulf^{pura}* alleles are epialleles. The paramutagenic SC148 allele, isolated from a different genetic background, gives rise to a phenotype intermediate between *sulf^{vag}* and *sulf^{pura}*. The extent of chlorophyll variegation directed by a specific paramutagenic *sulf* allele is reflected in its degree of paramutagenicity when heterozygous with the paramutable *sulf*⁺ allele. The paramutagenicity of *sulf^{vag}* allele is low (0–12% of heterozygous plants display variegated leaves), whereas that of the *sulf^{pura}* allele is generally high (varies between 0.5% and 100%). The paramutagenicity of the various alleles is influenced by the genetic background (Wisman *et al.*, 1983).

When *sulf*⁺ is paramutated, the resulting allele does not necessarily have the same epigenetic state as the paramutagenic allele: *sulf*⁺/*sulf^{pura}* plants can have a *sulf^vag* phenotype and yield *sulf^vag* offspring. It is hypothesized that *sulf^+* is paramutated in a stepwise manner, first from *sulf^+* to *sulf^vag* (partial inactivation) and subsequently to *sulf^pura* (complete inactivation).

The molecular mechanisms underlying *sulf* paramutation might be similar to those underlying position effect variegation (PEV), as both phenomena share some characteristics (Wisman *et al.*, 1983). PEV involves an X-ray-induced chromosomal rearrangement, positioning a euchromatic gene close to heterochromatin (Weiler & Wakimoto, 1995). This results in gene silencing in some cells but not in others; sectors of pigmented and unpigmented cells in *Drosophila* eyes are, for example, the consequence. Similar to PEV, the first paramutagenic *sulf* alleles were isolated using X-rays; the *sulf* locus maps close to heterochromatin, and the inactivated allele gives rise to variegated patterns in leaf color.

5.2.1.4   *Paramutation at the transgenic A1 locus in petunia*

The first example of paramutation involving transgenic plants concerns a transgene consisting of the maize *A1*-coding region driven by the constitutive CaMV-35S promoter (Meyer *et al.*, 1987, 1993). The *A1* gene is a structural gene encoding dihydroflavonol 4-reductase, an enzyme involved in anthocyanin biosynthesis. The introduction of this transgene into otherwise white flowering petunia plants resulted in brick-red pigmented flowers. Transgenic line #17 carries a single copy of the *A1* transgene and displays a metastable phenotype. Instead of fully pigmented flowers, due to spontaneous paramutation of the transgene, occasionally white, white-sectored or marbled flowers are observed (Prols & Meyer, 1992). When comparing red and white flowers, it turned out that the *A1* transgene was transcriptionally downregulated in the latter ones. The changed epigenetic state was called *17-W* (white flowers) and appeared paramutagenic when combined with the high-expressing state of the same allele (*17-R*; red flowers; Meyer *et al.*, 1993). Plants heterozygous for the *17-R* and *17-W* allele carried white or variable-colored flowers with a certain frequency. The *17-R* allele was heritably downregulated in these plants. Remarkably, the paramutation frequency showed a parent-of-origin effect. Paramutation was more pronounced when the 17-W line was used as a pollen donor than when the 17-R line was used (40% vs. 5% fully white flowers, respectively). A similar effect has also been observed for paramutation at the *nivea* locus of *Antirrhinum majus* (Harrison & Carpenter, 1973). The frequency of spontaneous *A1* paramutation appeared to depend on environmental effects: in field-grown plants, flowers developed early in the season were predominantly red, while flowers developed later in the season, when it was warmer and the light more bright, displayed less pigmentation (Meyer *et al.*, 1992). Endogenous factors, such as the age of the parental plant when crossed, also seemed to play a role. Although it has not been reported, it is very likely that the heritably downregulated *17-R* allele displays secondary paramutation.

A1 paramutation involves transcriptional silencing (Meyer *et al.*, 1993). The *17-R* state of the *A1* transgene is clearly transcribed, while the *17-W* state is transcriptionally downregulated. The transcriptional silencing correlated with increased DNA methylation and reduced nuclease sensitivity. Relative to the hypomethylated paramutable *17-R* state, the paramutagenic *17-W* state was hypermethylated in both symmetrical and non-symmetrical cytosines in and around the CaMV-35S promoter (Meyer *et al.*, 1992, 1993; Prols & Meyer, 1992). The cytosine methylation level at the promoter corresponded with the flower pigmentation level. White flowers showed more DNA methylation than marbled flowers, which in turn showed more methylation than fully red flowers. The inactive, paramutagenic *17-W* state of the *A1* gene was considerably less nuclease-sensitive than the active, paramutable *17-R* state. The regions flanking the *A1* gene showed no differences in DNA methylation level and nuclease sensitivity between 17-W and 17-R plants (Meyer & Heidmann, 1994; van Blokland *et al.*, 1997).

5.2.1.5   *Trans*-inactivation at the *PAI* loci in *Arabidopsis*

A paramutation-like phenomenon involving *trans*-interactions between non-allelic endogenous *phosphoribosylanthranilate isomerase* (*PAI*) genes has been observed in the Wassilewskija (WS) accession of *Arabidopsis*. The *PAI* genes encode enzymes active in the tryptophan biosynthesis pathway. An inverted repeat (IR) of two *PAI* genes heritably *trans*-inactivates unlinked homologous single-copy genes (Luff *et al.*, 1999).

WS contains four *PAI* genes located at three loci (Bender & Fink, 1995). One of these loci contains the *PAI1* and *PAI4* genes organized in a tail-to-tail IR. The other two loci contain the non-allelic *PAI2* and *PAI3* single copy genes. The sequences of *PAI1*, *PAI2* and *PAI4* are nearly 100% identical, whereas the sequence of *PAI3* is 90% identical to that of the other *PAI* genes (Bender & Fink, 1995; Melquist *et al.*, 1999). The majority of *PAI* transcripts are derived from the *PAI1* gene. *PAI2* and *PAI3* are transcriptionally silent, and *PAI4* lacks the upstream promoter sequences. Only *PAI1* and *PAI2* encode a functional enzyme. All *PAI* genes are heavily cytosine-methylated. Cytosine methylation at the *PAI2* and *PAI3* genes is present at the DNA sequences homologous to the *PAI1–PAI4* IR, including the promoter sequences.

The *PAI1–PAI4* IR can be considered paramutagenic and the single-copy *PAI* genes paramutable. The *PAI1–PAI4* locus heritably *trans*-inactivates (paramutates) the unmethylated single-copy *PAI* alleles derived from the Columbia (Col) accession of *Arabidopsis* (Luff *et al.*, 1999). This *trans*-inactivation is associated with *de novo* methylation of the Col single-copy genes. Unlike what is observed for most other paramutation phenomena, once paramutated, single-copy *PAI* genes do not become paramutagenic themselves (Luff *et al.*, 1999). The silenced state of single-copy *PAI* genes is relatively stable. Reversion of the silent state occurs in only 1–5% of the progeny of self-fertilized plants lacking the *PAI1–PAI4* IR (Bender & Fink, 1995). Once reactivated, the single-copy loci remain active in the absence of the *PAI1–PAI4* IR.

Double-stranded RNA (dsRNA) derived from the *PAI* IR is required for both the silencing of the single-copy *PAI* genes, and the maintenance of cytosine methylation at the IR (Melquist & Bender, 2003, 2004). RNA blot analysis indicated the production of *PAI1* 3′ read-through transcripts that include the palindromic *PAI4* sequence (Melquist & Bender, 2003). Severely reduced transcription of the IR resulted in decreased methylation at the single-copy *PAI* genes. A mutation blocking read-through transcription at the *PAI* IR reduced DNA methylation not only at the single-copy loci, but also at the IR itself (Melquist & Bender, 2004). Unlike observed for many cases of RNA silencing, the production of double-stranded *PAI* RNA does not result in detectable amounts of small RNAs (Melquist & Bender, 2003) or the posttranscriptional silencing of the *PAI1* gene. The authors suggest that either the dsRNA itself or levels of small RNAs below the detection limit serve as the *trans*-acting silencing signal.

5.2.2   *Paramutation in mammals and fungi*

More recently, paramutation has also been observed in organisms other than plants. In order to give a complete picture of paramutation, a few examples are discussed below.

5.2.2.1   *LoxP trans*-silencing in mice

Recently, two cases of paramutation in mice were described, each involving transgenic *loxP* sites (Rassoulzadegan *et al.*, 2002). In both cases, *loxP* sites became cytosine-methylated upon meiosis-specific expression of Cre recombinase. As a consequence they are no longer a substrate for Cre-mediated recombination. Once methylated, the *loxP* sites behave as paramutagenic alleles. If they are combined with an allele containing unmethylated *loxP* sites, the latter are methylated and heritably inactivated. Moreover, they become paramutagenic themselves (secondary paramutation).

The inactive, paramutagenic state spreads into neighboring endogenous sequences. The DNA methylation present at the *loxP* sites expands to flanking endogenous sequences up to several kilobases away within subsequent generations. The paramutagenic *loxP*-containing alleles are also able to paramutate homologous non-transgenic alleles, which in turn become paramutagenic: they are able to methylate and inactivate *loxP*-containing alleles. The nature of the epigenetic mark imposed on the non-transgenic alleles was not reported.

This *trans*-inactivation phenomenon was initially termed transvection, but has been renamed paramutation (M. Rassoulzadegan & F. Cuzin, personal communication) as paramutation involves a meiotically heritable change while transvection does not (Duncan, 2002).

5.2.2.2   *Trans*-nuclear inactivation of the *inf1* gene in *Phytophthora infestans*

In the plant pathogen *Phytophthora infestans*, a paramutation-like phenomenon was described involving *trans*-interactions between non-allelic *inf1 trans*- and endogenous genes (van West *et al.*, 1999). *P. infestans* is a diploid oomycete of which the mycelial cells can contain multiple, genetically different nuclei, resulting in heterokaryotic strains. The *inf1* gene is a highly expressed, single-locus gene encoding the secreted, easily detectable INF1 protein. INF1 is a member of the elicitin family inducing defense responses in plants. Transcriptionally silenced *inf1* transgenes behave as paramutagenic loci, heritably *trans*-inactivating the endogenous, paramutable *inf1* genes. This is true for transgenes containing the *inf1*-coding region in antisense or sense orientation, or without a promoter.

The *trans*-inactivation observed in *P. infestans* is *trans*-nuclear, suggesting the involvement of RNA in the *trans*-inactivation. In heterokaryons containing non-transgenic and transgenic *inf1* silenced nuclei, the endogenous *inf1* genes in the non-transgenic nuclei are heritably inactivated as well. These findings support a model involving the *trans*-nuclear transfer of sequence-specific silencing signals. Given recent data indicating the involvement of RNA in many silencing phenomena, RNA is likely to be involved in this case of *trans*-inactivation as well. Secondary paramutation has not been reported for the *inf1* gene.

5.2.2.3   Interchromosomal DNA methylation transfer in *Ascobolus immersus*

In the ascomycete fungus *Ascobolus immersus* a paramutation-like process is observed at the *b2* gene (Colot *et al.*, 1996). The *b2* gene is involved in spore pigmentation. A methylated, inactivated *b2* gene gives rise to a white spore phenotype (Colot & Rossingnol, 1995), whereas an unmethylated active *b2* gene gives rise to dark brown pigmented spores. The active *b2* gene is *trans*-inactivated by the meiotic transfer of DNA methylation. The methylation is transferred from the inactive (paramutagenic) donor allele to the active (paramutable) recipient allele, which now becomes methylated and inactive (paramutated).

The transfer of DNA methylation during meiosis is mechanistically related to recombination (Colot *et al.*, 1996). A strain carrying a methylated *b2* allele was fused to a strain carrying the unmethylated homolog. In *A. immersus*, once the two nuclei fuse, the resulting diploid cells immediately enter meiosis, giving rise to eight haploid ascospores (Zickler, 1973). Therefore, interactions between the methylated and unmethylated alleles leading to DNA methylation transfer are limited to meiosis. In ~9% of the cells a (partial) transfer of DNA methylation took place, visualized by the resulting spore colors. When tested, gene conversion (the non-reciprocal exchange of sequences from one sister chromatid to another) always went hand in hand with the transfer of DNA methylation, but not the other way around. Out of 100 asci displaying DNA methylation transfer, seven also showed gene conversion. The methylation transfer showed the same polarity (5′ to 3′ polarity) as gene conversion does. These results indicate that DNA methylation transfer might occur through an earlier intermediate in the recombination process.

Meiotic recombination is thought to involve DNA–DNA pairing inter-actions between the intact duplexes of homologous chromosomes (Kleckner, 1996). Another hypothesis we want to propose is the involvement of non-coding RNA in DNA methylation transfer and gene conversion. Intriguingly, a recent study reported a mouse recombination hot spot encoding a non-coding RNA (Nishant et al., 2004). The occurrence of transcriptional activity close to a recombination hot spot is supporting the hypothesis that chromatin accessi-bility is crucial to initiate recombination by DNA–DNA pairing. Alternatively, or in addition, the non-coding RNA itself is involved.

5.3 Paramutation models

Paramutation shares features with other epigenetic phenomena (see Section 5.4; Lippman & Martienssen, 2004; Matzke et al., 2004) and is therefore expected to involve similar mechanisms. Below, an RNA and physical pairing model are presented that might explain the various paramutation phenomena. The two models are not mutually exclusive; a combination of the two is presented as a third model.

5.3.1 RNA-based model

Given the recent evidence for a role of RNA in numerous epigenetic phenom-ena (reviewed by Cerutti, 2003; Grewal & Rice, 2004; Lecellier & Voinnet, 2004; Lippman & Martienssen, 2004), it is likely that RNA plays a role in various paramutation events as well. dsRNA and small interfering RNAs (siRNAs) are key players in RNA silencing. siRNAs are derived from dsRNA via cleavage by a dsRNA ribonuclease called Dicer, and are thought to trigger DNA methylation of homologous sequences (reviewed by Bender, 2004; Matzke et al., 2004). A second model for silencing triggered by RNA involves long single-stranded RNAs (Morey & Avner, 2004).

5.3.1.1   Silencing by dsRNA and siRNAs

The most efficient way to produce dsRNA is through transcription of DNA IRs. Although generally less efficient, directly repeated sequences (direct repeats, DRs), and even single-copy sequences can result in dsRNA production. dsRNA can be produced from the latter two by transcription in both sense and antisense orientation, or via antisense production by RNA-dependent RNA polymerase (RdRP; Figure 5.2A; Baulcombe, 2004; Lindbo *et al.*, 1993; Smith *et al.*, 1994). RdRP produces antisense RNA by using siRNAs as primers on sense templates, or by primer-independent action on a so-called 'aberrant' sense RNA template. It is still unclear which features make an 'aberrant' RNA a good target for RdRP.

RNA silencing triggered by IR-derived dsRNAs is easily sustained. The combined action of Dicer and RdRP results in a continuous multiplication of siRNAs, augmenting the silencing process (Figure 5.2A; Sijen *et al.*, 2001). Silencing by DRs is continuous once a rare antisense RNA is produced from the DR (Martienssen, 2003). Such an antisense RNA is sufficient to result in siRNA multiplication and thereby silencing. Due to the repeated structure, the DR-derived siRNAs can bind both upstream and downstream of the sequence where the siRNAs are derived from, supplying primers for RdRP (Figure 5.2A). Silencing triggered by single-copy sequences requires specific conditions. Single-copy sequences do not support a continuous multiplication of siRNAs by RdRP and Dicer, as RdRP activity has a 5' to 3' polarity (Martienssen, 2003; Sijen *et al.*, 2001). Therefore, silencing by single-copy sequences requires the continuous production of both sense and antisense RNA, or of good RdRP templates.

5.3.1.2   Silencing by long RNAs

Some paramutation phenomena might also involve long single-stranded RNAs rather than dsRNA and/or siRNAs (see Figure 5.2B). In this model, 'long' RNAs are involved in directing DNA methylation, histone modifications and the recruitment of chromatin proteins to the paramutable allele. There are several examples of long RNAs that result in chromatin silencing (Chow & Brown, 2003; Coady *et al.*, 1999; Geirsson *et al.*, 2003; Meller, 2003; Sleutels *et al.*, 2002). For example, the paternally expressed long, non-coding *Air* transcript overlaps the *Igf2r* gene in antisense direction and is required for the paternal repression of *Igf2r in cis* (Sleutels *et al.*, 2002).

5.3.1.3   RNA involvement in paramutation

RNA appears to play a role in at least two paramutation-like phenomena. A sequence-specific, diffusible factor mediates *trans*-inactivation in *P. infestans* (see Section 5.2.2.2; van West *et al.*, 1999), suggesting that RNA is involved. dsRNA produced from the *PAI1–PAI4* IR mediates DNA methylation and

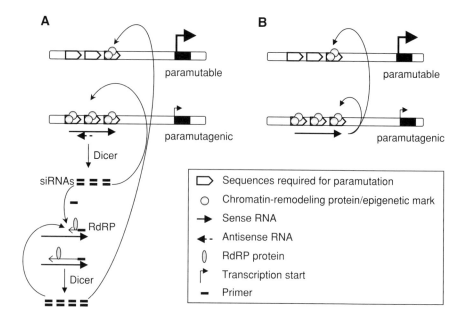

Figure 5.2   RNA model for paramutation. (A) Paramutation by small interfering RNAs (siRNAs). Double-stranded RNA (dsRNA) is generated from the sequences required for paramutation (e.g. direct repeats) by sense and rare antisense transcription. The Dicer enzyme cleaves the dsRNA into siRNAs. The siRNAs are hypothesized to trigger a cascade, resulting in chromatin silencing at the paramutable allele. In an amplification step, the primary formed siRNAs can bind to sense RNAs and act as a primer for RNA-dependent RNA polymerase (RdRP), which synthesizes antisense RNA. The resulting dsRNA starts the Dicer–RdRP cycle again. (B) Paramutation by long RNAs. In this model 'long' RNAs are involved in directing the DNA methylation, histone modifications and the recruitment of chromatin proteins to the paramutable allele.

trans-inactivation of the homologous, non-allelic single-copy *PAI* genes (see Section 5.2.1.5; Melquist & Bender, 2003, 2004). *PAI trans*-inactivation is unlike other RNA-silencing phenomena:

(1) full-length *PAI* transcripts are still detectable and result in PAI activity;
(2) small *PAI* RNAs cannot be detected;
(3) it takes several generations before the *trans*-inactivation of the single-copy genes is complete (Luff *et al.*, 1999).

The authors therefore suggest a mechanism in which either the dsRNA itself or undetectable levels of small RNAs are involved in the *trans*-inactivation.

5.3.2   Pairing-based model

A second model that could explain various paramutation phenomena hypothesizes physical *trans*-interactions between the paramutagenic and paramutable alleles (Figure 5.3). These interactions would enable the exchange of protein complexes, affecting the epigenetic state of the paramutable allele.

Physical pairing between homologous chromosomal regions has been detected in various organisms. In *Drosophila*, homologous chromosomes pair over their entire length in somatic cells (Cook, 1997), but in plants and vertebrates, pairing between whole chromosomes is in general restricted to meiosis and premeiotic stages (reviewed in McKee, 2004). Physical *trans*-interactions between homologous chromosomal regions have however been observed in interphase nuclei of various organisms, including plants and mammals (Abranches *et al.*, 2000; Aragon-Alcaide & Strunnikov, 2000; Csink & Henikoff, 1996; Dernburg *et al.*, 1996; Fransz *et al.*, 2002; Fuchs *et al.*, 2002; LaSalle & Lalande, 1996). For example, in *Arabidopsis* and yeast, transgenic repeats associate with each other in interphase nuclei based on sequence identity (Abranches *et al.*, 2000; Aragon-Alcaide & Strunnikov, 2000; Pecinka *et al.*, submitted).

Physical *trans*-interactions appear to play a role in various silencing phenomena (Bean *et al.*, 2004; Csink *et al.*, 2002; Dorer & Henikoff, 1997; Kassis, 2002; Lee *et al.*, 2004; Rossignol & Faugeron, 1995; Sage & Csink, 2003; Singer & Selker, 1995; Turner *et al.*, 2005). For example, in both *Neurospora* and *Ascobolus*, linked and unlinked repeated sequences are inactivated in pairs (Rossignol & Faugeron, 1995; Singer & Selker, 1995), suggesting the involvement of physical interaction. RNA-mediated silencing should not be limited to pairs. Furthermore, in *Neurospora*, pairing is required to prevent meiotic silencing. The presence of unpaired gene copies during meiosis triggers silencing of all copies of that gene (Lee *et al.*, 2004). Recently, similar observations have been made in other organisms, including mice (Bean *et al.*, 2004; Turner *et al.*, 2005). Pairing has been observed between differentially

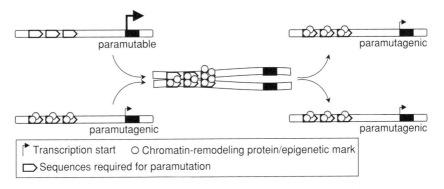

Figure 5.3 Pairing model for paramutation. Physical association between the paramutation sequences of the paramutable and paramutagenic allele results in an exchange of protein complexes, heritably changing the epigenetic state of the paramutable allele into a paramutagenic state.

imprinted chromosomal regions in wild-type human interphase nuclei and is suggested to be required for a correct pattern of parental imprinting (LaSalle & Lalande, 1996). Patients displaying a disturbed imprinting at a specific region lacked physical association.

The *trans*-interactions mediated by Polycomb group (PcG) proteins provide a good model of how to envisage pairing-induced paramutation (Bantignies *et al.*, 2003; Lavigne *et al.*, 2004). PcG proteins are involved in physical pairing between allelic and non-allelic chromosomal regions and also in pairing-dependent silencing in *Drosophila* (Bantignies *et al.*, 2003; Lavigne *et al.*, 2004; Pal-Bhadra *et al.*, 1997; Sigrist & Pirrotta, 1997). PcG proteins that are bound to one nucleosomal template can recruit and implement a repressed chromatin state on a second template (Lavigne *et al.*, 2004). PcG proteins act via regulatory Polycomb Response Elements (PREs). Pal-Bhadra *et al.* (1997) however showed that in case of repeat-induced silencing in *Drosophila*, PcG proteins can also bind to chromosomal sites lacking PREs. The Chromosome Conformation Capture (3C) method, successfully used to provide evidence for long-distance physical *in cis*-interactions (Dekker *et al.*, 2002; Murrell *et al.*, 2004; Tolhuis *et al.*, 2002), could be an excellent tool to examine if *trans*-interactions play a role in paramutation. The possible complication is that transient interactions, which will be difficult to detect, might be sufficient to establish the paramutated state.

5.3.3   Combined model

Paramutation could also involve both RNA and physical pairing. There are precedents for this hypothesis. Meiotic silencing, a silencing phenomenon observed in *Neurospora*, involves both pairing and RNA silencing (Lee *et al.*, 2004). Furthermore, both PcG proteins and components of the RNA-silencing pathway have been implicated in silencing of transgenic repeats in *Drosophila* and *Caenorhabditis elegans* (Lund & van Lohuizen, 2004). Zhang *et al.* (2004) showed that in *C. elegans* the RNA-binding domain of the PcG protein SOP-2 is essential for its localization and function.

5.4   Common features of paramutation phenomena

5.4.1   Involvement of repeats

Repeated sequences are a major trigger for the formation of silenced chromatin (Birchler *et al.*, 2000; Grewal & Rice, 2004; Henikoff, 1998; Lippman & Martienssen, 2004). Various transcriptionally downregulated regions in the genome, such as centromeres, telomeres and other heterochromatic regions are packed with repeated sequences (reviewed by Chan & Blackburn, 2004;

Dawe, 2003; Fransz *et al.*, 2003). The involvement of repeated sequences does not exclude any mechanistic model. Efficient RNA silencing and DNA pairing both require repeated sequences (see Section 5.3; Matzke *et al.*, 2004; Pecinka *et al.*, submitted). Transcription of repeats gives rise to dsRNA, a major trigger of RNA silencing, and repeated sequences physically pair more often than single-copy sequences. In *Arabidopsis*, transgenic *lac* operator arrays associate more often with each other than average euchromatic regions, and the same is observed for an inactive transgenic multi-copy *HPT* locus (Pecinka *et al.*, submitted). Furthermore, silenced repetitive sequences are prone to interact with heterochromatin (Csink & Henikoff, 1996; Dernburg *et al.*, 1996; Pecinka *et al.*, submitted), independent of sequence homology (Sage & Csink, 2003).

5.4.1.1 *Paramutation induced by repeats*

Although single-copy sequences can induce paramutation (Duvillie *et al.*, 1998; Meyer *et al.*, 1993; Qin & von Arnim, 2002; Qin *et al.*, 2003; Rassoul-zadegan *et al.*, 2002), in a considerable number of paramutation phenomena repeats are required for the induction of paramutation (English & Jones, 1998; Kermicle *et al.*, 1995; Luff *et al.*, 1999; Sidorenko & Peterson, 2001; Stam *et al.*, 2002a; van Houwelingen *et al.*, 1999; Walker & Panavas, 2001). Various types of repeats have been shown to induce paramutation: DRs, IRs and a combination of both in a complex structure.

Directly repeated sequences are required for *b1* and *r1* paramutation in maize and *SPT::Ac* paramutation in tobacco (English & Jones, 1998; Kermicle *et al.*, 1995; Stam *et al.*, 2002a). Multiple 853-bp direct repeats, situated ~100 kb upstream of the *b1*-coding region, are required for *b1* paramutagenicity and paramutability. The *r1* paramutagenic alleles contain at least two and at most five directly repeated *r1* genes. A stepwise decrease and increase in the *r1* copy number results in decreased and increased paramutagenicity, respectively (Kermicle *et al.*, 1995; Panavas *et al.*, 1999). Whereas the *b1* repeats are small, the repeated DNA fragments at the *r1* locus are at least 10 kb and contain the *r1*-coding region and flanking regions. Also for *SPT::Ac* paramutation in tobacco, DRs are required (English & Jones, 1998). The *SPT::Ac* loci carry two *streptomycine phosphotransferase* (*SPT*) genes in an IR, and zero to two copies of the transposable element *Activator* (*Ac*). Loci containing a functional *SPT* gene flanked on both sides by directly repeated *Ac* elements displayed efficient *in cis SPT* inactivation (Figure 5.4), and could *trans*-inactivate various active *SPT::Ac* alleles. *SPT::Ac* paramutation might involve physical interactions. The different *SPT::Ac* alleles offer sufficient possibilities for the production of dsRNA (Figure 5.4; English & Jones, 1998). Nevertheless, efficient silencing requires two *Ac* elements enclosing a functional *SPT* gene. We therefore hypothesize a mechanism involving physical interactions between the *Ac* elements, generating a silenced chromatin structure that is prone to *trans*-inactivate homologous alleles.

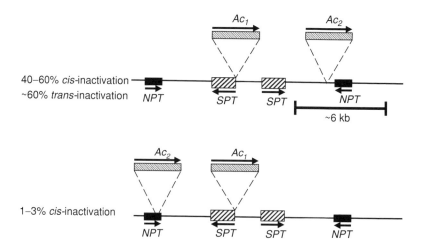

Figure 5.4   Organization of *SPT::Ac* alleles inducing *cis-* and *trans*-inactivation. Arrows show the direction of transcription of the *SPT* and *NPT* genes, and the *Ac* transposons. The thick bar shows the distance over which the second *Ac* can transpose without dramatically affecting the *incis* silencing efficiency. The allele organization at the top also induces *trans*-inactivation. The *trans*-inactivation capacity of the allele organization at the bottom has not been reported.

Some paramutation-like phenomena require IRs (Luff *et al.*, 1999; Melquist *et al.*, 1999; van Houwelingen *et al.*, 1999; Walker & Panavas, 2001). All paramutable *r1* alleles examined (16) contain two *r1* genes organized in an IR (Walker & Panavas, 2001). *PAI* *trans*-inactivation requires the *PAI1–PAI4* IR locus (see Sections 5.2.1.5 and 5.3.1; Melquist & Bender, 2003, 2004). The presence of two *dTph1* transposons organized in an IR is required for a paramutation-like *trans*-interaction, resulting in a novel transposition mechanism in petunia plants (van Houwelingen *et al.*, 1999).

Repeated sequences are also present in alleles of other paramutation systems, but their role in paramutation has not yet been reported (Bennett *et al.*, 1996; Mittelsten Scheid *et al.*, 2003; Stokes & Richards, 2002). *HPT* paramutation in tetraploid *Arabidopsis* plants involves a locus containing two CaMV-35S promoters in a direct orientation (Mittelsten Scheid *et al.*, 2003). The *Resistance*-like (*R*-like) gene cluster, of which two epigenetic variants show a paramutation-like *trans*-interaction (Stokes & Richards, 2002), contains several, highly homologous genes that are mostly organized in a direct orientation (E. Richards, personal communication). Furthermore, a variable number of direct repeats (VNTR) minisatellite, upstream of the insulin-coding region, is involved in human diabetes (Bell *et al.*, 1984; Bennett *et al.*, 1995). Class I alleles contain 26–63 repeats and predispose in a recessive way to type I diabetes, whereas class III alleles (140–209 repeats) generally protect against type I diabetes (Bennett *et al.*, 1996; Kelly *et al.*, 2003). Remarkably, if a father is heterozygous for a class III allele and a particular class I allele, and the offspring receives the paternal class I allele, then this allele does not predispose to the disease anymore (Bennett *et al.*, 1997; Stead *et al.*, 2000). Apparently, allelic interactions in the father between a specific class I and class III allele affect the epigenetic state of the class I allele in a meiotically heritable way. We like to speculate that the directly repeated sequences present at these various alleles are somehow involved in this paramutation process.

5.4.1.2   *Paramutation induced by single-copy sequences*

Repeated sequences are not required for all paramutation phenomena. Paramutation can also be triggered by single-copy sequences (*A1* and L91 transgenes in plants, *loxP* and recombinant *Ins2* alleles in mouse; Duvillie *et al.*, 1998; Meyer *et al.*, 1993; Qin & von Arnim, 2002; Rassoulzadegan *et al.*, 2002). These single-copy sequences might continuously produce RNAs triggering RNA silencing (see Section 5.3.1); alternatively, they are *trans*-inactivated via physical pairing mediated through specific protein-binding sites. For example, single-copy PREs are sufficient for PcG-dependent *trans*-inactivation (Sigrist & Pirrotta, 1997).

5.4.2   *Sequence requirements for paramutation*

Paramutation can require specific sequence elements. A transgenic approach, in which various distal *p1* promoter sequences were tested, demonstrated that only transgene loci containing the '1.2-kb fragment' were able to induce paramutation of the endogenous *P-rr* allele (Sidorenko & Peterson, 2001), suggesting this fragment contains special features. Remarkably, the 1.2-kb sequence is not only required for paramutagenicity, but it also acts as an enhancer at the *p1* locus (Sidorenko *et al.*, 1999, 2000). Similarly, the 853-bp repeats that are required for *b1* paramutagenicity are also required for *b1* enhancer activity (Stam *et al.*, 2002a).

How would enhancer sequences mediate *trans*-interactions? A good model is provided by Francastel *et al.* (1999). Their results suggest that a functional enhancer can avoid transgene silencing by influencing the subnuclear localization of a gene. It prevents the localization of the gene close to centromeric heterochromatin.

The presence of a specific sequence per se is however not necessarily sufficient to cause paramutation. In case of *p1* paramutation, repetition of the 1.2-kb sequence seems required. The *p1*-coding region of the *P1-rr* allele, an allele that can become paramutagenic, is flanked by, and partially overlapping with two 5.2-kb DRs (Figure 5.5; Athma *et al.*, 1992; Das & Messing, 1994). Each 5.2-kb repeat in turn contains two almost perfect 1.2-kb DRs. The *P1-wr* allele cannot be *trans*-inactivated (Figure 5.5; Sidorenko & Peterson,

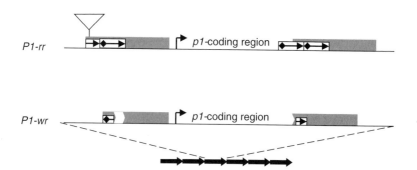

Figure 5.5 Map of the *P1-rr* and *P1-wr* allele. *P-rr* contains four 1.2-kb direct repeats, one of which is truncated and contains a 1.6-kb insertion (triangle). The *P1-wr* locus contains approximately six directly repeated copies of a fragment containing the *p1*-coding and flanking regions. Every copy contains only parts of the 1.2-kb repeats, and as a result truncated 5.2-kb repeats, represented by the split and truncated gray boxes. Gray boxes represent the 5.2-kb repeats flanking the *p1*-coding region; black arrows in white boxes represent the 1.2-kb repeats. The *p1* transcription start is indicated by a bent arrow.

2001). This allele contains approximately six directly repeated copies of a fragment containing both the *p1*-coding region and the flanking regulatory sequences. Each copy contains the 1.2-kb sequence only once, broken up in two separate parts (Chopra *et al.*, 1996, 1998). These data suggest that localized repetitiveness of the complete 1.2-kb sequence, as observed at the *P-rr* allele, might be required for *p1* paramutability.

5.4.3   Involvement of DNA methylation and chromatin structure

The silencing of genes and intergenic regions is generally correlated with DNA hypermethylation and specific chromatin structures (Bender, 2004; Lippman & Martienssen, 2004; Matzke *et al.*, 2004; Richards & Elgin, 2002). Similarly, in most paramutation systems, paramutagenicity correlates with hypermethylation, and *trans*-inactivation of the paramutable allele is associated with the acquisition of DNA methylation (Eggleston *et al.*, 1995; Forne *et al.*, 1997; Hatada *et al.*, 1997; Luff *et al.*, 1999; Meyer *et al.*, 1993; Mittelsten Scheid *et al.*, 2003; Rassoulzadegan *et al.*, 2002; Sidorenko & Peterson, 2001; Stam *et al.*, 2002a; Walker, 1998; Walker & Panavas, 2001). In case of *r1* paramutation, paramutagenic and neutral alleles have similar structural features, and cannot be distinguished based on those. They do however display different DNA methylation levels at specific regions; the paramutagenic alleles are hyper- and the neutral alleles hypomethylated in these regions (Walker & Panavas, 2001). Merely the presence of DNA methylation is however not sufficient for paramutation to occur. For example, the hypermethylated, inactivated *SUPERMAN* allele in *Arabidopsis* does not *trans*-inactivate its hypomethylated counterpart (Jacobsen & Meyerowitz, 1997).

What comes first, changes in DNA methylation or changes in chromatin structure? Or do they interconnect? In some cases, the rate of DNA methylation of the paramutable allele upon paramutation is very slow and does not reflect the extent of the phenotypic change, suggesting that chromatin-based silencing mechanisms precede DNA methylation. This applies, for example, for *HPT* paramutation (Mittelsten Scheid *et al.*, 2003). In addition there are cases where no correlation is observed between DNA methylation and paramutation (English & Jones, 1998; van West *et al.*, 1999). DNA methylation however probably plays an important role in the heritability of silenced epigenetic states (Dieguez *et al.*, 1998; Jones *et al.*, 2001; Kato *et al.*, 2003; Soppe *et al.*, 2002). This might explain why paramutation has not yet been described in organisms lacking extensive DNA methylation like *Drosophila, C. elegans, Schizosaccharomyces pombe* and *Saccharomyces cerevisiae.*

Up to now, data on the effect of paramutation on chromatin structure have been reported only for two paramutation systems (*b1* and *A1*; Chandler *et al.*, 2000; Stam *et al.*, 2002a; van Blokland *et al.*, 1997). In these systems, the inactive, paramutagenic states were clearly less accessible to nucleases than their active, paramutable counterparts. These differences in nuclease accessibility were mostly confined to the regions also displaying differences in DNA methylation (Meyer & Heidmann, 1994; Stam *et al.*, 2002a).

5.4.4   Secondary paramutation

Secondary paramutation refers to the ability of paramutable alleles to become paramutagenic once paramutated. Particular paramutable alleles do not display secondary paramutation because they lack the necessary features to become paramutagenic. Most paramutable alleles however show efficient secondary paramutation (see Bateson & Pellew, 1915; Hagemann & Berg, 1978; Hollick *et al.*, 1995; Meyer *et al.*, 1993; Patterson *et al.*, 1993; Rassoulzadegan *et al.*, 2002). These alleles generally have the exact same DNA sequence and sequence organization as their paramutagenic counterpart; their epigenetic state determines if they are paramutable or paramutagenic. Paramutable alleles that are homologous, but have a different sequence organization than the corresponding paramutagenic alleles, can lack the features required to become paramutagenic. In all paramutation systems not displaying secondary paramutation (Hatada *et al.*, 1997; Luff *et al.*, 1999; Park *et al.*, 1996), the paramutable alleles have a different sequence organization than their paramutagenic partner.

5.4.5   Stability of the epigenetic state

The epigenetic states of paramutation alleles can vary between very unstable and extremely stable, and this stability can be influenced, amongst others, by zygosity and environmental factors. The stability of the different epigenetic states has to be taken into account when thinking about models explaining the various phenomena.

The epigenetic state of some paramutable alleles is very stable (*P-rr, sulf, PAI2* and *PAI3*; Das & Messing, 1994; Hagemann, 1993; Melquist *et al.*, 1999), while that of various others is unstable ('ear rogue', *b1, A1, pl1,* and *SPT::Ac*; Bateson & Pellew, 1915; Coe, 1959; English & Jones, 1998; Hollick *et al.*, 1995; Meyer *et al.*, 1993). In the latter cases, the paramutable states spontaneously change into the paramutagenic state. This occurs with a specific frequency, which depends on the allele and other conditions. For example, the frequency with which derivatives of a particular *SPT::Ac* locus showed spontaneous *incis* silencing varied between 0% and 60% (English & Jones, 1998). This depended on the structure of the allele (see also Section 5.4.1 and Figure 5.4).

The stability of the paramutagenic state can also vary. The paramutagenic *A1, b1*, HPT, *pl* and *rl* states are very stable (Brink & Weyers, 1957; Coe, 1959; Mittelsten Scheid *et al.*, 2003; Sidorenko & Peterson, 2001), whereas the paramutagenic *Pl'* state is unstable; it reverts back to higher-expression states (Hollick *et al.*, 1995).

A number of paramutation alleles show a range of epigenetic states instead of one paramutable and one paramutagenic state (*sulf, A1, pl1* and *pl*; Hagemann, 1993; Hagemann & Berg, 1978; Hollick *et al.*, 1995; Meyer *et al.*, 1993; Sidorenko & Peterson, 2001).

The features that determine the stability of a paramutable or paramutagenic state are, amongst others, repetitiveness (see Section 5.4.1), specific sequence elements (see Section 5.4.2) and chromosomal location (Hagemann & Berg, 1978). For example, some alleles contain repeated sequences while others do not, and when present, repeated fragments vary in sequence, size and number. Given that repeats trigger the formation of silenced chromatin, it is to be expected that the more number of repeats are present, the more stable the paramutagenic state and the less stable the paramutable state.

Epigenetic stability can also be influenced by other circumstances, for example environmental and endogenous factors, the zygosity and ploidy. Environmental effects as well as the age of the plant have been shown to influence the frequency of spontaneous paramutation of the petunia *A1* transgene (discussed in Section 5.2.1.4; Meyer *et al.*, 1992). This might be explained by the observation that environmental factors like temperature influence DNA methylation levels and chromatin structure (Finnegan *et al.*, 2004). The stability of the epigenetic state can also be affected by the allele on the homologous chromosome (zygosity). The paramutable *b1, pl1, rl* and *SPT::Ac* alleles are less stable (more spontaneous paramutation) when they are in a homozygous state, than when they are heterozygous with a neutral allele or in a hemizygous situation (the allele on the homologous chromosome is deleted (Coe, 1966; English & Jones, 1998; Hollick *et al.*, 1995; Styles & Brink, 1966)). In contrast, the paramutagenic *Pl'* allele is stable when homozygous, but unstable when heterozygous with a neutral allele or when hemizyous (Hollick & Chandler, 1998). The effects of zygosity on the stability of the epigenetic state are more than twofold and therefore not due to a simple dosage effect of the affected alleles. Remarkably, the reversion of Pl' to a higher-expression state is only heritable when heterozygous with a neutral allele, not when hemizygous, suggesting allelic pairing might be involved in fixing the higher-expression state.

Paramutation can be influenced by and even be dependent on the ploidy level. For example, the paramutagenicity of the tomato *sulf* locus is reduced in tetraploid versus diploid plants (Hagemann & Berg, 1978), and *HPT* paramutation occurs in tetraploid, but not in diploid *Arabidopsis* plants (Mittelsten Scheid *et al.*, 2003). The *hygromycin phosphotransferase* (*HPT*) transgene locus confers a uniform hygromycin resistance in diploid *Arabidopsis* plants. Upon autotetraploidization, a number of hygromycin-sensitive plants were isolated in which the *HPT* transgene was transcriptionally silenced ('genotype' SSSS). In other siblings, the *HPT* genes were still active ('genotype' RRRR). When the S and R alleles were combined in a tetraploid background, the (paramutagenic) S alleles heritably *trans*-inactivated the (paramutable) R alleles. This paramutation-like phenomenon depends on the tetraploid state. After reduction in ploidy, the S alleles were still stably silenced, but no longer paramutagenic.

Upon polyploidization, a new balance has to be created between the different chromosomes, which can affect the epigenetic state of certain alleles. In autotetraploids for example, although most genes are expressed at a level proportional to the genome copy number, a number of genes show a higher or lower expression level per genome than observed in diploids (Guo *et al.*, 1996; Lee & Chen, 2001). In addition, limited chromosome rearrangements can be seen upon autotetraploidization, such as a rearrangement of the 45S rDNA locus (Weiss & Maluszynska, 2000).

5.4.6   Timing of paramutation

When studying the molecular mechanisms underlying paramutation, it is important to know when paramutation takes place. The change from a paramutable into a paramutagenic allele takes place after combining both alleles in one zygote, but does not necessarily occur immediately. It involves multiple events such as the change in epigenetic state of the paramutable allele, and the imposition of the heritable imprint onto the paramutable allele (epigenetic mark rendering the new epigenetic state heritable). These are not necessarily one and the same event. The change in epigenetic state could be followed by a series of events required to heritably secure the epigenetic state, but both processes might also go hand in hand.

The change in epigenetic state can occur during development of the organism. For example, when wild-type pea plants are crossed with plants carrying the paramutagenic 'ear rogue' allele, the lower nodes of the progeny plants look more or less wild-type, while the highest nodes look entirely rogue-like (narrower plant organs than the wild-type plant; Bateson and Pellew, 1915). Similarly, when the paramutable and paramutagenic tomato *sulf* alleles are combined, the cotyledons of the resulting progeny plants are green. Subsequent foliage leaves can be yellow speckled and later during development entirely yellow leaves can be formed (Hagemann and Berg, 1978, Hagemann, 1993). In a number of paramutation phenomena the genes affected are not expressed until relatively late during development (for example *b1*, *pll*, *pl*; Grotewold *et al.*, 1991, 1994; Hollick *et al.*, 1995; Patterson *et al.*, 1993) and, once expressed, the epigenetic change has already occured. In those cases it is not clear when during development the epigenetic change takes place.

The imposition of a heritable imprint can go hand in hand with or immediately follow the change in epigenetic state. For example, with the progeny plants of a cross between wild-type and rogue-like pea plants, the nodes that look more or less wild-type produce a relatively low percentage of rogue offspring, while the nodes that look entirely rogue-like produce exclusively rogues (Bateson and Pellew, 1915, 1920). With *b1* paramutation it is known that paramutable *B-I* and paramutagenic *B'* alleles can be present together in one nucleus up to the fourth or tenth leaf stage without *B-I* being heritably changed to *B'* (Coe, 1966).

Some paramutation examples do not show evidence of paramutation until the F2 generation (*r1*, *bal/cpr1-1*, *HPT*; Brink, 1973; Brink *et al.*, 1968; Mittelsten Scheid *et al.*, 2003; Stokes & Richards, 2002), suggesting paramutation occurs slowly, late during development, or needs meiosis to take place. The most extensively studied paramutable *r1* alleles are only expressed in seeds (Brink *et al.*, 1968; Brink, 1973). When combined in one zygote with a paramutagenic *r1* allèle, the pigment level of the resulting seed is not affected. The seed pigmentation level is not reduced until the next generation. This might suggest meiosis is needed for *r1* paramutation to occur. Alternatively, the seed tissue of the F1 zygote, which is made very early during development, is formed before *r1* paramutation takes place. In support of the latter hypothesis, paramutation of the *r1* allele called *R-d:Catspaw* allele, which can be scored in maize cotyledon and roots, is visible in the F1 (J. Kermicle, personal communication; Brink *et al.*, 1970). The *bal* and *cpr1-1* epigenetic variants of the *R*-like gene cluster *trans*-interact in the *bal/cpr1-1* F1 hybrid without visibly affecting the phenotype (Stokes & Richards, 2002). The effect of this interaction in the F1, the destabilization of the *cpr1-1* and/or *bal* epigenetic state, is only visible in the progeny of the *bal/cpr1-1* hybrids. *HPT* paramutation (see Section 5.4.5) also only becomes phenotypically apparent in the F2 generation (Mittelsten Scheid *et al.*, 2003). Genetic analyses however showed that the active *HPT* allele already becomes affected in the F1 hybrid. In case meiosis is essential for *HPT* paramutation, this requirement could be related to the fact that *HPT* paramutation is limited to tetraploid *Arabidopsis* plants. The arrangement and pairing of multiple homologous chromosomes during meiosis is more challenging in tetraploid versus diploids plants, and could somehow lead to *trans*-allelic interactions not occurring in diploid plants (Santos *et al.*, 2003; Weiss & Maluszynska, 2000).

5.5 *Trans*-acting mutations affecting paramutation

Paramutation is a complex epigenetic phenomenon with many features involved, and the underlying mechanisms are only starting to be unraveled. Isolation, cloning and characterization of mutations affecting various aspects of paramutation, establishment, maintenance or both, is crucial in order to uncover the mechanisms involved. Plants enable forward screens as well as reverse approaches to identify previously unknown and known factors playing a role in various aspects of paramutation, respectively. Below, mutations affecting paramutation and paramutation-like phenomena in maize and *Arabidopsis* are discussed. A strategy to isolate and characterize mutations affecting paramutation is illustrated using ethylmethanesulfonate (EMS) as the mutagen and the *mop1-1* mutation as an example.

5.5.1 *Maize mutations affecting paramutation*

Forward genetic screens using the maize *b1* and *pl1* paramutation systems led to the isolation of the recessive *mediator of paramutation* (*mop1-1*; Dorweiler *et al.*, 2000), *required to maintain repression* (*rmr1*, *rmr2*; Hollick & Chandler, 2001) and additional mutations (referred to in Hollick & Chandler, 2001; Lisch *et al.*, 2002). The *mop1-1* mutation affects various aspects of *b1*, *pl1* and *r1* paramutation. The *rmr* mutations affect *pl1* paramutation; the effects on the other systems have not yet been reported.

When isolating and characterizing mutations affecting paramutation, it is important to realize that they can affect various aspects of paramutation:

(1) the maintenance of the suppressed expression state (phenotype; Figure 5.6A);
(2) the maintenance of the paramutagenic state (the capability to cause paramutation; Figure 5.6B);
(3) the maintenance of the heritable imprint (epigenetic mark rendering the paramutagenic state heritable; Figure 5.6C).

To isolate mutations affecting the suppressed expression state, plants carrying a paramutagenic allele (*P'*) are fertilized by EMS-treated pollen carrying the same paramutagenic allele (*P'*; Figure 5.6A). The vast majority of the resulting

F1 generation consist of low-expressing P' plants. The presence of an exceptional, high-expressing F1 plant suggests the presence of a (semi)-dominant mutation (M) affecting the maintenance of the low-expression state. Subsequently, all F2 families, resulting from self-fertilized low-expressing F1 plants, have to be screened for families containing 25% of high-expressing

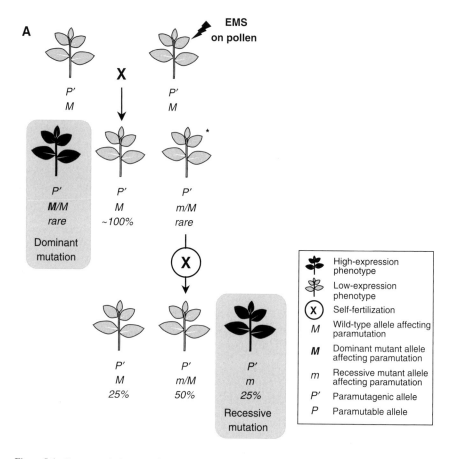

Figure 5.6 Strategy to isolate mutations affecting the maintenance of expression, paramutagenic state and/or heritable imprint. (A) Mutations affecting the maintenance of the low-expression state are isolated in the following type of screen: plants carrying a paramutagenic allele (P') and wild-type alleles of genes affecting paramutation (M) are fertilized by ethylmethanesulfonate (EMS)-treated pollen containing the same alleles. The resulting F1 generation mainly consists of wild-type, low-expressing paramutagenic plants ($P'M$). An exceptional, high-expressing F1 plant suggests the presence of a dominant mutation (M) affecting the maintenance of the low-expression state. To identify recessive mutations affecting paramutation (m), all low-expressing F1 plants are self-fertilized. All progeny of plants not carrying a recessive mutation are light-colored. The progeny of self-fertilized F1 plants carrying a recessive mutation (plant indicated with an asterisk) consist of 25% high-expressing (homozygous mutant, m) and 75% low-expressing plants (M and m/M).

(Continued)

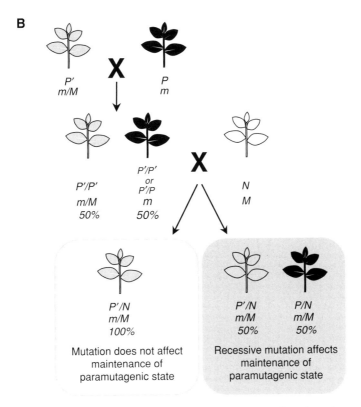

Figure 5.6 (*cont'd*) (B) In order to test if an identified recessive mutation affects the maintenance of the paramutagenic state, it has to be determined if a paramutagenic allele (*P'*) can paramutate its paramutable counterpart (*P*) in a homozygous mutant background (*m*). Therefore, a paramutagenic plant heterozygous for the mutation (*P'm/M*) is crossed with a paramutable plant homozygous for the mutation (*Pm*). Fifty percent of the progeny will be light-colored (*P'/P' m/M*) and 50% dark-colored. The latter are homozygous for the recessive mutation and either homozygous for *P'* (the mutation releases the maintenance of the low-expressed state, but not the maintenance of the paramutagenic state: *P'* is highly expressed, but can still paramutate *P*) or heterozygous for *P'* and *P* (the mutation releases the maintenance of both the low-expressed and the paramutagenic state: *P'* is highly expressed and cannot paramutate *P* in the mutant background). In order to distinguish the two possibilities, the dark, homozygous mutant plants are crossed with a plant carrying a neutral allele. If the mutation does not release the maintenance of the paramutagenic state, all progeny will be light-colored. If the mutation does release the maintenance of the paramutagenic state, 50% of the progeny will be dark and 50% light-colored. (*Continued*)

plants. Such a family could indicate a recessive mutation (*m*) affecting the maintenance of the suppressed-expression state.

The *mop1-1* allele was isolated as a recessive mutation, increasing the transcription rate of the paramutagenic *B'* allele (Dorweiler *et al.*, 2000). Heterozygous *B' Mop1/B' mop1-1* plants have a light pigmentation phenotype, while homozygous *B' mop1-1* mutant plants display a very dark pigmentation

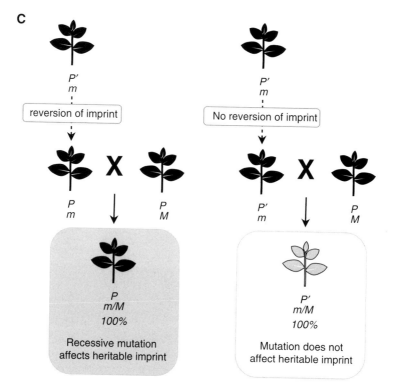

Figure 5.6 (*cont'd*) (C) A mutation affecting the maintenance of the low-expression and paramutagenic state and the heritable imprint will heritably revert the paramutagenic state (*P′*) to a paramutable state (*P*) during the lifespan of the plant. If that is the case (depicted on the left), crossing the homozygous mutant plant (*m*) with a homozygous wild-type plant carrying the paramutable allele (*P*) will result in 100% dark-colored heterozygous mutant progeny plants. If the heritable imprint is not affected (depicted on the right), all progeny will be light-colored. Key: *P′* = paramutagenic allele; *P* = paramutable allele; *N* = neutral allele; *M* = wild-type allele of gene affecting paramutation; *M* = dominant mutant allele affecting paramutation; *m* = recessive mutant allele affecting paramutation.

phenotype resembling *B-I* plants. These results indicate that the MOP1 protein plays a role in the maintenance of the repressed *B′* expression state. MOP1 is furthermore required for maintaining the low-expression state of *Pl′* but not that of a paramutated *r1* allele (J. Kermicle, personal communication). Like MOP1, the RMR1 and RMR2 proteins are also required to maintain the low *Pl′* expression level (Hollick & Chandler, 2001). The expression of neutral alleles is not affected by the *mop1* and *rmr* mutations (Dorweiler *et al.*, 2000; Hollick & Chandler, 2001).

To investigate whether a mutation influences the maintenance of the paramutagenic state, one has to test if a paramutagenic allele (*P′*) can paramutate its paramutable counterpart (*P*) in the mutant background (see Figure 5.6B).

The *mop1* mutation inhibits *b1*, *pl1* and *r1* paramutation, the *rmr* mutations *pl1* paramutation. In other words, the paramutagenic *b1*, *pl1* and *r1* states are no longer maintained in a *mop1* mutant background, and in addition the paramutagenic *pl1* state can be released in *rmr* mutant backgrounds. A mutation can also affect the maintenance of the heritable imprint, rendering the paramutagenic state heritable. In that case, the paramutagenic state heritably reverts to the paramutable state in the mutant background (Figure 5.6C). The *mop1-1* mutation does not alter the heritable imprint of the *B'* allele; once a wild-type *Mop1* allele is introduced, the light-colored, paramutagenic *B'* state is restored. In the case of *Pl'*, however, the *mop1-1* mutation can heritably revert the low-expressing paramutagenic *Pl'* state to the high-expressing paramutable *Pl-Rh* state. In other words, it can erase the heritable imprint associated with *Pl'*. The same holds for the *rmr* mutations. The reversion from *Pl'* to *Pl-Rh* however occurs at a higher frequency in *rmr1* than in *rmr2* mutant plants. In addition, the *Pl'* alleles that remained *Pl'* were in general less paramutagenic when derived from *rmr1* than when derived from *rmr2* mutants.

Homozygous *mop1-1* plants show serious pleiotropic developmental effects, indicating that the *mop1* mutation affects also loci other than the paramutation loci (Dorweiler *et al.*, 2000). *rmr1* and *rmr2* mutants do not display pleiotropic developmental abnormalities (Hollick and Chandler, 2001). The MOP1 protein is required for the transcriptional silencing of some, but not all silent transgenes tested (V. L. Chandler, K. McGinnis, Y. Lin, C. Springer, L. Sidorenko, C. Carey). MOP1 is also required for maintaining the *Mutator* (*Mu*) DNA methylation pattern that is correlated with transposon inactivity (Lisch *et al.*, 2002). In a *mop1-1* mutant, *Mu* element DNA methylation is decreased. The methylation level of certain other transposable elements, including one just upstream of the *B'* transcription start site, is however not changed. The *rmr* mutations caused a similar *Mu* element hypomethylation as the *mop1-1* mutation. Despite the immediate *Mu* hypomethylation, *Mu* elements only sporadically transpose again after multiple generations of continuous exposure to the *mop1-1* mutation. The silenced *Mu* state is apparently maintained independent of the examined methylation pattern. The effect of the *mop1-1* mutation is specific: *mop1-1* does not affect the level of DNA methylation of ribosomal and centromeric repeats, nor does it influence actin or ubiquitin RNA levels (Dorweiler *et al.*, 2000).

A functional MOP1 protein is required not only for three different paramutation systems, *b1*, *pl* and *r1*, but also for transgene silencing and *Mu* methylation, suggesting these phenomena share mechanistic features. At the same time, the unique effects on each of the systems suggest that the mechanisms involved are somewhat diverged. For example, the *mop1* mutation can affect the maintenance of the heritable imprint at the *pl1*, but not at the *b1* locus.

5.5.2 Arabidopsis mutations affecting trans-inactivation

To unravel the mechanisms underlying the paramutation-like *PAI* and *HPT* phenomena in *Arabidopsis* (discussed in Sections 5.2.1.5 and 5.4.5), both forward and reverse genetic approaches have been used. In combination with the *PAI trans*-inactivation system, the effect of the *ddm1*, *met1*, *cmt3* and *kyp/ suvh4* mutations has been studied. All four corresponding proteins are required for maintenance DNA methylation and *trans*-inactivation of the single-copy *PAI2* locus.

DDM1 is a SWI2/SNF2-like chromatin remodeling protein affecting CpG and non-CpG maintenance cytosine methylation, possibly via facilitating access of the methylation machinery to silenced chromatin (Jeddeloh *et al.*, 1999). MET1 is a CpG maintenance cytosine methyltransferase (Finnegan *et al.*, 1996). The *ddm1* and *met1* mutation cause demethylation of ribosomal and centromeric repeats, but only *ddm1* results in transposon mobilization (Hirochika *et al.*, 2000; Miura *et al.*, 2001; Singer *et al.*, 2001). CMT3 is a chromomethylase specialized in non-CpG maintenance methylation at specific genomic regions (Bartee *et al.*, 2001; Lindroth *et al.*, 2001). The fourth protein, KYP/SUVH4, is a SET-domain protein with histone H3Lys9 methyltransferase activity that indirectly affects maintenance DNA methylation, mainly in a non-CpG context (Jackson *et al.*, 2002; Johnson *et al.*, 2002; Malagnac *et al.*, 2002). KYP/SUVH4 possibly acts downstream of CpG methylation, reinforcing chromatin silencing (Jasencakova *et al.*, 2003; Soppe *et al.*, 2002; Tariq *et al.*, 2003).

All four proteins are required for maintenance DNA methylation at the single-copy *PAI* loci (Bartee & Bender, 2001; Bartee *et al.*, 2001; Malagnac *et al.*, 2002). MET1 and CMT3 are in addition required for maintenance DNA methylation of the *PAI* IR. The *ddm1* mutation has only a minor effect and *kyp/suvh4* does not affect maintenance DNA methylation at the *PAI* IR. Interestingly, DNA methylation at a nopaline synthase promoter (NOSpro) IR silencing locus is also MET-dependent and DDM1-independent (Aufsatz *et al.*, 2002). Like the *PAI* IR, the NOSpro IR transcriptionally *trans*-inactivates and methylates homologous sequences via the production of a dsRNA. This process is called RNA-dependent DNA methylation (RdDM). Mutations in *hda6* (encoding a putative histone deacetylase; Probst *et al.*, 2004) and *drd1* (encoding a putative SNF2-like chromatin-remodeling protein; Kanno *et al.*, 2004) affect RdDM as well. It would therefore be interesting to also study the effect of *hda6* and *drd1* mutations on the *PAI trans*-inactivation. The establishment of DNA methylation at the *PAI2* locus is independent of KYP/SUVH4. The data suggested that CMT3 is involved in establishing DNA methylation at the *PAI2* locus (Malagnac *et al.*, 2002). This effect could however be indirect, since the decreased methylation at the *PAI* IR locus in a *cmt3* mutant might weaken its *trans*-inactivating capacity. Various mutations in genes involved in the production and amplification of dsRNAs did not affect maintenance DNA methylation of the *PAI* loci (Melquist & Bender, 2003).

The data for the *PAI* system (Bartee & Bender, 2001; Bartee *et al.*, 2001; Malagnac *et al.*, 2002; Melquist & Bender, 2004) indicate that maintenance DNA methylation of the *PAI* IR depends on read-through transcription of the *PAI* IR, and on the MET1 and CMT3 maintenance methylases, and is independent of H3Lys9 histone methylation and the DDM1 protein. This suggests that the *PAI* IR locus has a typical chromatin conformation that can attract the methylation machinery independent of H3Lys9 methylation and DDM1.

The *trans*-inactivation of the *PAI* single-copy loci depends on the *PAI* IR-derived dsRNA, and is at least partially independent of H3Lys9 histone methylation (Malagnac *et al.*, 2002; Melquist & Bender, 2003, 2004). Once *trans*-inactivated, maintenance of the repressed state is again dependent on *PAI* dsRNA, and, in addition, on maintenance DNA methylation, H3Lys9 histone methylation and the DDM1 protein. It is however independent of a detectable level of small *PAI* RNAs and various genes involved in the production and amplification of dsRNAs (Melquist & Bender, 2003). This suggests that the *PAI* dsRNA itself might recruit the methylation machinery, either directly or via the production of undetectable levels of small RNAs.

The effect of *ddm1* and a mutation in the *mom1* gene were tested on *HPT* paramutation in *Arabidopsis*. MOM1 is a nuclear protein with only limited homology to the SWI2/SNF2 family. The *mom1* mutation releases transcriptional gene silencing without affecting DNA methylation (Amedeo *et al.*, 2000). Although *HPT* *trans*-inactivation is only observed in tetraploid plants, because of technical reasons the effect of the mutations on the maintenance of the silent state was tested in diploids (Mittelsten Scheid *et al.*, 2003). Whereas the *mom1* mutation had no effect on the maintenance of repression, the exposure to *ddm1* for multiple generations caused a slow demethylation and activation of the silent *HPT* transgenes.

More mutants are currently being isolated and tested for their effects on paramutation. A crucial step forward will be the cloning of the *trans*-acting maize mutations and characterization of their gene products. The maize *Mop1* gene is not the ortholog of the so far reported mutations affecting gene silencing in *Arabidopsis* (Chandler & Stam, 2004), so isolating mutations in the maize orthologs of these *Arabidopsis* silencing genes and studying their effect on paramutation in maize will remain interesting.

5.6   The possible roles and implications of paramutation

Paramutation could have various roles and evolutionary implications (see also Chandler & Stam, 2004; Chandler *et al.*, 2000). A few obvious roles are protection of the genome against invasive foreign DNA, and the stabilization of the genome upon hybridization and polyploidization. The existence of various epialleles might furthermore provide an organism an enhanced capacity to adapt to environmental circumstances. In order for paramutation to be of evolutionary significance, a reasonable number of alleles of different genes should be affected. Although paramutation is observed in a wide variety of organisms, only a limited number of genes are currently shown to participate. The majority of these examples are discovered because they affect a phenotype that is easily visible. If paramutation is more widespread than thus far suspected, whole transcriptome analyses should reveal numerous new examples.

Paramutation could be part of the cellular genome defense system that inactivates intrusive DNA. One of the main targets of the defense system is repetitive DNA. Recombination between repeated sequences can lead to potentially deleterious chromosomal rearrangements. An inactive epigenetic state is believed to prevent illegitimate recombination. Repeated sequences are involved in several paramutation phenomena (see Section 5.4.1) and paramutation generally leads to epigenetic inactivation, suggesting a link between paramutation and the defense system.

Paramutation may also play a role in the stabilization of the genome upon hybrid formation and polyploidization. Following an event of hybridization or polyploidization, especially allopolyploidization, the expression of many genes needs to find a new balance (Guo *et al.*, 1996; Lee & Chen, 2001; Liu & Wendel, 2003; Wang *et al.*, 2004). The effects of polyploidization on gene expression do not always have the same outcome in genetically identical polyploids. For example, upon polyploidization, the exact same allele can get silenced in one individual, and stay active in the other. This has also been observed for the *HPT* locus in *Arabidopsis* (Mittelsten Scheid *et al.*, 2003). Similar effects are true for hybrid formation (reviewed in Birchler *et al.*, 2003).

The existence of paramutagenic and paramutable alleles allows a relatively easy heritable adaptation to changes in environmental conditions without changes in DNA sequence being required. Consistent with this hypothesis, the epigenetic state of the *r1* and *A1* paramutation alleles is influenced by environmental factors (Meyer *et al.*, 1992; Mikula, 1967, 1995).

5.7 Concluding remarks and future directions

Paramutation has been discovered for various genes in a variety of organisms. All the different paramutation systems have in common that *trans*-interactions between homologous sequences result in heritable changes in epigenetic states. The various systems however also display unique features. Therefore it is likely that multiple mechanisms might be involved in paramutation. We hypothesize two models for paramutation, an RNA- and pairing-based model, which are not mutually exclusive.

To reveal more about the mechanisms underlying paramutation it is crucial to clone and characterize the gene (products) involved. Multiple *trans*-acting mutations affecting paramutation have been isolated and the screens are likely not saturated. None of the mutations isolated in the classical systems have however been assigned to a gene at the moment, but cloning is underway.

Paramutation involves changes in DNA methylation and chromatin structure. The role of chromatin structure is still a black box for most paramutation phenomena, however. Recently developed tools to examine sequence-specific and global chromatin structure changes should make it possible to reveal the role of chromatin structure changes in paramutation more easily.

At the moment, only a limited number of paramutation phenomena have been reported. Most of these affect a visible phenotype, facilitating their discovery. If paramutation is more widespread than thus far anticipated, techniques such as microarray analyses should be able to reveal several new examples. If that is the case, the role and evolutionary consequences of paramutation are much more extensive than currently appreciated.

Acknowledgments

We thank Roel van Driel for useful comments on the manuscript. M. Stam is funded by the Royal Netherlands Academy of Arts and Sciences (KNAW).

References

Abranches, R., Santos, A. P., Wegel, E., Williams, S., Castilho, A., Christou, P., Shaw, P. & Stoger, E. (2000) Widely separated multiple transgene integration sites in wheat chromosomes are brought together at interphase. *Plant Journal*, **24**(6), 713–23.

Amedeo, P., Habu, Y., Afsar, K., Scheid, O. M. & Paszkowski, J. (2000) Disruption of the plant gene *MOM* releases transcriptional silencing of methylated genes. *Nature*, **405**(6783), 203–206.

Aragon-Alcaide, L. & Strunnikov, A. V. (2000) Functional dissection of *in vivo* interchromosome association in *Saccharomyces cerevisiae*. *Nature Cell Biology*, **2**(11), 812–18.

Athma, P., Grotewold, E. & Peterson, T. (1992) Insertional mutagenesis of the maize *P* gene by intragenic transposition of *Ac*. *Genetics*, **131**(1), 199–209.

Aufsatz, W., Mette, M. F., van derWinden, J., Matzke, A. J. M. & Matzke, M. (2002) RNA-directed DNA methylation in *Arabidopsis*. *Proceedings of the National Academy of Sciences of the United States of America*, **99**, 16499–506.

Bantignies, F., Grimaud, C., Lavrov, S., Gabut, M. & Cavalli, G. (2003) Inheritance of Polycomb-dependent chromosomal interactions in *Drosophila*. *Genes & Development*, **17**, 2406–420.

Bartee, L. & Bender, J. (2001) Two *Arabidopsis* methylation-deficiency mutations confer only partial effects on a methylated endogenous gene family. *Nucleic Acids Research*, **29**, 2127–34.

Bartee, L., Malagnac, F. & Bender, J. (2001) *Arabidopsis* cmt3 chromomethylase mutations block non-CG methylation and silencing of an endogenous gene. *Genes & Development*, **15**(14), 1753–8.

Bateson, W. & Pellew, C. (1915) On the genetics of 'rogues' among culinary peas (*Pisum sativum*). *Journal of Genetics*, **5**, 15–36.

Bateson, W. & Pellew, C. (1920) The genetics of 'rogues' among culinary peas (*Pisum sativum*). *Proceedings of the Royal Society of London. Series B, Containing Papers of a Biological Character*, **91**(638), 186–95.

Baulcombe, D. (2004) RNA silencing in plants. *Nature*, **431**(7006), 356–63.

Bean, C. J., Schaner, C. E. & Kelly, W. G. (2004) Meiotic pairing and imprinted X chromatin assembly in *Caenorhabditis elegans. Nature Genetics*, **36**(1), 100–105.

Bell, G. I., Horita, S. & Karam, J. H. (1984) A polymorphic locus near the human insulin gene is associated with insulin-dependent diabetes mellitus. *Diabetes*, **33**(2), 176–83.

Bender, J. (2004) DNA methylation and epigenetics. *Annual Review of Plant Biology*, **55**, 41–68.

Bender, J. & Fink, G. R. (1995) Epigenetic control of an endogenous gene family is revealed by a novel blue fluorescent mutant of *Arabidopsis. Cell*, **83**(5), 725–34.

Bennett, S. T., Lucassen, A. M., Gough, S. C., Powell, E. E., Undlien, D. E., Pritchard, L. E., Merriman, M. E., Kawaguchi, Y., Dronsfield, M. J. & Pociot, F. *et al.* (1995) Susceptibility to human type 1 diabetes at IDDM2 is determined by tandem repeat variation at the insulin gene minisatellite locus. *Nature Genetics*, **9**(3), 284–92.

Bennett, S. T., Wilson, A. J., Cucca, F., Nerup, J., Pociot, F., McKinney, P. A., Barnett, A. H., Bain, S. C. & Todd, J. A. (1996) IDDM2-VNTR-encoded susceptibility to type 1 diabetes: dominant protection and parental transmission of alleles of the insulin gene-linked minisatellite locus. *Journal of Autoimmunity*, **9**(3), 415–21.

Bennett, S. T., Wilson, A. J., Esposito, L., Bouzekri, N., Undlien, D. E., Cucca, F., Nistico, L., Buzzetti, R., Bosi, E., Pociot, F., Nerup, J., Cambon-Thomsen, A., Pugliese, A., Shield, J. P., McKinney, P. A., Bain, S. C., Polychronakos, C. & Todd, J. A. (1997) Insulin VNTR allele-specific effect in type 1 diabetes depends on identity of untransmitted paternal allele. The IMDIAB Group. *Nature Genetics*, **17**(3), 350–52.

Birchler, J. A., Bhadra, M. P. & Bhadra, U. (2000) Making noise about silence: repression of repeated genes in animals. *Current Opinion in Genetics and Development*, **10**, 211–16.

Birchler, J. A., Auger, D. L. & Riddle, N. C. (2003) In search of the molecular basis of heterosis. *Plant Cell*, **15**(10), 2236–9.

Brink, R. A. (1956) A genetic change associated with the *R* locus in maize which is directed and potentially reversible. *Genetics*, **41**, 872–90.

Brink, R. A. (1958) Paramutation at the *R* locus. *Cold Spring Harbor Symposium on Quantitative Biology*, **23**, 379–91.

Brink, R. A. (1973) Paramutation. *Annual Review of Genetics*, **7**, 129–52.

Brink, R. A. & Weyers, W. H. (1957) Invariable genetic change in maize plants heterozygous for marbled aleurone. *Proceedings of the National Academy of Sciences of the United States of America*, **43**, 1053–1060.

Brink, R. A., Styles, E. D. & Axtell, J. D. (1968) Paramutation: directed genetic change. Paramutation occurs in somatic cells and heritably alters the functional state of a locus. *Science*, **159**(811), 161–70.

Brink, R. A., Kermicle, J. L. & Ziebur, N. K. (1970) Derepression in the female gametophyte in relation to paramutant *R* expression in maize endosperms, embryos, and seedlings. *Genetics*, **66**, 87–96.

Cerutti, H. (2003) RNA interference: traveling in the cell and gaining functions? *Trends in Genetics*, **19**, 39–46.

Chan, S. R. W. L. & Blackburn, E. H. (2004) Telomeres and telomerase. *Philosophical Transactions of the Royal Society of London. Series B, Biological Sciences*, **359**, 109–21.

Chandler, V. L. & Stam, M. (2004) Chromatin conversations: mechanisms and implications of paramutation. *Nature Reviews Genetics*, **5**(7), 532–44.

Chandler, V. L., Eggleston, W. B. & Dorweiler, J. E. (2000) Paramutation in maize. *Plant Molecular Biology*, **43**, 121–45.

Chopra, S., Athma, P. & Peterson, T. (1996) Alleles of the maize *P* gene with distinct tissue specificities encode Myb-homologous proteins with C-terminal replacements. *Plant Cell*, **8**(7), 1149–58.

Chopra, S., Athma, P., Li, X. G. & Peterson, T. (1998) A maize Myb homolog is encoded by a multicopy gene complex. *Molecular Genetics and Genomics*, **260**(4), 372–80.

Chow, J. C. & Brown, C. J. (2003) Forming facultative heterochromatin: silencing of an X chromosome in mammalian females. *Cellular and Molecular Life Sciences*, **60**, 2586–603.

Coady, M. A., Mandapati, D., Arunachalam, B., Jensen, K., Maher, S. E., Bothwell, A. L. & Hammond, G. L. (1999) Dominant negative suppression of major histocompatibility complex genes occurs in trophoblasts. *Transplantation*, **67**(11), 1461–7.

Coe, E. H. J. (1959) A regular and continuing conversion-type phenomenon at *b* locus in maize. *Maydica*, **24**, 49–58.

Coe, E. H. J. (1966) The properties, origin and mechanism of conversion-type inheritance at the *b* locus in maize. *Genetics*, **53**, 1035–63.

Colot, V. & Rossignol, J. -L. (1995) Isolation of the *Ascobolus immersus* spore color gene *b2* and study in single cells of gene silencing by methylation induced premeiotically. *Genetics*, **141**, 1299–314.

Colot, V., Maloisel, L. & Rossignol, J. -L. (1996) Interchromosomal transfer of epigenetic states in *Ascobolus*: transfer of DNA methylation is mechanistically related to homologous recombination. *Cell*, **86** (6), 855–64.

Cone, K. C., Cocciolone, S. M., Burr, F. A. & Burr, B. (1993) Maize anthocyanin regulatory gene *pl* is a duplicate of c1 that functions in the plant. *Plant Cell*, **5**(12), 1795–805.

Cook, P. R. (1997) The transcriptional basis of chromosome pairing. *Journal of Cell Science*, **110**, 1033–1040.

Csink, A. K. & Henikoff, S. (1996) Genetic modification of heterochromatic association and nuclear organization in *Drosophila*. *Nature*, **381** 529–31.

Csink, A. K., Bounoutas, A., Griffith, M. L., Sabl, J. F. & Sage, B. T. (2002) Differential gene silencing by *trans*-heterochromatin in *Drosophila melanogaster*. *Genetics*, **160**, 257–69.

Das, P. & Messing, J. (1994) Variegated phenotype and developmental methylation changes of a maize allele originating from epimutation. *Genetics*, **136**, 1121–41.

Dawe, R. K. (2003) RNA interference, transposons, and the centromere. *Plant Cell*, **15**, 297–301.

Dekker, J., Rippe, K., Dekker, M. & Kleckner, N. (2002) Capturing chromosome conformation. *Science*, **295**(5558), 1306–11.

Dernburg, A. F., Broman, K. W., Fung, J. C., Marshall, W. F., Phillips, J., Agard, D. A. & Sedat, J. W. (1996) Perturbation of nuclear architecture by long-distance chromosome interactions. *Cell*, **85**, 745–59.

Dieguez, M. J., Vaucheret, H., Paszkowski, J. & Mittelsten Scheid, O. (1998) Cytosine methylation at CG and CNG sites is not a prerequisite for the initiation of transcriptional gene silencing in plants, but it is required for its maintenance. *Molecular Genetics and Genomics*, **259**(2), 207–15.

Dorer, D. R. & Henikoff, S. (1997) Transgene repeat arrays interact with distant heterochromatin and cause silencing *in cis* and *trans*. *Genetics*, **147**(3), 1181–90.

Dorweiler, J. E., Carey, C. C., Kubo, K. M., Hollick, J. B., Kermicle, J. L. & Chandler, V. L. (2000) Mediator of paramutation1 is required for establishment and maintenance of paramutation at multiple maize loci. *Plant Cell*, **12**(11), 2101–18.

Duncan, I. W. (2002) Transvection effects in *Drosophila*. *Annual Review of Genetics*, **36**, 521–56.

Duvillie, B., Bucchini, D., Tang, T., Jami, J. & Paldi, A. (1998) Imprinting at the mouse *Ins2* locus: evidence for *cis*- and *trans*-allelic interactions. *Genomics*, **47**(1), 52–7.

Eggleston, W. B., Alleman, M. & Kermicle, J. L. (1995) Molecular organization and germinal instability of *R-stippled* maize. *Genetics*, **141**(1), 347–60.

English, J. J. & Jones, J. D. G. (1998) Epigenetic instability and *trans*-silencing interactions associated with an *SPT::Ac* T-DNA locus in tobacco. *Genetics*, **148**(1), 457–469.

Finnegan, E. J., Peacock, W. J. & Dennis, E. S. (1996) Reduced DNA methylation in *Arabidopsis thaliana* results in abnormal plant development. *Proceedings of the National Academy of Sciences of the United States of America*, **93**(16), 8449–54.

Finnegan, E. J., Sheldon, C. C., Jardinaud, F., Peacock, W. J. & Dennis, W. S. (2004) A cluster of *Arabidopsis* genes with a coordinate response to an environmental stimulus. *Current Biology*, **14**, 911–6.

Forne, T., Oswald, J., Dean, W., Saam, J. R., Bailleul, B., Dandolo, L., Tilghman, S. M., Walter, J. & Reik, W. (1997) Loss of the maternal *H19* gene induces changes in *Igf2* methylation in both *cis* and *trans. Proceedings of the National Academy of Sciences of the United States of America*, **94**(19), 10243–8.

Francastel, C., Walters, M. C., Groudine, M. & Martin, D. I. (1999) A functional enhancer suppresses silencing of a transgene and prevents its localization close to centometric heterochromatin. *Cell*, **99**, 259–69.

Fransz, P., De Jong, J. H., Lysak, M., Castiglione, M. R. & Schubert, I. (2002) Interphase chromosomes in *Arabidopsis* are organized as well defined chromocenters from which euchromatin loops emanate. *Proceedings of the National Academy of Sciences of the United States of America*, **99**(22), 14584–9.

Fransz, P., Soppe, W. & Schubert, I. (2003) Heterochromatin in interphase nuclei of *Arabidopsis thaliana. Chromosome Research*, **11**, 227–40.

Fuchs, J., Lorenz, A. & Loidl, J. (2002) Chromosome associations in budding yeast caused by integrated tandemly repeated transgenes. *Journal of Cell Science*, **115**, 1213–20.

Geirsson, A., Lynch, R. J., Paliwal, I., Bothwell, A. L. & Hammond, G. L. (2003) Human trophoblast noncoding RNA suppresses CIITA promoter III activity in murine B-lymphocytes. *Biochemical and Biophysical Research Communications*, **301**(3), 718–24.

Grewal, S. I. S. & Rice, J. C. (2004) Regulation of heterochromatin by histone methylation and small RNAs. *Current Opinion in Cell Biology*, **16**, 230–38.

Grotewold, E., Athma, P. & Peterson, T. (1991) Alternatively spliced products of the maize *P* gene encode proteins with homology to the DNA-binding domain of *myb*-like transcription factors. *Proceedings of the National Academy of Sciences of the United States of America*, **88**(11), 4587–91.

Grotewold, E., Drummond, B. J., Bowen, B. & Peterson, T. (1994) The myb-homologous *P* gene controls phlobaphene pigmentation in maize floral organs by directly activating a flavonoid biosynthetic gene subset. *Cell*, **76**(3), 543–53.

Guo, M., Davis, D. & Birchler, J. A. (1996) Dosage effects on gene expression in a maize ploidy series. *Genetics*, **142**(4), 1349–55.

Hagemann, R. (1958) Somatic conversion in *Lycopersicon esculentum* mill. *Zeitschrift für Vererbungslehre*, **89**(4), 587–613.

Hagemann, R. (1969) Somatic conversion (paramutation) at the *sulfurea* locus of *Lycopersicon esculentum* mill. III Studies with trisomics. *Canadian Journal of Genetics and Cytology*, **11**, 346–58.

Hagemann, R. (1993) Studies towards a genetic and molecular analysis of paramutation at the *sulfurea* locus of *Lycopersicon esculentum* mill. Technomic publishing company, Lancaster-Basel.

Hagemann, R. & Berg, W. (1978) Paramutation at the *sulfurea* locus of *Lycopersicon esculentum* mill. VII. Determination of the time of occurrence of paramutation by the quantitative evaluation of the variegation. *Theoretical and Applied Genetics*, **53**, 113–23.

Harrison, B. J. & Carpenter, R. (1973) A comparison of the instabilities at the *Nivea* and *Pallida* loci in *Antirrhinum majus. Heredity*, **31**, 309–23.

Hatada, I., Nabetani, A., Arai, Y., Ohishi, S., Suzuki, M., Miyabara, S., Nishimune, Y. & Mukai, T. (1997) Aberrant methylation of an imprinted gene *U2af1-rs1(SP2)* caused by its own transgene. *Journal of Biological Chemistry*, **272**(14), 9120–22.

Henikoff, S. (1998) Conspiracy of silence among repeated transgenes. *BioEssays*, **20**(7), 532–5.

Hirochika, H., Okamoto, H. & Kakutani, T. (2000) Silencing of retrotransposons in *Arabidopsis* and reactivation by the *ddm1* mutation. *Plant Cell*, **12**(3), 357–69.

Hollick, J. B. & Chandler, V. L. (1998) Epigenetic allelic states of a maize transcriptional regulatory locus exhibit overdominant gene action. *Genetics*, **150**(2), 891–7.

Hollick, J. B. & Chandler, V. L. (2001) Genetic factors required to maintain repression of a paramutagenic maize *pl1* allele. *Genetics*, **157**, 369–78.

Hollick, J. B., Patterson, G. I., Coe, E. H. Jr, Cone, K. C. & Chandler, V. L. (1995) Allelic interactions heritably alter the activity of a metastable maize *pl* allele. *Genetics*, **141**(2), 709–19.

Hollick, J. B., Patterson, G. I., Asmundsson, I. M. & Chandler, V. L. (2000) Paramutation alters regulatory control of the maize *pl* locus. *Genetics*, **154**(4), 1827–38.

Jackson, J. P., Lindroth, A. M., Cao, X. F. & Jacobsen, S. E. (2002) Control of CpNpG DNA methylation by the *KRYPTONITE* histone H3 methyltransferase. *Nature*, **416**, 556–60.

Jacobsen, S. E. & Meyerowitz, E. M. (1997) Hypermethylated *SUPERMAN* epigenetic alleles in *Arabidopsis*. *Science*, **277**(5329), 1100–103.

Jasencakova, Z., Soppe, W. J. J., Meister, A., Gernand, D., Turner, B. M. & Schubert, I. (2003) Histone modifications in *Arabidopsis* – high methylation of H3 lysine 9 is dispensable for constitutive heterochromatin. *Plant Journal*, **33**, 471–80.

Jeddeloh, J. A., Stokes, T. L. & Richards, E. J. (1999) Maintenance of genomic methylation requires a SWI2/SNF2-like protein. *Nature Genetics*, **22**(1), 94–7.

Johnson, L. M., Cao, X. F. & Jacobsen, S. E. (2002) Interplay between two epigenetic marks: DNA methylation and histone H3 lysine 9 methylation. *Current Biology*, **12**, 1360–67.

Jones, L., Ratcliff, F. & Baulcombe, D. F. (2001) RNA-directed transcriptional gene silencing in plants can be inherited independently of the RNA trigger and requires MET1 for maintenance. *Current Biology*, **11**, 747–57.

Kanno, T., Mette, M. F., Kreil, D. P., Aufsatz, W., Matzke, M. & Matzke, A. J. M. (2004) Involvement of putative SNF2 chromatin remodeling protein DRD1 in RNA-directed DNA methylation. *Current Biology*, **14**, 801–805.

Kassis, J. A. (2002) Pairing-sensitive silencing, Polycomb group response elements, and transposon homing in *Drosophila*. *Advances in Genetics*, **46**, 421–38.

Kato, M., Miura, A., Bender, J., Jacobsen, S. E. & Kakutani, T. (2003) Role of CG and non-CG methylation in immobilization of transposons in *Arabidopsis*. *Current Biology*, **13**, 421–6.

Kelly, M. A., Rayner, M. L., Mijovic, C. H. & Barnett, A. H. (2003) Molecular aspects of type 1 diabetes. *Molecular Pathology*, **56**(1), 1–10.

Kermicle, J. L., Eggleston, W. B. & Alleman, M. (1995) Organization of paramutagenicity in *R-stippled* maize. *Genetics*, **141**(1), 361–72.

Kleckner, N. (1996) Meiosis: how could it work? *Proceedings of the National Academy of Sciences of the United States of America*, **93**(16), 8167–74.

LaSalle, J. M. & Lalande, M. (1996) Homologous association of oppositely imprinted chromosomal domains. *Science*, **272**(5262), 725–8.

Lavigne, M., Francis, N. J., King, I. F. G. & Kingston, R. E. (2004) Propagation of silencing: recruitment and repression of naive chromatin – *in trans* by Polycomb repressed chromatin. *Molecular Cell*, **13**, 415–25.

Lecellier, C. H. & Voinnet, O. (2004) RNA silencing: no mercy for viruses? *Immunological Reviews*, **198**, 285–303.

Lee, H. S. & Chen, Z. J. (2001) Protein-coding genes are epigenetically regulated in *Arabidopsis* polyploids. *Proceedings of the National Academy of Sciences of the United States of America*, **98**, 6753–8.

Lee, D. W., Seong, K. Y., Pratt, R. J., Baker, K. & Aramayo, R. (2004) Properties of unpaired DNA required for efficient silencing in *Neurospora crassa*. *Genetics*, **167**(1), 131–50.

Lilienfeld, F. A. (1929) Vererbungsversuche mit schlitzblättrigen Sippen von *Malva parviflora*. I. Die *laciniata* Sippe. *Bibliotheca Genetica*, **13**, 214.

Lindbo, J. A., Silva-Rosales, L., Proebsting, W. M. & Dougherty, W. G. (1993) Induction of a highly specific antiviral state in transgenic plants: implications for regulation of gene expression and virus resistance. *Plant Cell*, **5**, 1749–59.

Lindroth, A. M., Cao, X. F., Jackson, J. P., Zilberman, D., McCallum, C. M., Henikoff, S. & Jacobsen, S. E. (2001) Requirement of CHROMOMETHYLASE3 for maintenance of CpXpG methylation. *Science*, **292**, 2077–80.

Lippman, Z. & Martienssen, R. (2004) The role of RNA interference in heterochromatic silencing. *Nature*, **431**(7006), 364–70.

Lisch, D., Carey, C. C., Dorweiler, J. E. & Chandler, V. L. (2002) A mutation that prevents paramutation in maize also reverses mutator transposon methylation and silencing. *Proceedings of the National Academy of Sciences of the United States of America*, **99**, 6130–35.

Liu, B. & Wendel, J. F. (2003) Epigenetic phenomena and the evolution of plant allopolyploids. *Molecular Phylogenetics and Evolution*, **29**(3), 365–79.

Luff, B., Pawlowski, L. & Bender, J. (1999) An inverted repeat triggers cytosine methylation of identical sequences in *Arabidopsis*. *Molecular Cell*, **3**(4), 505–11.

Lund, A. H. & van Lohuizen, M. (2004) Polycomb complexes and silencing mechanisms. *Current Opinion in Cell Biology*, **16**, 239–46.

Malagnac, F., Bartee, L. & Bender, J. (2002) An *Arabidopsis* SET domain protein required for maintenance but not establishment of DNA methylation. *EMBO Journal*, **21**, 6842–52.

Martienssen, R. A. (2003) Maintenance of heterochromatin by RNA interference of tandem repeats. *Nature Genetics*, **35**, 213–14.

Matzke, M., Aufsatz, W., Kanno, T., Daxinger, L., Papp, I., Mette, A. F. & Matzke, A. J. M. (2004) Genetic analysis of RNA-mediated transcriptional gene silencing. *Biochimica et Biophysica Acta*, **1677**, 129–41.

McKee, B. D. (2004) Homologous pairing and chromosome dynamics in meiosis and mitosis. *Biochimica et Biophysica Acta*, **1677**, 165–80.

Meller, V. H. (2003) Dosage compensation: making 1X equal 2X. *Trends in Cell Biology*, **10**, 54–9.

Melquist, S. & Bender, J. (2003) Transcription from an upstream promoter controls methylation signaling from an inverted repeat of endogenous genes in *Arabidopsis*. *Genes & Development*, **17**(16), 2036–47.

Melquist, S. & Bender, J. (2004) An internal rearrangement in an *Arabidopsis* inverted repeat locus impairs DNA methylation triggered by the locus. *Genetics*, **166**(1), 437–48.

Melquist, S., Luff, B. & Bender, J. (1999) *Arabidopsis PAI* gene arrangements, cytosine methylation and expression. *Genetics*, **153**(1), 401–13.

Meyer, P. & Heidmann, I. (1994) Epigenetic variants of a transgenic petunia line show hypermethylation in transgene DNA: an indication for specific recognition of foreign DNA in transgenic plants. *Molecular and General Genetics*, **243**(4), 390–99.

Meyer, P., Heidmann, I., Forkmann, G. & Saedler, H. (1987) A new petunia flower colour generated by transformation of a mutant with a maize gene. *Nature*, **330**(6149), 677–8.

Meyer, P., Linn, F., Heidmann, I., Meyer, H., Niedenhof, I. & Saedler, H. (1992) Endogenous and environmental factors influence 35S promoter methylation of a maize *A1* gene construct in transgenic petunia and its colour phenotype. *Molecular Genetics and Genomics*, **231**(3), 345–52.

Meyer, P., Heidmann, I. & Niedenhof, I. (1993) Differences in DNA-methylation are associated with a paramutation phenomenon in transgenic petunia. *Plant Journal*, **4**(1), 89–100.

Mikula, B. C. (1967) Heritable changes in *R*-locus expression in maize in response to environment. *Genetics*, **56**, 733–42.

Mikula, B. C. (1995) Environmental programming of heritable epigenetic changes in paramutant *r*-gene expression using temperature and light at a specific stage of early development in maize seedlings. *Genetics*, **140**(4), 1379–87.

Mittelsten Scheid, O., Afsar, K. & Paszkowski, J. (2003) Formation of stable epialleles and their paramutation-like interaction in tetraploid *Arabidopsis thaliana*. *Nature Genetics*, **34**(4), 450–54.

Miura, A., Yonebayashi, S., Watanabe, K., Toyama, T., Shimada, H. & Kakutani, T. (2001) Mobilization of transposons by a mutation abolishing full DNA methylation in *Arabidopsis*. *Nature*, **411**(6834), 212–14.

Morey, C. & Avner, P. (2004) Employment opportunities for non-coding RNAs. *FEBS Letters*, **567**, 27–34.

Murrell, A., Heeson, S. & Reik, W. (2004) Interaction between differentially methylated regions partitions the imprinted genes *Igf2* and *H19* into parent-specific chromatin loops. *Nature Genetics*, **36**(8), 889–93.

Nishant, K. T., Ravishankar, H. & Rao, M. R. (2004) Characterization of a mouse recombination hot spot locus encoding a novel non-protein-coding RNA. *Molecular Cell Biology*, **24**(12), 5620–34.

Pal-Bhadra, M., Bhadra, U. & Birchler, J. A. (1997) Cosuppression in *Drosophila*: gene silencing of alcohol dehydrogenase by white-Adh transgenes is Polycomb dependent. *Cell*, **90**(3), 479–90.

Panavas, T., Weir, J. & Walker, E. L. (1999) The structure and paramutagenicity of the *R-marbled* haplotype of *Zea mays*. *Genetics*, **153**(2), 979–91.

Park, Y. D., Papp, I., Moscone, E. A., Iglesias, V. A., Vaucheret, H., Matzke, A. J. M. & Matzke, M. A. (1996) Gene silencing mediated by promoter homology occurs at the level of transcription and results in meiotically heritable alterations in methylation and gene activity. *Plant Journal*, **9**(2), 183–94.

Patterson, G. I., Kubo, K. M., Shroyer, T. & Chandler, V. L. (1995) Sequences required for paramutation of the maize *b* gene map to a region containing the promoter and upstream sequences. *Genetics*, **140**, 1389–1406.

Patterson, G. I., Thorpe, C. J. & Chandler, V. L. (1993) Paramutation, an allelic interaction, is associated with a stable and heritable reduction of transcription of the maize *b* regulatory gene. *Genetics*, **135**(3), 881–94.

Pecinka, A., Kato, N., Meister, A., Probst, A. V., Schubert, I. & Lam, E. (submitted) Tandem repetitive transgenes and fluorescent chromatin tags alter the local interphase chromosome arrangement in *Arabidopsis thaliana*. *Journal of Cell Science*.

Probst, A. V., Fagard, M., Proux, F., Mourrain, P., Boutet, S., Earley, K., Lawrence, R. J., Pikaard, C. S., Murfett, J., Furner, I., Vaucheret, H. & Scheid, O. M. (2004) *Arabidopsis* histone deacetylase HDA6 is required for maintenance of transcriptional gene silencing and determines nuclear organization of rDNA repeats. *Plant Cell*, **16**, 1021–34.

Prols, F. & Meyer, P. (1992) The methylation patterns of chromosomal integration regions influence gene activity of transferred DNA in *Petunia hybrida*. *Plant Journal*, **2**, 465–75.

Qin, H. & von Arnim, A. G. (2002) Epigenetic history of an *Arabidopsis trans*-silencer locus and a test for relay of *trans*-silencing activity. *BMC Plant Biology*, **2**(1), 11.

Qin, H. X., Dong, Y. Z. & von Arnim, A. G. (2003) Epigenetic interactions between *Arabidopsis* transgenes: characterization in light of transgene integration sites. *Plant Molecular Biology*, **52**, 217–31.

Rassoulzadegan, M., Magliano, M. & Cuzin, F. (2002) Transvection effects involving DNA methylation during meiosis in the mouse. *EMBO Journal*, **21**,440–50.

Renner, O. (1959) Somatic conversion in the heredity of the *cruciata* character in *Oenothera*. *Heridity*, **13**, 283–8.

Richards, E. J. & Elgin, S. C. R. (2002) Epigenetic codes for heterochromatin formation and silencing: rounding up the usual suspects. *Cell*, **108**, 489–500.

Rossignol, J. L. & Faugeron, G. (1995) MIP: an epigenetic gene silencing process in *Ascobolus immersus*. *Current Topics in Microbiology and Immunology*, **197**, 179–91.

Sage, B. T. & Csink, A. K. (2003) Heterochromatic self-association, a determinant of nuclear organization, does not require sequence homology in *Drosophila*. *Genetics*, **165**, 1183–93.

Santos, J. L., Alfaro, D., Sanchez-Moran, E., Armstrong, S. J., Franklin, F. C. & Jones, G. H. (2003) Partial diploidization of meiosis in autotetraploid *Arabidopsis thaliana*. *Genetics*, **165**(3), 1533–40.

Selinger, D. A. & Chandler, V. L. (1999) Major recent and independent changes in levels and patterns of expression have occurred at the *b* gene, a regulatory locus in maize. *Proceedings of the National Academy of Sciences of the United States of America*, **96**(26), 15007–12.

Sidorenko, L. V. & Peterson, T. (2001) Transgene-induced silencing identifies sequences involved in the establishment of paramutation of the maize *p1* gene, *Plant Cell*, **13**(2), 319–35.

Sidorenko, L., Li, X., Tagliani, L., Bowen, B. & Peterson, T. (1999) Characterization of the regulatory elements of the maize *P-rr* gene by transient expression assays. *Plant Molecular Biology*, **39**(1), 11–19.

Sidorenko, L. V., Li, X., Cocciolone, S. M., Chopra, S., Tagliani, L., Bowen, B., Daniels, M. & Peterson, T. (2000) Complex structure of a maize *Myb* gene promoter: functional analysis in transgenic plants. *Plant Journal*, **22**(6), 471–82.

Sigrist, C. J. A. & Pirrotta, V. (1997) Chromatin insulator elements block the silencing of a target gene by the *Drosophila* Polycomb response element (PRE) but allow *trans* interactions between PREs on different chromosomes. *Genetics*, **147**(1), 209–21.

Sijen, T., Vijn, I., Rebocho, A., van Blokland, R., Roelofs, D., Mol, J. N. M. & Kooter, J. M. (2001) Transcriptional and posttranscriptional gene silencing are mechanistically related. *Current Biology*, **11**, 436–40.

Singer, M. J. & Selker, E. U. (1995) Genetic and epigenetic inactivation of repetitive sequences in *Neurospora crassa*: RIP, DNA methylation, and quelling. *Current Topics in Microbiology and Immunology*, **197**, 165–77.

Singer, T., Yordan, C. & Martienssen, R. A. (2001) Robertson's mutator transposons in *A. thaliana* are regulated by the chromatin-remodeling gene decrease in DNA methylation (*DDM1*). *Genes & Development*, **15**, 591–602.

Sleutels, F., Zwart, R. & Barlow, D. P. (2002) The non-coding Air RNA is required for silencing autosomal imprinted genes. *Nature*, **415**(6873), 810–13.

Smith, H. A., Swaney, S. L., Parks, T. D., Wernsman, E. A. & Dougherty, W. G. (1994) Transgenic plant virus resistance mediated by untranslatable sense RNAs: expression, regulation, and fate of nonessential RNAs. *Plant Cell*, **6**(10), 1441–53.

Soppe, W. J. J., Jasencakova, Z., Houben, A., Kakutani, T., Meister, A., Huang, M. S., Jacobsen, S. E., Schubert, I. & Fransz, P. F. (2002) DNA methylation controls histone H3 lysine 9 methylation and heterochromatin assembly in *Arabidopsis*. *EMBO Journal*, **21**, 6549–59.

Stam, M., Belele, C., Dorweiler, J. E. & Chandler, V. L. (2002a) Differential chromatin structure within a tandem array 100 kb upstream of the maize *b1* locus is associated with paramutation. *Genes & Development*, **16**, 1906–18.

Stam, M., Belele, C., Ramakrishna, W., Dorweiler, J. E., Bennetzen, J. L. & Chandler, V. L. (2002b) The regulatory regions required for *B'* paramutation and expression are located far upstream of the maize *b1* transcribed sequences. *Genetics*, **162**(2), 917–30.

Stead, J. D., Buard, J., Todd, J. A. & Jeffreys, A. J. (2000) Influence of allele lineage on the role of the insulin minisatellite in susceptibility to type 1 diabetes. *Human Molecular Genetics*, **9**(20), 2929–35.

Stokes, T. L. & Richards, E. J. (2002) Induced instability of two *Arabidopsis* constitutive pathogen-response alleles. *Proceedings of the National Academy of Sciences of the United States of America*, **99**(11), 7792–6.

Styles, E. D. & Brink, R. A. (1966) The metastable nature of paramutable *R* alleles in maize. I. Heritable enhancement in level of standard R^r action. *Genetics*, **54**, 433–9.

Tariq, M., Saze, H., Probst, A. V., Lichota, J., Habu, Y. & Paszkowski, J. (2003) Erasure of CpG methylation in *Arabidopsis* alters patterns of histone H3 methylation in heterochromatin. *Proceedings of the National Academy of Sciences of the United States of America*, **100**, 8823–7.

Tolhuis, B., Palstra, R. J., Splinter, E., Grosveld, F. & deLaat, W. (2002) Looping and interaction between hypersensitive sites in the active beta-globin locus. *Molecular Cell*, **10**, 1453–65.

Turner, J. M., Mahadevaiah, S. K., Fernandez-Capetillo, O., Nussenzweig, A., Xu, X., Deng, C. X. & Burgoyne, P. S. (2005) Silencing of unsynapsed meiotic chromosomes in the mouse. *Nature Genetics*, **37**(1), 41–7.

van Blokland, R., tenLohuis, M. & Meyer, P. (1997) Condensation of chromatin in transcriptional regions of an inactivated plant transgene: evidence for an active role of transcription in gene silencing. *Molecular and General Genetics*, **257**(1), 1–13.

van Houwelingen, A., Souer, E., Mol, J. & Koes, R. (1999) Epigenetic interactions among three *dTph1* transposons in two homologous chromosomes activate a new excision-repair mechanism in petunia. *Plant Cell*, **11**(7), 1319–36.

van West, P., Kamoun, S., van 't Klooster, J. W. & Govers, F. (1999) Internuclear gene silencing in *Phytophthora infestans*. *Molecular Cell*, **3**(3), 339–48.

Walker, E. L. (1998) Paramutation of the *r1* locus of maize is associated with increased cytosine methylation. *Genetics*, **148**(4), 1973–81.

Walker, E. L. & Panavas, T. (2001) Structural features and methylation patterns associated with paramutation at the *r1* locus of *Zea mays*. *Genetics*, **159**(3), 1201–15.

Wang, J., Tian, L., Madlung, A., Lee, H. S., Chen, M., Lee, J. J., Watson, B., Kagochi, T., Comai, L. & Chen, Z. J. (2004) Stochastic and epigenetic changes of gene expression in *Arabidopsis* polyploids. *Genetics*, **167**(4), 1961–73.

Weiler, K. S. & Wakimoto, B. T. (1995) Heterochromatin and gene expression in *Drosophila. Annual Review of Genetics*, **29**, 577–605.

Weiss, H. & Maluszynska, J. (2000) Chromosomal rearrangement in autotetraploid plants of *Arabidopsis thaliana. Hereditas*, **133**(3), 255–61.

Wisman, S., Ramanna, M. S. & Koornneef, M. (1983) Isolation of a new paramutagenic allele of the *sulfurea* locus in the tomato cultivar Moneymaker following *in vitro* culture. *Theoretical and Applied Genetics*, **87**, 289–94.

Zhang, H., Christoforou, A., Aravind, L., Emmons, S. W., vandenHeuvel, S. & Haber, D. A. (2004) The *C. elegans Polycomb* gene *sop-2* encodes an RNA binding protein. *Molecular Cell*, **14**, 841–7.

Zickler, D. (1973) Fine structure of chromosome pairing in ten Ascomycetes: meiotic and premeiotic (mitotic) synaptonemal complexes. *Chromosoma*, **40**(4), 401–16.

6 Genomic imprinting in plants: a predominantly maternal affair

Ueli Grossniklaus

6.1 Introduction

Genomic imprinting refers to an epigenetic modification of maternally and paternally inherited alleles that leads to their differential expression in a parent-of-origin-dependent manner. Due to genomic imprinting, maternal and paternal genomes are functionally non-equivalent. Consequently, imprinted genes are expected to display parent-of-origin effects when mutated. Although Kermicle (1970) first described genomic imprinting of a specific gene in maize, the phenomenon has been studied predominantly in mammals after its discovery in mice (McGrath & Solter, 1984; Surani *et al.*, 1984). Therefore, most of what we know about the mechanisms of genomic imprinting comes from studies in mice (reviewed in Delaval & Feil, 2004; Ferguson-Smith *et al.*, 2003).

Over the last few years, however, genomic imprinting in plants has attracted renewed interest after the discovery of *MEDEA*, the first imprinted plant gene essential to development (Grossniklaus *et al.*, 1998b; Kinoshita *et al.*, 1999; Vielle-Calzada *et al.*, 1999). Recent work on genomic imprinting in plants has revealed interesting parallels between the underlying mechanisms in mammals and plants despite the independent evolution of imprinting in the two kingdoms (Baroux *et al.*, 2002b; Köhler *et al.*, 2005; Vielle-Calzada *et al.*, 1999). Several recent reviews have dealt with the occurrence, evolution and function of genomic imprinting in plants (Alleman & Doctor, 2000; Baroux *et al.*, 2002b; Gehring *et al.*, 2004; Grossniklaus *et al.*, 2001; Gutierrez-Marcos *et al.*, 2003; Kermicle & Alleman, 1990; Köhler & Grossniklaus, 2005; Messing & Grossniklaus, 1999; Scott & Spielman, 2004; Spillane *et al.*, 2002; Vinkenoog *et al.*, 2003). This chapter summarizes the background of genomic imprinting and focuses on recent findings related to the regulation of genomic imprinting in plants, which involves DNA methylation and chromatin modification, both of which also play a role in regulating imprinting in mammals.

6.2 Plant reproduction

In both animals and plants, many events crucial to genomic imprinting occur during reproduction. Plant reproduction is peculiar in that it involves a complex interplay between the two generations of the plant life cycle, the haploid gametophyte and the diploid sporophyte. It is during gametogenesis and seed development where genomic imprints are established and imprinted genes play crucial roles, respectively. To provide the background for genomic imprinting, plant reproduction is summarized.

6.2.1 Gametogenesis and double fertilization

Unlike in animals, where meiotic products differentiate directly into gametes, mitotic divisions follow meiosis to produce multicellular gametophytes (Grossniklaus & Schneitz, 1998). The gametophytes, in turn, lead to the production of the gametes that participate in fertilization. In flowering plants, the male gametophyte (pollen) usually consists of a large vegetative cell that contains two sperm cells (McCormick, 2004). The most widely distributed type of female gametophyte is a seven-celled embryo sac, containing the two female gametes (Yadegari & Drews, 2004). Both of these, the egg and central cell, fuse with one sperm cell during the process of double fertilization (Weterings & Russell, 2004). In most species, male and female gametophytes are derived from a single meiotic product, such that their constituent cells are genetically identical. The postmeiotic divisions that form the mature, multi-cellular gametophytes may, however, lead to epigenetic differences between gametes (Messing & Grossniklaus, 1999). The unusual segregation of epigenetic effects in a *methyltransferase1* (*met1*) mutant background in *Arabidopsis* provides evidence for epigenetic differences among gametophytic cells (Saze *et al.*, 2003).

6.2.2 Seed development

Double fertilization initiates seed development, which involves a complex interplay of the maternal sporophyte in which the embryo sac is embedded and the two fertilization products (Figure 6.1). The fertilized egg cell produces the embryo, the next sporophytic generation, while the fertilized central cell leads to the formation of the endosperm, a terminally differentiated tissue that is thought to provide nutrition to the developing embryo and may be viewed analogous to the mammalian placenta (Haig & Westoby, 1989). Because the central cell contains two polar nuclei that fuse with the sperm nucleus, the endosperm is usually triploid, although not in all taxa (Baroux *et al.*, 2002a). The endosperm of many species, e.g. the grasses, is permanent and crucial to successful germination and survival of the seedling. In other species including *Arabidopsis*, however, the endosperm is transient and is consumed during seed development, such that only one cell layer is left in the mature seed (Olsen, 2004). The sporophytic tissues that surround the embryo sac develop into the seed coat, and the seeds are usually enclosed in the fruit.

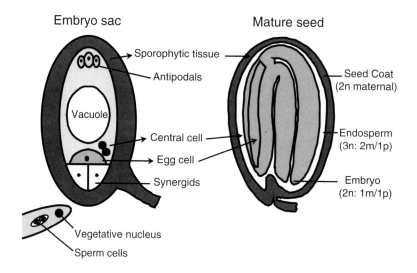

Figure 6.1 Schematic representation of an ovule at fertilization and a mature seed. The ovule contains the embryo sac with its four differentiated cell types. Fertilization of the egg cell by one of the sperm cells delivered by the pollen tube leads to the production of the embryo, while the fusion of the second sperm with the central cells initiates endosperm development. The endosperm of *Arabidopsis* is transient and gets consumed during seed development such that only one cell layer is left in the mature seed. The sporophytic tissue of the ovule develops into the seed coat after fertilization such that the mature seed consist of three parts with a distinct genetic constitution: a diploid maternal seed coat, a diploid embryo with one paternal (p) and one maternal (m) genome and an endosperm with two maternal and one paternal genome.

From a developmental perspective, a variety of influences can affect the formation of the seed. On the one hand, the mother plant or sporophyte provides nutrients and possibly developmental signals to the developing seed. Indeed, sporophytic maternal effects on germination (Gualberti *et al.*, 2002), embryogenesis (Ray *et al.*, 1996) and endosperm development (Colombo *et al.*, 1997; Felker *et al.*, 1985) have been reported. On the other hand, the maternal (or paternal) gametophyte may exert gametophytic parental effects on seed development. These could rely either on a specialized gametic cytoplasm or on epigenetic changes, i.e. a specialized gametic epigenome. Indeed, several gametophytic maternal effect mutants were discovered over the last few years in *Arabidopsis* and maize (Chaudhury *et al.*, 1997; Evans & Kermicle, 2001; Grini *et al.*, 2002; Grossniklaus *et al.*, 1998b; Guitton *et al.*, 2004; Köhler *et al.*, 2003; Moore, 2002; Ohad *et al.*, 1996). Thus, seed development involves complex regulatory interactions of the zygotic products with maternal influences of both sporophytic and gametophytic origin.

6.3   The nature of genomic imprinting

Genes that are regulated by genomic imprinting show differential expression in the zygotic products, with the activity of an allele depending on its parental origin. Thus, genetically identical alleles differ with respect to transcriptional activity despite them being in the same nucleus. This implies that paternal and maternal alleles differ in their epigenetic make-up. Their altered expression states are mitotically heritable but not related to changes in DNA sequence. Given that paternally inherited alleles can be passed on to the progeny by a female and vice versa, the epigenetic change must be reversible, i.e. reset in each generation according to the sex of the individual. In genetic terms, the state of an imprinted allele depends on its parent but not on its grandparent. Of course, this definition makes little sense for a terminally differentiated tissue such as the endosperm but it is relevant to the embryo. Mutations in imprinted genes are expected to show parent-of-origin (maternal or paternal) effects. Because such effects can result from a variety of mechanisms it is often difficult to demonstrate that a gene is regulated by genomic imprinting. As described below, regulation by genomic imprinting has only been unambiguously shown for three *Arabidopsis* and three maize genes, although many others are likely to represent imprinted loci.

6.3.1   Parental effects and the discovery of genomic imprinting

As early as the 1920s Metz described chromosome loss during male spermatogenesis in sciarid flies, a phenomenon that also occurs at other stages of the life cycle, e.g. during embryogenesis (reviewed in Goday & Esteban, 2001; Herrick & Seger, 1999). When Crouse (1960) discovered that the eliminated chromosome set was invariably of paternal origin, he coined the term 'imprinting' to describe that paternal chromosomes must carry some sort of mark, distinguishing them from the maternal chromosome set. Imprinting in sciarid flies and some other taxa among the arthropods and nematodes affects entire genomes or chromosomes. While a related phenomenon occurs with non-random X chromosome inactivation in mammals, genomic imprinting in plants and mammals can affect individual genes or small gene clusters.

The first case of genomic imprinting at a specific gene rather than an entire chromosome or genome was reported by Kermicle (1970) in maize. Dominant alleles of the *red1* (*r1*) locus confer anthocyanin pigmentation to the aleurone layer of the endosperm in maize kernels (reviewed in Kermicle, 1996). The *R-r:standard* allele (designated *R*) was found to show a parent-of-origin-dependent phenotype with respect to aleurone pigmentation. If *R* was crossed as the female parent to the colorless *r-g* allele (designated *r*) the resulting kernels with triploid endosperm of the genotype *R/R/r* were fully pigmented.

In the reciprocal cross the endosperm of the genotype $r/r/R$ showed a mottled pigmentation where most aleurone cells do not produce anthocyanin (Plate 6.1).

This difference in phenotype depending on the direction of the cross is the hallmark of a maternal effect mutation. However, the underlying nature of such a maternal effect can vary. Firstly, a maternal effect could be caused by haplo-insufficiency, where a single dose of R may not provide enough activity to produce sufficient pigment. Secondly, cytoplasmic factors in the female gametes may be missing. In principle, such a cytoplasmic effect could be due to defective plastids or mitochondria, which are usually inherited maternally in plants, or caused by the lack of a maternal gene product that is produced before fertilization but only required after fertilization during seed development. Finally, maternal and paternal alleles of the maternal effect locus may have differential activity in the zygotic products, i.e. be regulated by genomic imprinting.

In an elegant series of genetic experiments, Kermicle (1970) could exclude gene dosage and cytoplasmic effects and concluded that the activity of R depended on its parental origin (reviewed in Alleman & Doctor, 2000; Baroux et al., 2002b). Thus, R represents the first single gene for which regulation by genomic imprinting was unambiguously demonstrated. Almost 15 years later, nuclear transfer experiments in mice showed that paternal and maternal genomes are not equivalent in mammals (Barton et al., 1984; McGrath & Solter, 1984; Surani et al., 1984). Genetic experiments showed that imprinted genes in mice are not scattered throughout the genome but often occur in clusters on specific chromosomes (Cattanach & Kirk, 1985). The cloning of the first imprinted mouse genes in the early 1990s (e.g. Barlow et al., 1991; Bartolomei et al., 1991; DeChiara et al., 1991) initiated intensive research on the molecular mechanisms of genomic imprinting in mammals.

6.3.2   Genomic imprinting and gene dosage effects

As outlined above, the underlying nature of parental effects is diverse and only a subset of paternal or maternal effect phenotypes is due to the disruption of an imprinted locus. Unfortunately, parental effects have often been equated to genomic imprinting in the literature, although other possible causes have not been excluded. For instance, interploidy crosses cause developmental aberrations in endosperm development that lead to seed abortion in maize and affect seed size in *Arabidopsis* (Lin, 1984; Redei, 1964; Scott et al., 1998). While these findings have been taken as evidence for the role of genomic imprinting in seed development, they could equally well be due to other mechanisms, in particular gene dosage effects (Baroux et al., 2002b; Birchler, 1993). In maize, genetic studies using translocation stocks have shown that specific chromosomal regions carry factors controlling endosperm size (*Efs*): if a particular

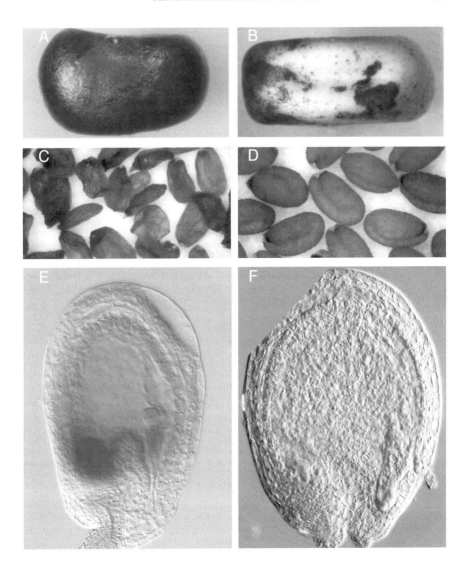

Plate 6.1 Examples of imprinted genes in plants. (A) and (B) show kernels with a maternally and a paternally inherited *R* allele of maize, respectively. The *R* allele was exposed to *R-st* in the previous generation, which strengthens the effect. (C) and (D) show *Arabidopsis* seeds with a maternally and a paternally inherited mutant *medea* allele, respectively. Maternal inheritance of a mutant *medea* allele causes seed abortion irrespective of the paternal contribution. (E) and (F) show a maternally and a paternally inherited *MEDEA-promoter::GUS* fusion construct, respectively. Only the maternally inherited allele is active and produces β-galactosidase (GUS), which is detected as a blue stain in a cytochemical reaction. Some diffusion of the blue product into the surrounding sporophytic tissue is visible. See also color plate section for color version of this figure.

region containing an *Ef* on a paternally inherited chromosome is missing, endosperm size is reduced. Lin (1982) showed that the introduction of a maternal *Ef* copy of the same chromosomal region could not rescue the seed size defect and concluded that the maternal *Ef* was inactive and, thus, that the *Efs* represent paternally active imprinted genes. While this is certainly an attractive possibility, such genetic experiments do not exclude all gene dosage effects and additional experiments indeed suggested that at least some mater- nally derived *Ef* alleles are functional (Birchler, 1993).

Gene dosage effects can stem from disturbances in the relative amount of maternal and paternal gene activities in the zygotic products after fertilization; this may be particularly important in the endosperm where a ratio of 2 maternal (m) to 1 paternal (p) genomes (2m:1p) is crucial for normal devel- opment. In addition, such effects may also result from imbalances in the dosage of maternal products produced before fertilization in relation to zygotic products; this ratio of female gametophytic (g) to zygotic (z) genomes (em- bryo 1g:2z; endosperm 2g:3z) is also altered in interploidy crosses and affects both fertilization products. An in-depth discussion of dosage effects and their relation to imprinting can be found elsewhere (Birchler, 1993; von Wangen- heim & Peterson, 2004).

6.3.3   *Genomic imprinting and asymmetry of parental gene activity*

As is apparent from the discussion above, an unequivocal demonstration of genomic imprinting remains difficult. In maize, sophisticated genetic experi- ments have been used to show that *R* is regulated by genomic imprinting (Kermicle, 1970). In *Arabidopsis*, similar experiments are currently not pos- sible and a demonstration of genomic imprinting relies on molecular ap- proaches. Most often, differential expression of maternal and paternal alleles is taken as evidence for imprinting. Transcripts derived from maternal and paternal alleles are detected by polymorphisms assayed in RNA gel blots or allele-specific reverse transcription polymerase chain reaction (RT-PCR) assays. Alternatively, the promoters of potentially imprinted genes were fused to a reporter gene and their activity assayed in reciprocal crosses. Provided a gene is not expressed in the gametes but only in the zygotic products after fertilization, differential levels of maternally and paternally derived transcripts provide clear evidence for regulation by genomic imprinting. Purely zygotic, allele-specific transcription was reported for three maternally expressed genes in maize, *ZmFie1* (Danilevskaya *et al.*, 2003), *no apical meristem related protein1* (*nrp1*, Guo *et al.*, 2003), and *maternally expressed gene1* (*meg1*, Gutierrez-Marcos *et al.*, 2004), and recently for the first paternally expressed gene in *Arabidopsis*, *PHERES1* (*PHE1*) (Köhler *et al.*, 2005).

Many genes that show differential expression during seed development, however, are also expressed in the female gametes before fertilization. Thus, it cannot be unambiguously determined whether the origin of maternally derived transcripts is pre- or postfertilization. In these cases maternally derived transcripts may be produced before fertilization and stored in the gametes (gametophytic expression), be derived from imprinted gene expression after fertilization (zygotic expression) or represent a combination of the two (Figure 6.2). Only if active transcription in the zygotic products is demonstrated, for instance by detecting nascent transcripts using RNA-FISH (Braidotti, 2001), can regulation by genomic imprinting be concluded. Although this is routine in mammalian systems, active transcription during seed development has only been demonstrated for one plant gene, *MEDEA*, where endosperm nuclei carry two active and one silent allele (Vielle-Calzada *et al.*, 1999).

Over the last years, maternal expression was shown for a large number of endogenous genes, lines carrying promoter–reporter gene fusion constructs or enhancer detector elements (e.g. Baroux *et al.*, 2001; Golden *et al.*, 2002; Luo *et al.*, 2000; Springer *et al.*, 2000; Sorensen *et al.*, 2001; Vielle-Calzada *et al.*, 2000; Yadegari *et al.*, 2000). For most *Arabidopsis* loci studied, paternal

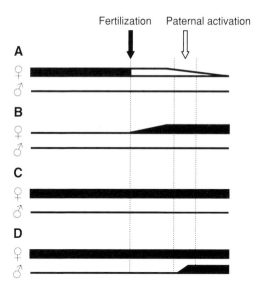

Figure 6.2 Selected possibilities for expression of maternally and paternally inherited alleles that result in differential gene activity. (A) shows a gametophytically expressed gene with a cytoplasmic effect: its product is detectable after fertilization although the gene is not actively transcribed any more. (B) shows a zygotically expressed gene that is regulated by genomic imprinting. (C) shows a combination of (A) and (B) where the gene is expressed only from the maternal allele but both before and after fertilization. (D) shows one of many paternally inherited alleles that become active a few days after fertilization, although the exact time point depends on the specific gene analyzed. In combination with any of the cases shown in (A)–(C) (here illustrated with (C)), the period between fertilization and paternal activation will be exclusively under maternal control. The observed differential expression does not reflect regulation by genomic imprinting in all cases.

expression during early seed development was undetectable, a finding that was recently confirmed for maize (Danilevskaya *et al.*, 2003; Guo *et al.*, 2003; Gutierrez-Marcos *et al.*, 2004; Grimanelli *et al.*, 2005). A few genes with early paternal expression have been reported, but even these show much higher levels of maternally derived transcripts (Weijers *et al.*, 2001). This is consistent with the findings of Baroux *et al.* (2001) who found low basal levels of paternal activity early during seed development when using a highly sensitive assay based on barnase but not with a less sensitive reporter gene. While the widespread absence of paternal activity clearly illustrates that early seed development is to a large extent under maternal control (Vielle-Calzada *et al.*, 2000), it cannot be distinguished whether these maternal transcripts are of gametophytic or zygotic origin because most are already expressed before fertilization. It is likely that maternal transcripts during early seed development represent a combination of all possibilities depicted in Figure 6.2. For a few of the loci studied by Vielle-Calzada *et al.*, however, there is evidence for regulation by genomic imprinting. For instance, enhancer detector line ET1811 is not expressed in the central cell and gets induced in the endosperm after fertilization but only if inherited maternally. Another enhancer detector line is expressed in the egg cell before fertilization but gets restricted to the apical cell lineage after fertilization indicating active postfertilization transcription only in these cells (R. Baskar & U. Grossniklaus, unpublished data). Furthermore, enhancer detector line KS117 is expressed in the endosperm only after fertilization if inherited maternally (Sorensen *et al.*, 2001). Thus, these three lines likely reflect imprinted genes active in endosperm and embryo, respectively.

6.4   Imprinted genes in *Zea mays* and *Arabidopsis thaliana*

To date, imprinted genes have only been identified in maize and *Arabidopsis*. However, parent-of-origin effects in crosses, which may be related to genomic imprinting, have been observed in many species such that imprinting is likely to be widespread among the flowering plants. For many years, genomic imprinting was studied only in maize until the discovery of the *Arabidopsis* maternal effect mutants (reviewed in Grossniklaus *et al.*, 2001) renewed the interest in this field, which has since been studied intensely in *Arabidopsis*.

6.4.1   Imprinted genes and potentially imprinted genes in maize

After the initial description of genomic imprinting at the *r1* locus, the identification of further imprinted loci had to await molecular techniques. Recently, several new imprinted genes have been identified using novel approaches. Genome-wide RNA profiling led to the discovery of the imprinted locus *nrp1* (Guo *et al.*, 2003), and allelic message display to the isolation of *meg1*, whose maternally inherited alleles are expressed specifically in the transfer layer of the endosperm (Gutierrez-Marcos *et al.*, 2004). As both these genes are not expressed in the female gametes, they are clearly regulated by genomic imprinting. However, while mono-allelically expressed early in seed development, the paternal allele also becomes active at around 10 days after pollination. Thus, *meg1* represents a stage-specific case of imprinted gene expression. *ZmFie1*, one of the two maize genes with similarity to the *Arabidopsis* gene *FERTILIZATION-INDEPENDENT ENDOSPERM* (*FIE*, see below), starts maternal allele-specific expression at about six days after pollination and is expressed mono-allelically throughout seed development (Danilevskaya *et al.*, 2003).

There are several other maize genes whose expression was proposed to be regulated by genomic imprinting (reviewed in Alleman & Doctor, 2000; Baroux *et al.*, 2002b; Gehring *et al.*, 2004). These include *dzr1*, a locus showing a parent-of-origin-dependent effect on the posttranscriptional regulation of 10-kDa zein (Chaudhuri & Messing, 1994), and the *α-tubulin* and *α-zein* genes (Lund *et al.*, 1995a, b), which show high levels of the maternally derived transcripts during kernel development. As the expression profile of these genes before fertilization is not known, their imprinting status has not been unambiguously determined. The second *FIE* gene in maize, *ZmFie2*, is expressed already before fertilization but only maternally derived transcripts are detectable until five days after pollination, when the paternal allele is activated (Danilevskaya *et al.*, 2003). Whether this differential expression level is due to cytoplasmically stored mRNA, genomic imprinting or a combination of the two is not known.

Importantly, only specific alleles of the *r1*, *dzr1*, *α-tubulin* and *α-zein* genes show parent-of-origin effects whereas others do not. This contrasts with imprinting in mammals, where usually all alleles of a given locus are subject to imprinting. Also, none of these imprinted alleles plays a crucial role in seed development. The function of the newly described *npr1*, *meg1*, *ZmFie1* and *ZmFie2* genes is not known as mutants in these genes have yet to be isolated. It is possible that these loci show locus- rather than allele-specific imprinting as the two (*ZmFie1*, *ZmFie2* and *meg1*) and four alleles (*npr1*) tested, respectively, showed the same parent-of-origin effects.

6.4.2   *The FIS class of genes in* Arabidopsis

The maize genes introduced above either do not play a role in seed development or their function is currently unknown. In contrast, the genes of the *FIS* class in *Arabidopsis* are essential to development. Currently, four members of this class, which are all required maternally for normal seed development, are known. The *medea* (*mea*) mutant was isolated in a screen for gametophytic maternal effect mutations (Grossniklaus *et al.*, 1998b), while *fie* and the three *fertilization-independent seed* mutants (*fis1* allelic to *mea*, *fis2*, *fis3* allelic to *fie*) were identified in a screen for silique elongation in the absence of fertilization (Chaudhury *et al.*, 1997; Ohad *et al.*, 1996; Peacock *et al.*, 1995). The fourth member of the *fis* class, *msi1*, was characterized by a reverse genetics approach (Köhler *et al.*, 2003). Additional alleles of all four members and possibly a fifth one were identified in a screen for mutants deregulating an endosperm marker (Guitton *et al.*, 2004), which had previously been shown to be affected in *fis* class mutants (Sorensen *et al.*, 2001).

In general, the *fis* class mutants share the following phenotypes, although there is some variation between mutants and different genetic backgrounds: (1) seeds derived from a *fis* mutant gametophyte abort irrespective of the paternal contribution, i.e. they are gametophytic maternal effect embryo lethal; (2) embryo and endosperm derived from a *fis* mutant gametophyte show delayed development and aberrant cell proliferation; (3) *fis* mutant central cells initiate endosperm development in the absence of fertilization.

How pre- (3) and postfertilization (1, 2) phenotypes are interrelated is currently not well understood but collectively these phenotypes suggest a crucial role of the *FIS* class genes in the control of cell proliferation (reviewed in Chaudhury & Berger, 2001; Grossniklaus *et al.*, 2001).

6.4.3   The MEA–FIE Polycomb group complex

The molecular basis of the common phenotypes of the *fis* class mutants became clear when their molecular identity was determined. All four *FIS* class genes encode subunits of a conserved mutli-protein *Polycomb* group (PcG) complex, which is referred to as Enhancer of zeste–Extra sex combs (E(z)–Esc) complex or *Polycomb* repressive complex 2 (PRC2) in *Drosophila melanogaster* (reviewed in Levine *et al.*, 2004; Ringrose & Paro, 2004). All versions of E(z)–Esc and its human counterpart contain the core proteins E(z), Esc, and Supressor of zeste12 (Su(z)12) as well as some accessory proteins, depending on the particular complex and purification procedure used. Among these, the protein p55, a homolog of the human retinoblastoma-associated proteins RbAp46 and RbAp48 and the histone deacetylase Rpd3 play a prominent role (reviewed in Levine *et al.*, 2004).

PcG proteins are best known for their role in maintaining the genes of the HOX clusters in a repressed state (reviewed in Orlando, 2003; Ringrose & Paro, 2004). PcG complexes were proposed to be epigenetic regulators that form the basis of a cellular memory mechanism maintaining gene expression patterns over main cell divisions. The molecular basis for this function was elucidated when it was shown that SET domains (Jones & Gelbart, 1993; Tschiersch *et al.*, 1994), also present in E(z), confer histone methyltransferase (HMT) activity (Rea *et al.*, 2000). While purified SET-domain proteins of the Su(var)3–9 class have HMT activity on their own in several organisms

(Jackson *et al.*, 2002; Peters *et al.*, 2002; Rea *et al.*, 2000), HMT activity of SET-domain proteins of the E(z) class has so far only been demonstrated in partially assembled/purified complexes that methylate histone 3 at lysine 9 and/or lysine 27 (Czermin *et al.*, 2002; Kuzmichev *et al.*, 2002; Müller *et al.*, 2002). Both these histone methylation marks are associated with transcriptionally inactive chromatin (reviewed in Wang *et al.*, 2004).

The molecular characterization of the *FIS* class genes showed that they encode proteins similar to the subunits of the E(z)–Esc complex: MEA was found to encode a protein similar to E(z) (Grossniklaus *et al.*, 1998b), FIE corresponds to Esc (Ohad *et al.*, 1999), and FIS2 was shown to have a zinc-finger (Luo *et al.*, 1999) and was later found to be similar to Su(z)12 (Birve *et al.*, 2001). As its *Drosophila* and human counterparts MEA and FIE interact physically in yeast (Luo *et al.*, 2000; Spillane *et al.*, 2000; Yadegari *et al.*, 2000), as well as *in vitro* (Spillane *et al.*, 2000), and they form a large molecular weight complex that also contains MSI1, an *Arabidopsis* protein with similarity to p55 of *Drosophila* (Köhler *et al.*, 2003). The MEA–FIE complex is thought to repress target genes that prevent endosperm proliferation before fertilization and also regulates target genes controlling cell proliferation after fertilization in the developing seed. Two target genes, *PHE1* and *MEIDOS*, were recently identified and the MEA–FIE complex was shown to physically interact with the *PHE1* promoter region (Köhler *et al.*, 2003). In *Arabidopsis*, several other versions of PRC2-like complexes exist. The role of these complexes in plant development has been reviewed extensively and will not be discussed further (Goodrich & Tweedie, 2002; He & Amasino, 2005; Hsieh *et al.*, 2003; Köhler & Grossniklaus, 2002; Reyes & Grossniklaus, 2003).

6.4.4   *Imprinted genes and potentially imprinted genes in* Arabidopsis

As outlined above, the gametophytic maternal effect of the *fis* class mutants could be caused by a variety of underlying mechanisms, including genomic imprinting. Differential expression of maternally and paternally inherited alleles during seed development was indeed found for *MEA*, *FIS2* and *FIE* using promoter–reporter gene fusions or allele-specific RT-PCR (Kinoshita *et al.*, 1999; Luo *et al.*, 2000; Vielle-Calzada *et al.*, 1999; Yadegari *et al.*, 2000). However, all these genes are also expressed before fertilization such that a distinction between maternally deposited gene products and imprinted expression requires that active transcription of the maternally inherited allele be demonstrated. This has, so far, only been done for *MEA*, where two of the three alleles were found to be transcribed in the early endosperm after fertilization (Vielle-Calzada *et al.*, 1999). Together with allele-specific RT-PCR experiments, this indicates that only the maternally inherited *MEA* alleles are active during seed development.

Recently, parental alleles of *FWA* were shown to be differentially expressed during seed development (Kinoshita *et al.*, 2004). The *fwa-1* mutant was initially isolated as a late flowering mutant (Koornneef *et al.*, 1991). Molecular analysis showed that the mutant phenotype was not associated with a nucleotide change in the *FWA* gene but rather with ectopic expression due to a permanent loss of DNA methylation (Soppe *et al.*, 2000). In the wild type, *FWA* is expressed in the female gametophyte before fertilization and in the developing endosperm. Maternal expression is correlated with a loss of DNA methylation in the 5′ region of the *FWA* gene, which only occurs in the tissue where *FWA* expression has been detected (Kinoshita *et al.*, 2004). *FWA* is expressed in the endosperm but its role in seed development is currently unknown. Due to the genetic interactions of *FWA* with regulators of genomic imprinting (see below), it is a likely candidate gene for being imprinted. However, unambiguous proof is currently missing as the gene is already active before fertilization and transcriptional activity during endosperm development has not yet been demonstrated.

Until recently, all imprinted and potentially imprinted genes described in maize and *Arabidopsis* were expressed from the maternally inherited allele only, at least during certain stages of seed development. None of them showed mono-allelic expression from the paternally inherited allele, whereas in mice, maternally and paternally expressed imprinted genes are about equally common (see Morison *et al.*, 2001). Expression studies on the *MEA* target gene *PHE1* have recently shown that *PHE1* is preferentially expressed from the paternally inherited allele during seed development (Köhler *et al.*, 2005). The maternally inherited allele is also expressed, but at a much lower level. Differential expression of imprinted genes that does not follow an all-or-nothing pattern is not uncommon in mice, where imprinting can also be stage- or tissue-specific (e.g. Lewis *et al.*, 2004). Thus, given that parental *PHE1* alleles are differentially expressed and *PHE1* is only activated after fertilization, it represents the first paternally-expressed imprinted gene in plants. In summary, although there are several candidate genes likely to be imprinted in *Arabidopsis*, regulation by genomic imprinting has unambiguously been demonstrated only for *MEA* and *PHE1*.

6.4.5   *Genomic imprinting in embryo and endosperm*

Traditionally, genomic imprinting in plants is viewed as specific to the endosperm (reviewed in Gehring *et al.*, 2004; Haig & Westoby, 1991; Kermicle & Alleman, 1990). On the one hand, this is based on the observations that interploidy crosses and the *Ef*s, which have often been implicated in imprinting, mainly affect the development of the endosperm. On the other hand, plant embryos can form in culture from single sporophytic cells or microspores (reviewed in Mordhorst *et al.*, 1997), and haploids with either only paternal or only maternal genomes have been recovered in many species (Kermicle, 1969; Kimber & Riley, 1963; Sarkar & Coe, 1966). Thus, embryo development does not appear to depend on having either both a paternal and a maternal genome or a specialized gametic cytoplasm. Consequently, maternal effects on embryogenesis, be it through imprinting or cytoplasmic mechanisms, are not expected to affect embryo development. However, the occurrence of haploids in nature is very rare and somatic embryogenesis in culture may represent a rather specialized case under unusual circumstances. How these situations relate to normal zygotic embryogenesis is unknown. The recent discovery that maternal effects play a crucial role in seed development indicates that somatic and zygotic embryogenesis may have less in common than originally thought. In fact, even endosperm can develop in the absence of a paternal genome in some apomicts (Haig & Westoby, 1989; Koltunow & Grossniklaus, 2003; Nogler, 1984) and the conventional view may have to be revisited.

Imprinted and potentially imprinted maize genes described above are expressed in the endosperm only, *ZmFie2* being an exception. Because until a few years ago plant imprinting was only studied in maize, this finding strengthened the view that imprinting in plants is specific to the endosperm. The fact that more imprinted genes seem to be expressed in the endosperm than the embryo may indicate that imprinting first evolved for the endosperm. Reik and Lewis (2005) recently proposed that imprinting evolved for the placenta in mammals, the tissue playing an analogous role to the endosperm of plants. However, with the exception of *FWA*, which seems to be expressed specifically in the endosperm (Kinoshita *et al.*, 2004), all imprinted and potentially imprinted genes in *Arabidopsis* are expressed in both products of fertilization (Köhler *et al.*, 2003, 2005; Luo *et al.*, 2000; Spillane *et al.*, 2000; Vielle-Calzada *et al.*, 1999; Yadegari *et al.*, 2000). Experiments to determine allele-specific expression specifically in the embryo and endosperm were performed for *MEA* and *FWA* at later stages of seed development, from 6 days after pollination onwards, when a dissection of embryo and endosperm becomes feasible (Kinoshita *et al.*, 1999, 2004).

Based on RT-PCR experiments on RNA from isolated embryo and endosperm tissue, *MEA* was reported to be expressed mono-allelically in the endosperm but bi-allelically in the embryo (Kinoshita *et al.*, 1999). At earlier stages, RT-PCR experiments on RNA from developing fruits, containing both embryo and endosperm, *MEA* was found to be expressed only from the maternally inherited allele, suggesting that it is regulated by imprinting in both the embryo and the endosperm (Vielle-Calzada *et al.*, 1999). Mono-allelic expression during early seed development may reflect the widespread lack of detectable paternal expression, which was found subsequently and was discussed above (Vielle-Calzada *et al.*, 2000). However, recent experiments using allele-specific probes in highly sensitive quantitative RT-PCR have failed to detect above-background levels of paternally derived transcripts throughout seed development until 10 days postpollination (Page, 2004). Thus, these results confirm the view that *MEA* is expressed mono-allelically from the maternally inherited allele in both the embryo and the endosperm. The discrepancy between reported results may lie in technical differences or in the sensitivity of *MEA* expression to the different genetic backgrounds used. Indeed, a screen of over 50 different *Arabidopsis* accessions showed that about half the accessions have some modifying activity on the *mea* seed abortion phenotype (Spillane *et al.*, 2002; C. Spillane & U. Grossniklaus, unpublished data). The expressivity of various aspects of the *mea* phenotype has also been reported to depend on genetic background (Kiyosue *et al.*, 1999; Luo *et al.*, 2000; Vielle-Calzada *et al.*, 1999).

6.5   Molecular mechanisms of genomic imprinting

While the molecular mechanisms of imprinting in mammals have been studied extensively (reviewed in Sleutels & Barlow, 2002), the study of the regulation of imprinting in plants has just begun. In principle, the imprinted state could be either the active or the inactive one, with the alternative state representing the default. Alternatively, both states may require specific regulation by genomic imprinting. It is clear that a number of *cis*-acting elements and *trans*-acting factors are involved in the regulation of imprinted genes. Here, recent findings on *trans*-acting factors involved in imprinting as well as the *cis*-regulatory elements required for imprinting in plants are reviewed briefly. More detailed information can be found in several recent reviews (Gehring *et al.*, 2004; Köhler & Grossniklaus, 2005; Spillane *et al.*, 2004a).

6.5.1   Trans-*acting factors affecting imprinting*

The first *trans*-acting factor involved in imprinting was described by Kermicle (1978) in maize. In the absence of activity from the *maternal derepression of r1* (*Mdr1*) locus, a maternally inherited *R* allele is not expressed in all aleurone cells as expected when maternally transmitted, but rather produces a mottled phenotype as if it was inherited paternally. As the *mdr1* mutant was found to be recessive, full expression of the maternally inherited allele requires *Mdr1* activity and the low-expression level of paternally inherited alleles appears to be the default state.

Recently, a factor with an activity similar to *Mdr1* was described in *Arabidopsis*: the *DEMETER* (*DME*) gene is required for the expression of maternally inherited *MEA* alleles (Choi *et al.*, 2002). *DME* was found to be expressed in the embryo sac before fertilization, with *DME* RNA decreasing rapidly after fertilization. This suggests that *DME* activates the maternal *MEA* allele before fertilization and ensures mono-allelic expression afterwards through some kind of lasting epigenetic modification. *DME* was isolated as a mutant with a phenotype similar to that of *MEA*, consistent with its function as an activator of *MEA* (Choi *et al.*, 2002; Guitton *et al.*, 2004). In addition, a screen for second-site modifiers aimed at the identification of regulators of *MEA* identified three additional alleles of *DME* (S. Tahar-Chaouch, C. Baroux & U. Grossniklaus, unpublished data). *DME* was also found to activate the *FWA* gene in the central cell before fertilization and seems to play a general role in the regulation of imprinted genes (Kinoshita *et al.*, 2004).

DME encodes a putative DNA glycosylase (Choi *et al.*, 2002) and a residue conserved in all glycosylases was found to be essential for the activation of *MEA* (Choi *et al.*, 2004). DNA glycosylases function in DNA repair and excise mismatched, altered or damaged bases. How the activity of *DME* leads to the activation of *MEA* and *FWA* is currently unknown but important hints come from a screen for second-site suppressors of *dme* that identified several new alleles of *met1* (Xiao *et al.*, 2003). *MET1* is the major maintenance DNA methyltrasferase in *Arabidopsis* and *met1* mutants show reduced DNA methylation, both at repeated- and at single-copy sequences (Kankel *et al.*, 2003). A lack of *MET1* during gametogenesis leads to the formation gametes that differ in their epigenetic make-up through passive demethylation (Saze *et al.*, 2003). Genetically, *MET1* was reported to act antagonistically to *DME* in the control of maternal *MEA* expression (Xiao *et al.*, 2003). It has been proposed that *DME* may remove methylated cytosines to explain the antagonistic effect with *MET1*. Indeed, DNA glycosylases have been reported to act as demethylases (Jost *et al.*, 1995) but do this reaction with extremely low activity (Scharer & Jiricny, 2001; Zhu *et al.*, 2000). The finding that *REPRESSOR OF SILENCING1* (*ROS1*), a factor involved in transcriptional gene silencing, removes 5-methylcytosine *in vitro* also suggests that DNA glycosylases may be involved in the regulation of DNA methylation and that *DME* may indeed activate *MEA* through demethylation. An unresolved problem, however, is whether the *MEA* locus is methylated at all. Conflicting data have been reported with respect to methylation of the promoter region of *MEA* (Choi *et al.*, 2002; Xiao *et al.*, 2003) and no differences in methylation status in wild-type versus *dme* mutant plants have been reported to date.

The fact that DNA or chromatin modifications involved in imprinting are expected to occur in the gametophytes, consisting of a very small number of cells, makes it extremely difficult to directly address whether such mechanisms are involved. An essay of DNA methylation is only informative if performed in the relevant cells and that was not possible up to now due to their inaccessibility. Therefore, most of the evidence suggesting the involvement of transacting factors in imprinting is based on genetic data where it is difficult to distinguish direct from indirect effects.

In plants, evidence for a possible involvement of DNA methylation in imprinting, which is prevalent in mammals (reviewed in Bestor, 2003;

Shermer & Razin, 1996), has come from genetic and molecular investigations. In maize, maternally expressed imprinted alleles of *r1*, *α-zein* and *α-tubulin* were found to be hypomethylated in comparison with the paternal allele, correlating with increased activity (Kermicle, 1996; Lund *et al.*, 1995a, b). In *Arabidopsis*, reciprocal crosses between the wild type and plants with a reduced level of *MET1* (*MET1a/s*; Finnegan *et al.*, 1996) showed effects on seed size similar to those observed in interploidy crosses (Adams *et al.*, 2000; Scott *et al.*, 1998). Although the basis of the effect in interploidy crosses is not clear (see above), it might, in part, be related to genomic imprinting. *MET1 a/s* plants were also found to be able to rescue the embryo lethality of *fis* class mutants when used as a pollen donor (Luo *et al.*, 2000; Vinkenoog *et al.*, 2000). However, *MET1a/s* did not alter the expression of a promoter–reporter gene fusion with *MEA* and *FIS2*, suggesting that the rescue effect is indirect, possibly working via target genes of the MEA–FIE complex. In contrast, repression of a paternally inherited *FWA* allele seems to depend on *MET1* activity (Kinoshita *et al.*, 2004).

The complexity of the interactions is also reflected by the effect of mutations in *decreased DNA methylation1* (*DDM1*), which reduces genome-wide methylation levels by about 70% (Vongs *et al.*, 1993) and encodes a SWI2/SNF2-related chromatin remodeling protein (Brzeski & Jerzmanowski, 2003; Jeddeloh *et al.*, 1999). *ddm1* mutants show a progressive accumulation of phenotypes with increased inbreeding. If a *ddm1* allele from a normally methylated plant is introduced into a *mea* background, seeds derived from a *mea* mutant embryo sac can survive if also homozygous for *ddm1* (Vielle-Calzada *et al.*, 1999). However, if the *ddm1* strain has been inbred and carries a demethylated genome, the introduction of the *ddm1* mutant through the male is sufficient to rescue *mea* embryo lethality (Yadegari *et al.*, 2000). While genetic evidence suggests that *DDM1* acts as a zygotic repressor of the paternal *MEA* allele (Vielle-Calzada *et al.*, 1999), inbred *ddm1* strains have certainly other effects. Among them is the loss of activity of the *MEA* target gene *PHE1* in a *ddm1*-inbred background (Köhler *et al.*, 2003). Due to the complex inheritance characteristics of epimutations induced by *ddm1*, which are stably maintained after *ddm1* is segregated away, it is not easy to distinguish direct and indirect effects. A better definition of the role of *DDM1* in the regulation of genomic imprinting will require additional molecular experiments.

Also the exact role of DNA methylation in the regulation of imprinted genes is not yet clear, DNA methylation has been implicated in imprinting in various ways in plants as it has in mammals. Another *trans*-acting factor that has recently been reported to play a role in mono-allelic expression in mammals is the PRC2 complex. The mouse Eed protein, which is similar to FIE, is required for the mono-allelic expression in random and imprinted X chromosome inactivation and also in the regulation of some autosomal imprinted genes (Mager *et al.*, 2003; Silva *et al.*, 2003; Wang *et al.*, 2001). Although genomic imprinting has evolved independently in mammals and plants, it was found that the MEA–FIE complex is also involved in the regulation of imprinted genes in plants. Imprinted expression of *PHE1* depends on *MEA* activity: in a *mea* mutant, the maternally inherited *PHE1* allele is strongly derepressed while the paternally inherited one is not derepressed significantly (Köhler *et al.*, 2005). The recruitment of a similar PcG complex to regulate imprinted gene expression in these distinct lineages constitutes an interesting case of convergent evolution as does the involvement of DNA methylation, although its exact role in imprinting in plants has yet to be determined.

6.5.2   Cis-*acting elements involved in imprinting*

Genetic experiments have identified several *trans*-acting factors of imprinting in plants. Far less is known about the *cis*-regulatory regions required for imprinting. This is surprising, given the wealth of data we have on this aspect in mice (reviewed in Sleutels & Barlow, 2002). Transposons and direct repeats have been proposed to play a role in several epigenetic phenomena (Martienssen, 1996; Matzke *et al.*, 1996). Recently, the role of a transposon in imprinting was demonstrated in plants for the *FWA* locus (Lippman *et al.*, 2004). An *AtSINE3* element inserted 980 bp upstream of the *FWA* start codon contributes the first two exons to *FWA*. This brings *FWA* under the control of *DDM1*, which is a major factor controlling transposons and related repeats (Lippman *et al.*, 2004). The presence of small interfering RNAs associated with the transposon in the *FWA* promoter suggests that it is responsible for the epigenetic regulation of *FWA*.

The *MEA* locus has also been analyzed for the presence of transposons and repeat sequences and their potential role in genomic imprinting was investigated (Spillane *et al.*, 2004a). At 4363 bp upstream of the *MEA* start codon an *AtREP2* helitron transposable element was found, followed by a large tandem duplication encompassing over 3 kb in total. In the 3′ region of the gene, seven 182-bp repeats (Cao & Jacobsen, 2002) and a shorter trinucleotide repeat were found (Spillane *et al.*, 2004a). By exploiting natural alleles of the *MEA* locus that do not contain either the *AtREP2* helitron or the downstream repeats, it could be shown that these elements play no role in regulating the imprinted expression of the *MEA* locus. The results were confirmed using promoter–reporter gene fusions (Plate 6.1) that lack these elements but still show parent-of-origin-dependent expression (Spillane *et al.*, 2004b). Thus, distinct mechanisms seem to operate in the regulation of *FWA* and *MEA*. This is reminiscent of the situation in mammals, where a multitude of different mechanisms controlling imprinted expression have been reported (see Sleutels & Barlow, 2002).

6.6   Role of imprinting in plant development and evolution

There are many theories for the evolution of imprinting that have been reviewed extensively and will be covered very shortly here (see Baroux *et al.*, 2002b; Gutierrez-Marcos *et al.*, 2003; Haig & Westoby, 1989; Hurst & McVean, 1998; Messing & Grossniklaus, 1999; Moore, 2001; Moore & Haig, 1991; Spielman *et al.*, 2001; Tilghman, 1999; Wilkins & Haig, 2003).

The parental conflict theory proposed by Haig and Westoby (1989, 1991) has attracted a lot of attention as it provides a theoretical basis for growth-related defects that have been reported for many mutants affecting imprinted genes in both plants and mammals. The parallels between the phenotype of *mea* and that of a large number of mouse mutants with respect to cell proliferation and growth suggest that similar selective pressures led to the recruitment of growth-controlling genes for regulation by genomic imprinting (Baroux *et al.*, 2002b; Vielle-Calzada *et al.*, 1999). These similarities could be based on the fact that both mammals and plant have a placental habit, where all nutrients supporting the developing embryo are provided by the mother. In polygamous organisms, this is expected to result in a conflict between maternal and paternal interests over the allocation of nutrients from mother to offspring as stated in David Haig's parental conflict theory.

While *mea* fits the parental conflict paradigm very nicely, it has to be stated that most imprinted and potentially imprinted genes in plants have either no essential role in development or their phenotype is presently unknown. There are also several imprinted mouse genes that do not have obvious effects on growth, such that alternative scenarios should be considered (Hurst & McVean, 1998). In plants, the view that interploidy crosses and translocation studies provide evidence for genomic imprinting has been challenged (Birchler, 1993; von Wangenheim & Peterson, 2004) and more emphasis has been put on dosage effects, which are expected for any gene product whose function is dosage-sensitive, particularly those working in multi-subunit regulatory complexes (Birchler *et al.*, 2001). In this respect, not only the maternal-to-paternal genome ratio but also the gametophytic-to-zygotic genome ratio plays a role in dosage sensitivity. It is clear that imprinting can be viewed as a way to fine-tune gene expression levels, as either maternal or paternal alleles of imprinted genes are inactive. This adds an additional possibility to adjust the expression level of a gene: while for non-imprinted genes expressed in the endosperm only the relative levels of $3\times$ and $0\times$ (expressed or not expressed) are possible, imprinted genes could also be expressed at a relative level of $1\times$ or $2\times$, depending on whether the maternal or paternal allele is inactive (Kermicle & Alleman, 1990). While the expression level of many genes may indeed have to be fine-tuned precisely to ensure normal development, this does not exclude a parental conflict as the underlying basis for the evolution of imprinted loci. The diversity of imprinting mechanisms and phenotypes of imprinted genes may indicate that diverse selective pressures have played a role in the evolution of imprinted loci.

Because plants have a haploid, postmeiotic phase that includes the formation of several differentiated cell types, imprinting may also play a role in the epigenetic differentiation of egg and central cell, which follow very different developmental pathways after fertilization (Messing & Grossniklaus, 1999). Imprinting may differ between these cell types and contribute to the distinct development of the fertilization products. Such an effect would not be subject to a parental conflict as the female gametes are thought to become fertilized at random by the two sperm cells (Messing & Grossniklaus, 1999). Due to the difficulties in isolating these secluded cell types, the hypothesis that egg and central cell have a distinct epigenome has not yet been extensively tested. But the fact that quite a few imprinted genes are specific to the endosperm suggests that this is indeed the case.

In summary, imprinting in plants seems to play several distinct roles and may have evolved due to a variety of selective pressures and advantages. The isolation of additional imprinted loci and ongoing investigations into the molecular mechanisms of genomic imprinting will certainly shed more light onto the function and evolutionary role of imprinted genes in plants.

Acknowledgments

I am indebted to Juan-Miguel Escobar-Restrepo for his help with the bibliography and to Amal Johnston for help with the figures. Célia Baroux provided the pictures shown in Plate 6.1C and D and Jerry Kermicle provided the kernels photographed in Plate 6.1A and B. I thank Peter Meyer, the editor of this volume, for his enormous patience. Our work on imprinting is funded by the University of Zürich, the Roche Research Foundation, the Swiss National Science Foundation and the EPIGENOME Network of Excellence of the European Union.

References

Adams, S., Vinkenoog, R., Spielman, M., Dickinson, H. G. & Scott, R. J. (2000) Parent-of-origin effects on seed development in *Arabidopsis thaliana* require DNA methylation. *Development*, **127**(11), 2493–502.

Alleman, M. & Doctor, J. (2000) Genomic imprinting in plants: observations and evolutionary implications. *Plant Molecular Biology*, **43**, 147–61.

Barlow, D. P., Stoger, R., Herrmann, B. G., Saito, K. & Schweifer, N. (1991) The mouse insulin-like growth factor type-2 receptor is imprinted and closely linked to the Tme locus. *Nature*, **349**(6304), 84–7.

Baroux, C., Blanvillain, R. & Gallois, P. (2001) Paternally inherited transgenes are down-regulated but retain low activity during early embryogenesis in *Arabidopsis*. *FEBS Letters*, **509**(1), 11–6.

Baroux, C., Spillane, C. & Grossniklaus, U. (2002a) Evolutionary origins of the endosperm in flowering plants. *Genome Biology*, **3**(9), 1026. Review.

Baroux, C., Spillane, C. & Grossniklaus, U. (2002b) Genomic imprinting during seed development. *Advances in Genetics*, **46**, 165–214. Review.

Bartolomei, M. S., Zemel, S. & Tilghman, S. M. (1991) Parental imprinting of the mouse H19 gene. *Nature*, **351**(6322), 153–5.

Barton, S. C., Surani, M. A. & Norris, M. L. (1984) Role of paternal and maternal genomes in mouse development. *Nature*, **311**(5984), 374–6.

Bestor, T. H. (2003) Cytosine methylation mediates sexual conflict. *Trends in Genetics*, **19**(4), 185–90.

Birchler, J. A. (1993) Dosage analysis of maize endosperm development. *Annual Review of Genetics*, **27**, 181–204.

Birchler, J. A., Bhadra, U., Bhadra, M. P. & Auger, D. L. (2001) Dosage-dependent gene regulation in multicellular eukaryotes: implications for dosage compensation, aneuploid syndromes, and quantitative traits. *Developmental Biology*, **234**(2), 275–88.

Birve, A., Sengupta, A. K., Beuchle, D., Larsson, J., Kennison, J. A., Rasmuson-Lestander, A. & Müller, J. (2001) *Su(z)12*, a novel *Drosophila Polycomb* group gene that is conserved in vertebrates and plants. *Development*, **128**(17), 3371–9.

Braidotti, G. (2001) RNA-FISH to analyze allele-specific expression. *Methods in Molecular Biology*, **181**, 169–80.

Brzeski, J. & Jerzmanowski, A. (2003) Deficient in DNA methylation 1 (DDM1) defines a novel family of chromatin-remodeling factors. *Journal of Biological Chemistry*, **278**(2), 823–8.

Cao, X. & Jacobsen, S. E. (2002) Role of the *Arabidopsis* DRM methyltransferases in *de novo* DNA methylation and gene silencing. *Current Biology*, **12**(13), 1138–44.

Cattanach, B. M. & Kirk, M. (1985) Differential activity of maternally and paternally derived chromosome regions in mice. *Nature*, **315**(6019), 496–8.

Chaudhuri, S. & Messing, J. (1994) Allele-specific parental imprinting of *dzr1*, a post-transcriptional regulator of zein accumulation. *Proceedings of the National Academy of Sciences of the United States of America*, **91**(11), 4867–71.

Chaudhury, A. M. & Berger, F. (2001) Maternal control of seed development. *Seminars in Cell and Developmental Biology*, **12**(5), 381–6. Review.

Chaudhury, A. M., Ming, L., Miller, C., Craig, S., Dennis, E. S. & Peacock, W. J. (1997) Fertilisation-independent seed development in *Arabidopsis thaliana*. *Proceedings of the National Academy of Sciences of the United States of America*, **94**(8), 4223–8.

Choi, Y., Gehring, M., Johnson, L., Hannon, M., Harada, J. J., Goldberg, R. B., Jacobsen, S. E. & Fischer, R. L. (2002) DEMETER, a DNA glycosylase domain protein, is required for endosperm gene imprinting and seed viability in *Arabidopsis*. *Cell*, **110**(1), 33–42.

Choi, Y., Harada, J. J., Goldberg, R. B. & Fischer, R. L. (2004) An invariant aspartic acid in the DNA glycosylase domain of DEMETER is necessary for transcriptional activation of the imprinted MEDEA gene. *Proceedings of the National Academy of Sciences of the United States of America*, **101**(19), 7481–6.

Colombo, L., Franken, J., van der Krol, A. R., Wittich, P. E., Dons, H. J & Angenent, G. C. (1997) Downregulation of ovule-specific MADS box genes from petunia results in maternally controlled defects in seed development. *Plant Cell*, **9**(5), 703–15.

Crouse, H. V. (1960) The controlling element in sex chromosome behavior in *Sciara*. *Genetics*, **45**, 1429–43.

Czermin, B., Melfi, R., McCabe, D., Seitz, V., Imhof, A. & Pirrotta V. (2002) *Drosophila* Enhancer of zeste/ESC complexes have a histone H3 methyltransferase activity that marks chromosomal *Polycomb* sites. *Cell*, **111**(2), 185–96.

Danilevskaya, O. N., Hermon, P., Hantke, S., Muszynski, M. G., Kollipara, K. & Ananiev, E. V. (2003) Duplicated *fie* genes in maize: expression pattern and imprinting suggest distinct functions. *Plant Cell*, **15**(2), 425–38.

DeChiara, T. M., Robertson, E. J. & Efstratiadis, A. (1991) Parental imprinting of the mouse insulin-like growth factor II gene. *Cell*, **64**(4), 849–59.

Delaval, K. & Feil, R. (2004) Epigenetic regulation of mammalian genomic imprinting, *Current Opinion in Genetics and Development*, **14**(2), 188–95.

Evans, M. M. & Kermicle, J. L. (2001) Interaction between maternal effect and zygotic effect mutations during maize seed development. *Genetics*, **159**(1), 303–15.

Felker, C., Peterson, D. M. & Nelson, O. E. (1985) Anatomy of immature grains of eight maternal effect shrunken endosperm barley mutants. *American Journal of Botany*, **72**, 248–56.

Ferguson-Smith, A., Lin, S. P., Tsai, C. E., Youngson, N. & Tevendale, M. (2003) Genomic imprinting – insights from studies in mice. *Seminars in Cell and Developmental Biology*, **14**(1), 43–9.

Finnegan, E. J., Peacock, W. J. & Dennis, E. S. (1996) Reduced DNA methylation in *Arabidopsis thaliana* results in abnormal plant development. *Proceedings of the National Academy of Sciences of the United States of America*, **93**, 8449–54.

Gehring, M., Choi, Y. & Fischer, R. L. (2004) Imprinting and seed development. *Plant Cell*, **16**(Suppl), S203–13.

Goday, C. & Esteban, M. R. (2001) Chromosome elimination in sciarid flies. *BioEssays*, **23**(3), 242–50.

Golden, T. A., Schauer, S. E., Lang, J. D., Pien, S., Mushegian, A. R., Grossniklaus, U., Meinke, D. W. & Ray, A. (2002) *SHORT INTEGUMENTS1/SUSPENSOR1/CARPEL FACTORY*, a Dicer homolog, is a maternal effect gene required for embryo development in *Arabidopsis*. *Plant Physiology*, **130**, 808–22.

Goodrich, J. & Tweedie, S. (2002) Remembrance of things past: chromatin remodeling in plant development. *Annual Review of Cell and Developmental Biology*, **18**, 707–46. Review.

Grimanelli D., Perotti, E., Ramirez, J. & Leblanc, O. (2005) Timing of the maternal-to-zygotic transition during early seed development in maize. *Plant Cell*, **17**, 1061–72.

Grini, P. E., Jürgens, G. & Hülskamp, M. (2002) Embryo and endosperm development is disrupted in the female gametophytic *capulet* mutants of *Arabidopsis*. *Genetics*, **162**(4), 1911–25.

Grossniklaus, U. & Schneitz, K. (1998) The molecular and genetic basis of ovule and megagametophyte development. *Seminars in Cell and Developmental Biology*, **9**(2), 227–38.

Grossniklaus, U., Moore, J. M. & Gagliano, W. B. (1998a) Molecular and genetic approaches to understanding and engineering apomixis: *Arabidopsis* as a powerful tool. In S. S. Virmani, E. A. Siddiq & K. Muralidharan (eds) *Advances in Hybrid Rice Technology*. Proceedings of the 3rd International Symposium on Hybrid Rice 1996, International Rice Research Institute, Manila, Philippines, pp. 187–211.

Grossniklaus, U., Vielle-Calzada, J. P., Hoeppner, M. A. & Gagliano, W. B. (1998b) Maternal control of embryogenesis by *MEDEA*, a *Polycomb* group gene in *Arabidopsis*. *Science*, **280**(5362), 46–450.

Grossniklaus, U., Spillane, C., Page, D. R. & Köhler, C. (2001) Genomic imprinting and seed development: endosperm formation with and without sex. *Current Opinion in Plant Biology*, **4**(1), 21–7.

Gualberti, G., Papi, M., Bellucci, L., Ricci, I., Bouchez, D., Camilleri, C., Costantino, P. & Vittorioso, P. (2002) Mutations in the *Dof* zinc finger genes *DAG2* and *DAG1* influence with opposite effects the germination of *Arabidopsis* seeds. *Plant Cell*, **14**(6), 1253–63.

Guitton, A. E, Page, D. R, Chambrier, P., Lionnet, C., Faure, J. E., Grossniklaus, U. & Berger, F. (2004) Identification of new members of *fertilisation independent seed Polycomb* group pathway involved in the control of seed development in *Arabidopsis thaliana*. *Development*, **131**(12), 2971–8.

Guo, M., Rupe, M. A., Danilevskaya, O. N, Yang, X. & Hu, Z. (2003) Genome-wide mRNA profiling reveals heterochronic allelic variation and a new imprinted gene in hybrid maize endosperm. *Plant Journal*, **36**(1), 30–44.

Gutierrez-Marcos, J. F., Pennington, P. D., Costa, L. M. & Dickinson, H. G. (2003) Imprinting in the endosperm: a possible role in preventing wide hybridisation. *Philosophical Transactions of the Royal Society of London. Series B, Biological Sciences*, **358**(1434), 1105–1111. Review.

Gutierrez-Marcos, J. F., Costa, L. M., Biderre-Petit, C., Khbaya, B., O'Sullivan, D. M., Wormald, M., Perez, P. & Dickinson, H. G. (2004) *Maternally expressed gene1* is a novel maize endosperm transfer cell-specific gene with a maternal parent-of-origin pattern of expression. *Plant Cell*, **16**(5), 1288–301.

Haig, D. & Westoby, M. (1989) Parent specific gene expression and the triploid endosperm. *American Naturalist*, **134**, 147–55.

Haig, D. & Westoby, M. (1991) Genomic imprinting in endosperm: its effect on seed development in crosses between species, and between different ploidies of the same species, and its implications for the evolution of apomixes. *Philosophical Transactions of the Royal Society of London. Series B, Biological Sciences*, **333**, 1–13.

He, Y. & Amasino, R. M. (2005) Role of chromatin modification in flowering-time control. *Trends in Plant Science*, **10**(1), 30–35.

Herrick, G. & Seger, J. (1999) Imprinting and paternal elimination in insects. *Results and Problems in Cell Differentiation*, **25**, 41–71.

Hsieh, T. F., Hakim, O., Ohad, N. & Fischer, R. L. (2003) From flour to flower: how *Polycomb* group proteins influence multiple aspects of plant development. *Trends in Plant Science*, **8**(9), 439–45. Review.

Hurst, L. D. & McVean, G. T. (1998) Do we understand the evolution of genomic imprinting? *Current Opinion in Genetics and Development*, **8**(6), 701–708.

Jackson, J. P., Lindroth, A. M., Cao, X. & Jacobsen, S. E. (2002) Control of CpNpG DNA methylation by the KRYPTONITE histone H3 methyltransferase. *Nature*, **416**, 556–60.

Jeddeloh, J. A., Stokes, T. L. & Richards, E. J. (1999) Maintenance of genomic methylation requires a SWI2/SNF2-like protein. *Nature Genetics*, **22**(1), 94–7.

Jones, R. S. & Gelbart, W. M. (1993) The *Drosophila Polycomb*-group gene *Enhancer of zeste* contains a region with sequence similarity to *trithorax*. *Molecular and Cellular Biology*, **13**(10), 6357–66.

Jost, J. P., Siegmann, M., Sun, L. & Leung, R. (1995) Mechanisms of DNA demethylation in chicken embryos. Purification and properties of a 5-methylcytosine-DNA glycosylase. *The Journal of Biological Chemistry*, **270**(17), 9734–9.

Kankel, M. W., Ramsey, D. E., Stokes, T. L., Flowers, S. K., Haag, J. R., Jeddeloh, J. A., Riddle, N. C., Verbsky, M. L. & Richards, E. J. (2003) *Arabidopsis* MET1 cytosine methyltransferase mutants. *Genetics*, **163**(3), 1109–22.

Kermicle, J. L. (1969) Androgenesis conditioned by a mutation in maize. *Science*, **166**(3911), 1422–4.

Kermicle, J. L. (1970) Dependance of the *R*-mottled aleurone phenotype in maize on mode of sexual transmission. *Genetics*, **66**, 69–85.

Kermicle, J. L. (1978) Imprinting of gene action in maize endosperm. In D. B. Walden (ed.) *Maize Breeding and Genetics*. Wiley, New York, pp. 357–71.

Kermicle, J. L. (1996) Epigenetic silencing and activation of a maize *r* gene. In V. E. A. Russo, R. A. Martiensson & A. D Riggs (eds) *Epigenetic Mechanisms of Gene Regulation*. Cold Spring Harbor Press, Cold Spring Harbour, USA, pp. 267–88.

Kermicle, J. L. & Alleman, M. (1990) Genomic imprinting in maize in relation to the angiosperm life cycle. *Development*, (Suppl), 9–14.

Kimber, G. & Riley, R. (1963) Haploid angiosperms. *The Botanical Review*, **29**, 480–531.

Kinoshita, T., Yadegari, R., Harada, J. H., Goldberg, R. B. & Fisher, R. L (1999) Imprinting of the *MEDEA Polycomb* gene in the *Arabidopsis* endosperm. *Plant Cell*, **11**(10), 1945–52.

Kinoshita, T., Miura, A., Choi, Y., Kinoshita, Y., Cao, X., Jacobsen, S. E., Fischer, R. L. & Kakutani, T. (2004) One-way control of *FWA* imprinting in *Arabidopsis* endosperm by DNA methylation. *Science*, **303**(5657), 521–3.

Kiyosue, T., Ohad, N., Yadegri, E., Hannon, M., Dinnery, J., Wells, D., Katz, A., Margossian, L., Harada, J., Goldberg, R. B. & Fisher, R.L (1999) Control of fertilisation-independent endosperm development by the *MEDEA Polycomb* gene in *Arabidopsis*. *Proceedings of the National Academy of Sciences of the United States of America*, **96**(7), 4186–91.

Köhler, C. & Grossniklaus, U. (2002) Epigenetic inheritance of expression states in plant development: the role of *Polycomb* group proteins. *Current Opinion in Cell Biology*, **14**(6), 773–9. Review.

Köhler, C. & Grossniklaus, U. (2005) Seed development and genomic imprinting in plants. *Progress in Molecular and Subcellular Biology*, **38**, 237–62. Review.

Köhler, C., Hennig, L., Bouveret, R., Gheyselinck, J., Grossniklaus, U. & Gruissem, W. (2003) *Arabidopsis* MSI1 is a component of the MEA/FIE *Polycomb* group complex and required for seed development. *EMBO Journal*, **22**(18), 4804–14.

Köhler, C., Page, D. R., Gagliardini, V. & Grossniklaus, U. (2005) The *Arabidopsis thaliana* MEDEA *Polycomb* group protein controls expression of *PHERES1* by parental imprinting. *Nature Genetics*, **37**(1), 28–30.

Koltunow, A. M. & Grossniklaus, U. (2003) Apomixis: a developmental perspective. *Annual Reviews of Plant Biology*, **54**, 547–74. Review.

Koornneef, M., Hanhart, C. J. & van der Veen, J. H. (1991) A genetic and physiological analysis of late flowering mutants in *Arabidopsis thaliana*. *Molecular and General Genetics*, **229**(1), 57–66.

Kuzmichev, A., Nishioka, K., Erdjument-Bromage, H., Tempst, P. & Reinberg, D. (2002) Histone methyl-transferase activity associated with a human multiprotein complex containing the Enhancer of zeste protein. *Genes and Development*, **16**, 2893–905.

Levine, S. S., King, I. F. & Kingston, R. E. (2004) Division of labor in *Polycomb* group repression. *Trends in Biochemical Sciences*, **29**(9), 478–85. Review.

Lewis, A., Mitsuya, K., Umlauf, D., Smith, P., Dean, W., Walter, J., Higgins, M., Feil, R. & Reik, W. (2004) Imprinting on distal chromosome 7 in the placenta involves repressive histone methylation independent of DNA methylation. *Nature Genetics*, **36**(12), 1291–5.

Lin, B. -Y. (1982) Association of endosperm reduction with parental imprinting in maize. *Genetics*, **100**, 475–86.

Lin, B. -Y. (1984) Ploidy barrier to endosperm development in maize. *Genetics*, **107**, 103–15.

Lippman, Z., Gendrel, A. V., Black, M., Vaughn, M. W., Dedhia, N., McCombie, W. R., Lavine, K., Mittal, V., May, B., Kasschau, K. D., Carrington, J. C., Doerge, R. W., Colot, V. & Martienssen, R. (2004) Role of transposable elements in heterochromatin and epigenetic control. *Nature*, **430**(6998), 471–6.

Lund, G., Ciceri, P. & Viotti, A. (1995a) Maternal-specific demethylation and expression of specific alleles of zein genes in the endosperm of *Zea mays* L. *Plant Journal*, **8**(4), 571–81.

Lund, G., Messing, J. & Viotti, A. (1995b) Endosperm-specific demethylation and activation of specific alleles of alpha-tubulin genes of *Zea mays* L. *Molecular and General Genetics*, **246**(6), 716–22.

Luo, M., Bilodeau, P., Koltunow, A., Dennis, E. S., Peacock, W. J. & Chaudhury, A. (1999) Genes controlling fertilization-independent seed development in *Arabidopsis thaliana*. *Proceedings of the National Academy of Sciences of the United States of America*, **96**(1), 296–301.

Luo, M., Bilodeau, P., Dennis, E. S., Peacock, J. & Chaudhury, A. (2000) Expression and parent-of-origin effects for *FIS2*, *MEA* and *FIE* in the endosperm of developing *Arabidopsis* seeds. *Proceedings of the National Academy of Sciences of the United States of America*, **97**(19), 10637–42.

Mager, J., Montgomery, N. D., de Villena, F. P. & Magnuson, T. (2003) Genome imprinting regulated by the mouse *Polycomb* group protein Eed. *Nature Genetics*, **33**(4), 502–507.

Martienssen, R. (1996) Epigenetic phenomena: paramutation and gene silencing in plants. *Current Biology*, **6**(7), 810–13. Review.

Matzke, M. A., Matzke A. J. M. & Eggleston, W. B. (1996) Paramutation and transgene silencing: a common response to invasive DNA? *Trends in Plant Science*, **1**, 382–8.

McCormick S. (2004) Control of male gametophyte development. *Plant Cell*, **16**(Suppl), S142–53.

McGrath, J. & Solter, D. (1984) Completion of mouse embryogenesis requires both the maternal and paternal genomes. *Cell*, **37**(1), 179–83.

Messing, J. & Grossniklaus, U. (1999) Genomic imprinting in plants. *Results and Problems in Cell Differentiation*, **25**, 23–40. Review.

Moore, M. M. (2002) Isolation and characterization of female gametophytic mutants in *Arabidopsis thaliana*. PhD thesis, State University of New York, Stony Brook.

Moore, T. (2001) Genetic conflict, genomic imprinting and establishment of the epigenotype in relation to growth. *Reproduction*, **122**(2), 185–93. Review.

Moore, T. & Haig, D. (1991) Genomic imprinting in mammalian development: a parental tug-of-war. *Trends in Genetics*, **7**, 45–9.

Mordhorst, A. P., Toonen M. A. J. &. de Vries S. C. (1997) Plant embryogenesis. *Critical Reviews in Plant Science*, **16**, 535–76.

Morison, I. M., Paton, C. J. & Cleverley, S. D. (2001) The imprinted gene and parent-of-origin effect database. *Nucleic Acids Research*, **29**(1), 275–6.

Müller, J., Hart, C. M., Francis, N. J., Vargas, M. L., Sengupta, A., Wild, B., Miller, E. L., O'Connor, M. B., Kingston, R. E. & Simon, J. A. (2002) Histone methyltransferase activity of a *Drosophila Polycomb* group repressor complex. *Cell*, **111**(2), 197–208.

Nogler, G. A. (1984) Gametophytic apomixis. In B. M. Johri (ed.) *Embryology of Angiosperms*. Springer, Berlin Heidelberg New York, pp. 475–518.

Ohad, N., Margossian, L., Hsu, Y. C., Williams, C., Repetti, P. & Fisher, R. L. (1996) A mutation that allows endosperm development without fertilisation. *Proceedings of the National Academy of Sciences of the United States of America*, **93**(11), 5319–24.

Ohad, N., Yadegari, R., Margossian, L., Hannon, M., Michaeli, D., Harada, J. J., Goldberg, R. B. & Fischer, R. L. (1999) Mutations in *FIE*, a WD *Polycomb* group gene, allow endosperm development without fertilization. *Plant Cell*, **11**(3), 407–16.

Olsen, O. A. (2004) Nuclear endosperm development in cereals and *Arabidopsis thaliana*. *Plant Cell*, **16**(Suppl), S214–27.

Orlando, V. (2003) *Polycomb*, epigenomes, and control of cell identity. *Cell*, **112**(5), 599–606. Review.

Page, D. R. (2004) Maternal effects during seed development: an expression analysis of the *FERTILIZATION-INDEPENDENT SEED (FIS)* class gene *MEDEA* and a search for new *fis* class mutants. PhD thesis, University of Zürich, Switzerland.

Peacock, J., Luo, M., Craig, S., Dennis, E. & Chaudhury, A. (1995) A mutagenesis programme for apomixis genes in *Arabidopsis*. In *Induced Mutations and Molecular Techniques for Crop Improvement*. International Atomic Energy Agency, Vienna, pp. 117–25.

Peters, A. H., Mermoud, J. E., O'Carroll, D., Pagani, M., Schweizer, D., Brockdorff, N. & Jenuwein, T. (2002) Histone H3 lysine 9 methylation is an epigenetic imprint of facultative heterochromatin. *Nature Genetics*, **30**, 77–80.

Ray, S., Golden, T. & Ray, A. (1996) Maternal effects of the short integument mutation on embryo development in *Arabidopsis*. *Developmental Biology*, **180**(1), 365–9.

Rea, S., Eisenhaber, F., O'Carroll, D., Strahl, B. D., Sun, Z. W., Schmid, M., Opravil, S., Mechtler, K., Ponting, C. P., Allis, C. D. & Jenuwein, T. (2000) Regulation of chromatin structure by site-specific histone H3 methyltransferases. *Nature*, **406**, 593–9.

Redei, G. P. (1964) Crossing experiences with polyploids. *http://www.arabidopsis.org/ais/1964/redei-1964-aagkf.html*

Reik, W. & Lewis, A. (2005) Co-evolution of X-chromosome inactivation and imprinting in mammals. *Nature Reviews. Genetics*, **6**(5), 403–10.

Reyes, J. C. & Grossniklaus, U. (2003) Diverse functions of *Polycomb* group proteins during plant development. *Seminars in Cell and Developmental Biology*, **14**(1), 77–84. Review.

Ringrose, L. & Paro, R. (2004) Epigenetic regulation of cellular memory by the *Polycomb* and *trithorax* group proteins. *Annual Review of Genetics*, **38**, 413–43.

Sarkar, K. R. & Coe, E. H. (1966) A genetic analysis of origin of maternal haploids in maize. *Genetics*, **54**, 453.

Saze, H., Scheid, O. M. & Paszkowski, J. (2003) Maintenance of CpG methylation is essential for epigenetic inheritance during plant gametogenesis. *Nature Genetics*, **34**(1), 65–9.

Scharer, O. D. & Jiricny, J. (2001) Recent progress in the biology, chemistry and structural biology of DNA glycosylases. *BioEssays*, **23**(3), 270–81.

Shermer, R. & Razin, A. (1996) Establishment of imprinted methylation patterns during development. In V. E. A. Russo, R. A. Martienssen & A. D. Riggs (eds) Cold Spring Harbor Press, Cold Spring Harbour, USA, pp. 215–30.

Scott, R. J. & Spielman, M. (2004) Epigenetics: imprinting in plants and mammals—the same but different? *Current Biology*, **14**(5), R201–203. Review.

Scott, R. J., Spielman, M., Bailey J. & Dickinson H. G. (1998) Parent-of-origin effects on seed development in *Arabidopsis thaliana*. *Development*, **125**(17), 3329–41.

Silva, J., Mak, W., Zvetkova, I., Appanah, R., Nesterova, T. B., Webster, Z., Peters, A. H., Jenuwein, T., Otte, A. P. & Brockdorff, N. (2003) Establishment of histone h3 methylation on the inactive X chromosome requires transient recruitment of Eed-Enx1 Polycomb group complexes. *Developmental Cell*, **4**(4), 481–95.

Sleutels, F. & Barlow, D. (2002) The origins of genomic imprinting in mammals. *Advances in Genetics*, **46**, 119–63. Review.

Soppe, W. J., Jacobsen, S. E., Alonso-Blanco, C., Jackson, J. P., Kakutani, T., Koornneef, M. & Peeters, A. J. (2000) The late flowering phenotype of *fwa* mutants is caused by gain-of-function epigenetic alleles of a homeodomain gene. *Molecular Cell*, **6**(4), 791–802.

Sorensen, M. B., Chaudhury, A. M., Robert, H., Bancharel, E. & Berger, F. (2001) *Polycomb* group genes control pattern formation in plant seed. *Current Biology*, **11**(4), 277–81.

Spielman, M., Vinkenoog, R., Dickinson, H. G. & Scott, R. J. (2001) The epigenetic basis of gender in flowering plants and mammals. *Trends in Genetics*, **17**(12), 705–11.

Spillane, C., MacDougall, C., Stock, C., Köhler, C., Vielle-Calzada, J. P., Nunes, S. M., Grossniklaus, U. & Goodrich, J. (2000) Interaction of the *Arabidopsis Polycomb* group proteins FIE and MEA mediates their common phenotypes. *Current Biology*, **10**(23), 1535–8.

Spillane, C., Vielle-Calzada, J. P. & Grossniklaus, U. (2002) Parent-of-origin effects and seed development: genetics and epigenetics. In Y. H. Hui, G. G. Khachatourians, W. K. Nip & R. Scorza (eds) *Handbook of Transgenic Food*. Marcel Dekker, New York, pp. 109–36.

Spillane, C., Baroux, C., Escobar-Restrepo, J.-M., Page, D. R., Laoueille, S. & Grossniklaus, U. (2004a) Transposons and tandem repeats are not involved in the control of genomic imprinting at the *MEDEA* locus in *Arabidopsis*. *Cold Spring Harbor Symposia on Quantitative Biology*, **69** (in press).

Spillane, C., Curtis, M. D. & Grossniklaus, U. (2004b) Apomixis technology development-virgin births in farmers' fields? *Nature Biotechnology*, **22**(6), 687–91.

Springer, P. S., Holding, D. R., Groover, A., Yordan, C. & Martienssen, R. A. (2000) The essential Mcm7 protein PROLIFERA is localized to the nucleus of dividing cells during the G(1) phase and is required maternally for early *Arabidopsis* development. *Development*, **127**(9), 1815–22.

Surani, M. A. H., Barton, S. C. & Norris, M. L. (1984) Development of reconstituted mouse eggs suggests imprinting of the genome during gametogenesis. *Nature*, **308**(5959), 548–50.

Tilghman, S. M. (1999) The sins of the fathers and mothers: genomic imprinting in mammalian development. *Cell*, **96**, 185–93.

Tschiersch, B., Hofmann, A., Krauss, V., Dorn, R., Korge, G. & Reuter, G. (1994) The protein encoded by the *Drosophila* position-effect variegation suppressor gene *Su(var)3–9* combines domains of antagonistic regulators of homeotic gene complexes. *EMBO Journal*, **13**(16), 3822–31.

Vielle-Calzada, J. P., Thomas, J., Spillane, C., Coluccio, A., Hoeppner, M. A. & Grossniklaus, U. (1999) Maintenance of genomic imprinting at the *Arabidopsis medea* locus requires zygotic *DDM1* activity. *Genes & Development*, **13**(22), 2971–82.

Vielle-Calzada, J. P., Baskar, R. & Grossniklaus, U. (2000) Delayed activation of the paternal genome during seed development. *Nature*, **404**(6773), 91–4.

Vinkenoog, R., Spielman, M., Adams, S., Fisher, R. L., Dickinson, H. G. & Scott, R. J. S. (2000) Hypo-methylation promotes autonomous endosperm development and rescues postfertilization lethality in *fie* mutants. *Plant Cell*, **12**(11), 2271–82.

Vinkenoog, R., Bushell, C., Spielman, M., Adams, S., Dickinson, H. G. & Scott, R. J. (2003) Genomic imprinting and endosperm development in flowering plants. *Molecular Biotechnology*, **25**(2), 149–84. Review.

Vongs, A., Kakutani, T., Martienssen, R. A. & Richards, E. J. (1993) *Arabidopsis thaliana* DNA methylation mutants. *Science*, **260**(5116), 1926–8.

von Wangenheim, K. H. & Peterson, H. P. (2004) Aberrant endosperm development in interploidy crosses reveals a timer of differentiation. *Developmental Biology*, **270**(2), 277–89.

Wang, J., Mager, J., Chen, Y., Schneider, E., Cross, J. C., Nagy, A. & Magnuson, T. (2001) Imprinted X inactivation maintained by a mouse Polycomb group gene. *Nature Genetics*, **28**(4), 371–5.

Wang, Y., Fischle, W., Cheung, W., Jacobs, S., Khorasanizadeh, S. & Allis, C. D. (2004) Beyond the double helix: writing and reading the histone code. *Novartis Foundation Symposium*, **259**, 3–17, discussion 17–21, 163–9. Review.

Weijers, D., Geldner, N., Offringa, R., & Jurgens, G. (2001) Seed development: early paternal gene activity in *Arabidopsis*. *Nature*, **414**(6965), 709–710.

Weterings, K. & Russell, S. D. (2004) Experimental analysis of the fertilization process. *Plant Cell*, **16**(Suppl), S107–118.

Wilkins, J. F. & Haig, D. (2003) What good is genomic imprinting: the function of parent-specific gene expression. *Nature Reviews. Genetics*, **4**(5), 359–68. Review.

Xiao, W., Gehring, M., Choi, Y., Margossian, L., Pu, H., Harada, J. J., Goldberg, R. B., Pennell, R. I. & Fischer, R. L. (2003) Imprinting of the MEA Polycomb gene is controlled by antagonism between MET1 methyltransferase and DME glycosylase. *Developmental Cell*, **5**(6), 891–901.

Yadegari, R. & Drews, G. N. (2004) Female gametophyte development. *Plant Cell*, **16**(Suppl), S133–41. Review.

Yadegari, R., Kinoshita, T., Lotan, O., Cohen, G., Katz, A., Choi, Y., Katz, A., Nakashima, K., Harada, J. J., Goldberg, R. B., Fischer, R. L. & Ohad, N. (2000) Mutations in the *FIE* and *MEA* genes that encode interacting *Polycomb* proteins cause parent-of-origin effects on seed development by distinct mechanisms. *Plant Cell*, **12**(12), 2367–81.

Zhu, B., Zheng, Y., Angliker, H., Schwarz, S., Thiry, S., Siegmann, M. & Jost J, P. (2000) 5-Methylcytosine DNA glycosylase activity is also present in the human MBD4 (G/T mismatch glycosylase) and in a related avian sequence. *Nucleic Acids Research*, **28**(21), 4157–65.

7 Nucleolar dominance and rRNA gene dosage control: a paradigm for transcriptional regulation via an epigenetic on/off switch

Nuno Neves, Wanda Viegas and Craig S. Pikaard

7.1 Introduction

In eukaryotes, four ribosomal RNAs (rRNAs) and ~85 proteins are needed to build ribosomes, the molecular machines responsible for all protein synthesis. The regulation of ribosome biogenesis is therefore crucial for regulating cell growth and is closely tied to the mechanisms that regulate proliferation (Moss & Stefanovsky, 2002; Warner, 1999). Building a eukaryotic ribosome is a complicated enterprise, requiring the coordination of all three nuclear DNA-dependent RNA polymerase transcription systems, namely RNA polymerases I, II and III. RNA polymerase I is responsible for transcribing a large family of essentially identical genes, each of which encodes a 45S (S = Svedberg unit; describes sedimentation behavior upon ultracentrifugation) primary transcript that is subsequently processed into three of the four rRNAs that form the catalytic core of the ribosome, namely rRNAs of 18S, 5.8S and 25/28S (25S in plants and yeast; 28S in mammals; Pederson & Politz, 2000) (Grummt, 1999). By convention, the genes transcribed by RNA polymerase I are known as rRNA genes or, collectively, as rDNA. RNA polymerase II is responsible for the transcription of the ribosomal protein genes. RNA polymerase III transcribes the repetitive genes encoding the fourth structural RNA of the ribosomes, which is 5S in size; thus these genes are known by convention as 5S genes. This chapter focuses on the regulation of the rRNA genes (rDNA) transcribed by RNA polymerase I.

rRNA genes are present in hundreds to thousands of copies in eukaryotic cells, organized as head-to-tail repeats at one or more chromosomal loci that span millions of base pairs (Figure 7.1A). Transcription of the rRNA genes by RNA polymerase I gives rise to the nucleolus, the most conspicuous compartment of the nucleus and the cell's factory for ribosome assembly (Hernandez-Verdun et al., 2002; Leung & Lamond, 2003; Schwarzacher & Mosgoeller, 2000; Shaw & Jordan, 1995) (Figure 7.1B). As a result, the loci where rRNA genes are localized are termed nucleolar organizing regions, or NORs (McClintock, 1934). Several recent reviews have discussed the organization and transcriptional features of rRNA genes and their central role in cellular metabolism (Grummt, 2003; Moss & Stefanovsky, 2002; Neves et al., 2005;

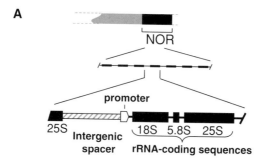

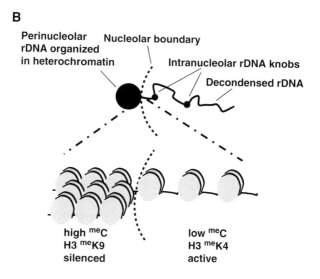

Figure 7.1 Molecular and cytological organization of rRNA gene arrays. (A) At a nucleolus organizer region (NOR), rRNA genes are clustered in long tandem arrays of head-to-tail repeats. Each transcription unit, which encodes a precursor processed into 18S, 5.8S and 25S rRNA, is separated from the next transcription unit by an intergenic spacer. (B) In interphase cells, most of the rRNA genes are condensed in heterochromatin at the periphery of the nucleolus. This condensed perinucleolar rDNA is made up of inactive rRNA genes. Within the nucleolus, active rRNA gene chromatin is decondensed. However, intranucleolar condensed rDNA knobs occur in some species, suggesting that contiguous stretches of active rRNA genes can be punctuated by inactive genes assembled into heterochromatin.

Pikaard, 2002). Likewise, we refer the interested reader to several excellent reviews concerning the organization and functions of the nucleolus (Hernandez-Verdun & Roussel, 2003; Hernandez-Verdun *et al.*, 2002; Leung & Lamond, 2003; Scheer & Weisenberger, 1994; Shaw & Jordan, 1995). These topics are beyond the scope of this chapter, which focuses on the epigenetic regulation of rRNA gene transcription.

7.2   Ribosomal RNA gene dosage control

An unexplained observation is that eukaryotic cells appear to have more rRNA genes than they need, such that only a subset of the available rRNA genes are active at any given time. This appears to be the case from yeast to humans, which have ~140 to ~400 rRNA genes, respectively, and is especially true in plants, which typically have thousands of rRNA genes (Flavell, 1986; Gerbi, 1985; Reeder, 1974; Rogers & Bendich, 1987). In resting cells, rRNA transcripts are produced at very low levels, probably due to the fact that protein synthesis is not occurring at a rapid rate, leading to a lesser need for ribosomes. Conversely, in active growing cells rRNA gene transcription can account for as much as 80% of total gene expression (Jacob, 1995; Moss & Stefanovsky, 2002). However, even in proliferating cells, 50% or more of the rRNA genes are not active. This conclusion is apparent from multiple lines of evidence (reviewed in Grummt & Pikaard, 2003). For instance, chicken cells with two, three or four NORs synthesize the same amounts of rRNA despite the changes in rRNA gene dosage (Muscarella *et al.*, 1985, 1987). A cytological study using antibodies against RNA polymerase I estimated that the number of transcriptionally active rRNA genes in vertebrates ranges from 45 to 145, depending on the species, although each cell possesses many hundreds of rRNA genes (Haaf *et al.*, 1991). This estimate agrees with a study of growing HeLa cells in which BrUTP incorporation suggested that only about 120 rRNA genes are being actively transcribed (Jackson *et al.*, 1998). In plants, three-dimensional electron microscopy reconstructions of BrUTP-labeled *Pisum sativum* nuclei in proliferating meristematic cells found that the number of rDNA transcription sites is between 250 and 300 (González-Melendi *et al.*, 2001), corresponding to fewer than 5% of the total number of rRNA genes in this species. Other elegant studies in proliferating yeast and murine cells, using accessibility to psoralen cross-linking as a means for discriminating active from inactive genes, showed that only one-half to one-third of the rDNA units are in a psoralen-accessible conformation that reflects active transcription (Conconi *et al.*, 1989; Dammann *et al.*, 1993, 1995). Yet another line of evidence comes from examination of rRNA gene arrays by electron microscopy (McKnight & Miller, 1976; Miller & Beatty, 1969), which reveals that only a subset of the rRNA genes are active and engaged by transcribing RNA polymerase I elongation complexes.

Whereas active and inactive rRNA genes are interspersed with one another in yeast (Dammann *et al.*, 1995; French *et al.*, 2003; Sandmeier *et al.*, 2002), the active and inactive subsets of rRNA genes are likely to be more compartmentalized within the NORs of multi-cellular eukaryotes, such as plants. Molecular cytogenetic studies in diploid rye show that only the distal part of each NOR (the portion furthest from the centromere) contains active rRNA genes (Caperta *et al.*, 2002). At metaphase, these genes associate with proteins that stain intensely with silver and are reliable markers of transcribed, but not silent, rRNA genes (Goodpasture & Bloom, 1975; Jimenez *et al.*, 1988). By contrast, centromere-proximal rRNA genes within the rye NORs are silver-negative and are condensed within heterochromatic knobs that persist through-out interphase and that localize at the periphery of the nucleolus. A similar topological organization of rRNA genes is observed in many plants. For instance, pea, wheat and *Arabidopsis* all have conspicuous perinucleolar knobs of heterochromatin that contain rDNA (Figure 7.1B) (Leitch *et al.*, 1992; Pontes *et al.*, 2003; Shaw *et al.*, 1993). These rDNA knobs display a compact chromatin conformation that is presumably inaccessible to the tran-scription machinery, and no BrUTP is incorporated into nascent RNA transcripts at the knobs, unlike decondensed portions of the NOR (M. Silva, N. Neves and W. Viegas, unpublished data). Interestingly, Shaw *et al.* (1993) demonstrated in pea that rDNA perinucleolar knobs could undergo decondensation and incorporation into the central region of the nucleolus, leading to increased nucleolar volume, reflecting an increased transcriptional activity of rRNA genes. Similarly, chemical or genetic changes that are known to dere-press silenced rRNA genes in *Arabidopsis* can be correlated with rDNA decondensation (Pontes *et al.*, 2003) (unpublished data of the Viegas and Pikaard laboratories).

Collectively, the available evidence suggests that eukaryotes tend to have large excesses of rRNA genes relative to the number of genes that are required for ribosome biogenesis and that cells are somehow able to regulate the effective dosage of their rRNA genes. Hypothetically, the cell could achieve such modulation of rRNA synthesis by controlling the number of transcripts made from a constant subset of rRNA genes. Alternatively, because there are hundreds of rRNA genes (thousands in plants), genes might be turned 'on' or 'off', as needed. There is evidence that both strategies are employed. Collectively, the available evidence favors regulation at two levels, first by controlling the number of rRNA genes in the 'on' or 'off' states ('dosage control') and subsequently by modulating the frequency of pol I initiation among the active subset (for a review, see Grummt & Pikaard, 2003). The latter fine-tuning of pol I initiation frequency is most likely accomplished by growth-regulated modifications of pol I tran-scription factors (Cavanaugh *et al.*, 2002; Hannan *et al.*, 2003; Milkereit & Tschochner, 1998; Pelletier *et al.*, 2000; Peyroche *et al.*, 2000; Stefanovsky *et al.*, 2001). The mechanisms responsible for switching rRNA genes on or off within an NOR are only recently coming to light, with studies of nucleolar dominance in plants playing an important role in elucidating the switch mechanisms.

7.3 Nucleolar dominance

Nucleolar dominance is a striking phenomenon that occurs in genetic hybrids of both plants and animals, but was first discovered in plants. Navashin (1934) noticed that specific *Crepis* chromosomes display secondary constrictions at metaphase but that this distinctive cytogenetic feature was often missing from the appropriate chromosome of one parent in hybrids resulting from crossing two related species. Navashin noted that the secondary constriction was always lost from chromosomes of the same species, regardless of whether that species served as the maternal or the paternal parent in the cross. Navashin also found that by backcrossing the *Crepis* hybrid with the underdominant parent (the species whose chromosomes lack secondary constriction in the hybrid), the underdominant chromosomes contributed by the hybrid would once again form secondary constrictions in the backcross progeny. These results indicated that hybridization had not caused a permanent loss and/or damage of the chromosome and indicated that the loss of secondary constriction formation was a reversible phenomenon.

Navashin's contemporary, Heitz (1931) showed that secondary constrictions occur at or very near the sites where nucleoli form. McClintock (1934) later proved this to be the case by showing that a chromosome break at the single maize secondary constriction, followed by translocation of the two chromosome pieces to other chromosomes, caused the novel formation of two nucleoli and two secondary constrictions on the recombinant chromosomes. McClintock realized that the original *locus* must have contained redundant information in order to be divisible and she was the first to call the *locus* the nucleolar organizer. Decades later, it became clear that NORs are the *loci* where rRNA genes are located in long tandem arrays, that secondary constrictions correspond to sites of active rRNA gene transcription and that nucleolar dominance is due to the transcription of only one parental set of rRNA genes (Honjo & Reeder, 1973; Phillips *et al.*, 1971; Ritossa & Spiegelman, 1965; Wallace & Birnstiel, 1966; Wallace & Langridge, 1971).

The mechanisms responsible for NOR choice and differential activity in nucleolar dominance are unclear. Because nucleolar dominance occurs independent of maternal or paternal effects, this suggests that gametic imprinting (Brannan & Bartolomei, 1999; Huynh & Lee, 2001; Sleutels *et al.*, 2000) does not dictate NOR choice in a hybrid. Likewise, NOR inactivation is not random, as is the case for X chromosome inactivation in somatic cells of female mammals (Heard *et al.*, 1997; Lee & Jaenisch, 1997). Instead, the same parental set of rRNA genes is always silenced. Nucleolar dominance is also not dependent simply on the number of rRNA genes contributed by each parent (Chen & Pikaard, 1997a; Flavell, 1986; Flavell & O'Dell, 1979; Navashin, 1934). An early hypothesis stemming from the rapid co-evolution of rRNA gene sequences and transcription factors was that inactivation of a species-specific transcription factor in a hybrid could inactivate the matching set of rRNA genes (Grummt *et al.*, 1982; Miesfeld & Arnheim, 1984; Miesfeld *et al.*, 1984; Learned *et al.*, 1985). Support for this hypothesis came from cell cultures derived from mouse–human cell fusions that expressed the mouse or human rRNA genes, but not both, and from *in vitro* transcription experiments that defined at least one species-specific pol I transcription factor. However, the species-specific transcription factor hypothesis fails to explain nucleolar dominance in natural hybrids. Specifically, underdominant *Brassica* or *Arabidopsis* rRNA gene promoters were shown to be fully functional when transfected into protoplasts of the dominant species, showing that rRNA genes of either parent can use the same transcription machinery (Chen *et al.*, 1998; Frieman *et al.*, 1999). Another hypothesis, supported by oocyte injection experiments in *Xenopus*, suggested that dominant genes might have superior binding affinities for transcription factors present in limiting amounts (Caudy & Pikaard, 2002; Reeder, 1985; Reeder & Roan, 1984). However, direct tests in plants have failed to reveal any differences in the competitive strength of dominant and underdominant genes for transcription factors, either *in vivo* or *in vitro* (Chen *et al.*, 1998; Frieman *et al.*, 1999).

Instead of dominant genes being preferentially activated, it now appears that underdominant genes are preferentially silenced. This change in thinking stems from the demonstration that silenced rRNA genes and NORs can be derepressed by chemical inhibitors of cytosine methylation (Amado *et al.*, 1997; Chen & Pikaard, 1997b) or histone deacetylation (Chen & Pikaard, 1997b). Importantly, derepression using both types of chemicals together is no more effective than with either compound alone, suggesting that cytosine methylation and histone deacetylation are partners in the same gene-silencing pathway (Chen & Pikaard, 1997b). It is now clear that blocking cytosine methylation or histone deacetylation induces the same set of chromatin modifications, coincident with transcriptional derepression (Lawrence *et al.*, 2004). Furthermore, it is apparent that similar mechanisms silence rRNA genes in non-hybrids and hybrids alike, leading to the hypothesis that dosage control and nucleolar dominance are aspects of the same regulatory mechanisms. These ideas are discussed in more detail in the sections that follow.

7.4   DNA methylation and rRNA gene regulation

In those eukaryotes that methylate their DNA, cytosine methylation plays an important role in the genome. This is illustrated by the fact that defects in the cytosine methylation machinery can result in embryonic lethality in mammals or pleiotropic developmental defects in plants. Cytosine methylation is thought to provide defense against transposable elements and retroviruses by inactivating these elements (Bestor, 1998; Bird, 2002; Martienssen & Colot, 2001; Richards, 1997) and plays an important role in heterochromatin maintenance. Methylation-induced transcriptional silencing can result from inhibiting activator protein–DNA interactions and/or from recruitment of methylcytosine-binding domain proteins (MBDs) that associate with, or re-cruit, protein complexes that establish dense heterochromatic structures that are incompatible with transcription.

Correlations between cytosine methylation, chromatin accessibility and rRNA gene activity have been noted for two decades. For instance, in hex-aploid bread wheat, whose genome was formed by the merger of three distinct ancestral diploid genomes, rRNA genes within the most active NORs were shown to be less methylated and more accessible to DNase I than were rRNA genes at inactive NORs (Flavell *et al.*, 1988; Sardana *et al.*, 1993; Thompson & Flavell, 1988). Similarly in maize, an undermethylated subset of rRNA genes was shown to be preferentially expressed and more accessible to DNase I than the subset of hypermethylated genes (Jupe & Zimmer, 1993). Likewise, in wheat × rye hybrids that display nucleolar dominance, the underdominant rye rRNA genes are more densely methylated within the rDNA inter-genic spacer than are active wheat rRNA genes (Houchins *et al.*, 1997; Neves *et al.*, 1995).

More definitive evidence that cytosine methylation plays an important role in rRNA gene silencing awaited the demonstration that treatment with 5-aza-2′-deoxycytosine (aza-dC), a potent inhibitor of cytosine methylation, will reactivate silent NORs and/or rRNA genes in wheat–rye hybrids (Amado *et al.*, 1997) or in allotetraploid hybrids in the genera *Brassica* (Chen & Pikaard, 1997b) or *Arabidopsis* (Lawrence *et al.*, 2004). In the case of *Arabidopsis suecica*, reactivation of silent rRNA gene arrays in *A. suecica* following a aza-dC treatment has been shown to be accompanied by strong NOR decondensation (O. Pontes, W. Viegas and C. S. Pikaard, unpublished data). Methylated and inactive rDNA repeats can also be reactivated in mouse cells following aza-dC treatment (Santoro & Grummt, 2001).

Recently, Lawrence *et al.* (2004) mapped the positions of methylated cytosines in *Arabidopsis thaliana* rRNA gene promoters by bisulfite-mediated DNA sequencing (Lawrence *et al.*, 2004). Importantly, the majority of pro-moters were found to be extensively methylated whereas a less abundant class of promoters was lightly methylated overall and completely unmethylated in the minimal promoter region. Chromatin immunoprecipitation (ChIP) using an antibody specific for RNA polymerase I revealed that only the hypomethy-lated rRNA genes in *A. thaliana* associate with the polymerase and are thus the transcriptionally active subset (Lawrence *et al.*, 2004). Analysis of nucleolar dominance in *A. suecica*, the natural allotetraploid hybrid of *A. thaliana* and *A. arenosa* confirmed and extended these results. In *A. suecica*, *A. arenosa–* derived rRNA genes are dominant and active whereas *A. thaliana*–derived rRNA genes are silent (Chen *et al.*, 1998). Importantly, the *A. thaliana* genes are found exclusively in a hypermethylated state whereas the dominant *A. arenosa*–derived rRNA genes are partitioned into hypomethylated (active) and hypermethylated (inactive) classes (Lawrence *et al.*, 2004). The fact that not all of the dominant *A. arenosa* rRNA genes are hypomethylated suggests that the dominant class of genes is still in excess over the number of rRNA genes needed, and that these excess genes are dosage-controlled, as in non-hybrids, by a mechanism involving cytosine hypermethylation.

7.5   Histone modifications and rRNA gene regulation

7.5.1   *Histone acetylation*

Eukaryotic DNA does not exist in the nucleus as naked DNA but instead is assembled by proteins into chromatin. The most basic unit of chromatin is the nucleosome core particle, which consists of ~146 bp of DNA wrapped around a histone octamer that is composed of two molecules each of histones H2A, H2B, H3 and H4 (Luger *et al.*, 1999; Richmond & Davey, 2003). Nucleosomal histones are subject to a variety of posttranslational modifications including acetylation, methylation, phosphorylation, glycosylation, ubiquitylation, ADP-ribosylation (Berger, 2002; Felsenfeld & Groudine, 2003; Jenuwein & Allis, 2001; Strahl & Allis, 2000) and sumoylation (Shiio & Eisenman, 2003). Collectively, these modifications are thought to comprise a 'histone code' that coordinates the activity of activator or repressor complexes that interact with chromatin (Berger, 2002; Felsenfeld & Groudine, 2003; Jenuwein & Allis, 2001; Rice & Allis, 2001; Richards & Elgin, 2002; Strahl & Allis, 2000). Posttranslational modifications of histone N-terminal tails are particularly important aspects of the code. For instance, the acetylation or methylation of specific lysine (K) residues in the tails of histones H3 and H4 has been shown to be important marks of active euchromatin or silent heterochromatin (for reviews, see Felsenfeld & Groudine, 2003; Richards & Elgin, 2002; Vermaak *et al.*, 2003).

The degree to which histones are acetylated reflects a steady-state balance resulting from the opposing actions of two types of enzymes, namely *h*istone *a*cetyl*t*ransferases (HATs) and *h*istone *dea*cetylases (HDACs) (Kuo & Allis, 1998; Richards & Elgin, 2002; Vogelauer *et al.*, 2000). As a general rule, transcriptionally active genes are associated with histones that are hyperacetylated, whereas inactive genes tend to be hypoacetylated (Csordas, 1990; Hebbes *et al.*, 1988; Turner, 1991, 2000). Acetylation is thought to exert a positive effect on gene activity by inducing a decondensation of the nucleosome structure, as suggested by *in vitro* studies that demonstrated an increased binding affinity for pol II or pol III transcription factors to nucleosomes assembled using hyper-acetylated histones (Lee *et al.*, 1993; Vettese-Dadey *et al.*, 1996). Many

transcriptional activators possess intrinsic HAT activity or recruit protein complexes that include one or more HAT activities (Struhl, 1998). Conversely,
histone hypoacetylation is generally associated with transcriptional silencing
and transcriptionally silent heterochromatic domains are generally underacety-
lated in comparison with euchromatin (Dobosy & Selker, 2001; Grewal, 2000;
Ng & Bird, 2000; Struhl, 1998). Examples include the inactive X chromosome
in somatic cells of female mammals (Jeppesen & Turner, 1993), yeast telomeric
heterochromatin (Moazed, 2001) and heterochromatic regions of plant chromosomes (Buzek *et al.*, 1998; Houben *et al.*, 1996).

Evidence that histone deacetylation plays a role in rRNA gene silencing
came initially from the demonstration that silenced rRNA genes subjected
to nucleolar dominance in *Brassica* allotetraploids could be derepressed by
growing seedlings on medium containing sodium butyrate or trichostatin
A (TSA), both of which are HDAC inhibitors (Chen & Pikaard, 1997b).
Corroborating evidence was found using mammalian cultured cells, in which
TSA treatments resulted in a dose-dependent increase in transcription from
an rRNA reporter gene and from endogenous rRNA genes (Hirschler-
Laszkiewicz *et al.*, 2001). Silenced *A. thaliana*–derived rRNA genes subjected
to nucleolar dominance in *A. suecica* seedlings are also derepressed by growth
on medium containing TSA (Lawrence *et al.*, 2004). Interestingly, TSA
treatment not only causes the reactivation of *A. thaliana* NORs but also causes
a two- to threefold upregulation of the dominant *A. arenosa*–derived rRNA
genes (Lawrence *et al.*, 2004), further suggesting that the same mechanisms
responsible for silencing the underdominant genes are also responsible for
dosage control among the dominant set.

7.5.2 *Histone methylation*

The fact that amino acids within the histone tails are methylated has been known
for almost four decades. However, the significance of this modification for
epigenetic regulation has been unraveled only recently (Kouzarides, 2002;
Rea *et al.*, 2000; Richards & Elgin, 2002). Specific histone methylation marks
carry important information concerning the partitioning of the genome into
silenced, heterochromatic regions and active euchromatic regions. For instance,
methylation of histone H3 lysine 9 (H3mK9) and H3 lysine 4 (H3mK4) has
been found to discriminate different chromatin territories (Bernstein *et al.*,
2002; Litt *et al.*, 2001; Noma *et al.*, 2001), with the number of methyl groups
added to the lysines also being important (Dutnall, 2003). For example, in
mammals, trimethylated H3K9 is preferentially located in pericentromeric
heterochromatin, whereas the mono- and dimethylated forms are found within
euchromatic domains (Peters *et al.*, 2003; Rice *et al.*, 2003). In contrast, both
mono- and dimethylated H3K9 are found at silent *loci* in *Arabidopsis*, with the
trimethylated form being undetectable (Jackson *et al.*, 2004).

The relationship between H3K9 methylation, chromatin structure and gene silencing is increasingly clear. An important initial insight came from the discovery that the mammalian homolog of *Drosophila* suppressor of variegation 3–9 (Su(var)3–9), a protein required for heterochromatic silencing, is a histone H3K9 methylase (Rea *et al.*, 2000). Likewise, H3K9 methylase genes of fission yeast, *Neurospora* and *Arabidopsis* were identified based on loss-of-function mutations that disrupted transcriptional silencing (Jackson *et al.*, 2002; Nakayama *et al.*, 2001; Tamaru & Selker, 2001). Importantly, histone demethylases have not yet been identified, suggesting that histone methylation, unlike histone deacetylation, is not a reversible process, possibly allowing methylation marks to 'lock in' a given transcriptional state. Therefore, changes in histone methylation in a given chromosomal interval presumably require that methylated histones be displaced by transcription (Ahmad & Henikoff, 2002), by complete proteolysis or tail clipping (Bannister *et al.*, 2002) or by conversion of methylated amino acids to non-conventional amino acids (e.g. conversion of methylarginine to citrulline; Cuthbert *et al.*, 2004). Another important point is that H3K9 can be acetylated, or methylated, but not both at the same time (Nakayama *et al.*, 2001). In euchromatin, H3K9 is typically acetylated, whereas in heterochromatin it is not acetylated, but methylated. Therefore, H3K9 is thought to be an important site at which deacetylation is a prerequisite for methylation, with the irreversibility of the latter modification serving as a more durable mark.

A clear correlation between histone methylation states and epigenetic regulation of rRNA genes has recently been demonstrated both for dosage control in non-hybrid *A. thaliana* and for nucleolar dominance in the allotetraploid, *A. suecica* (Lawrence *et al.*, 2004). In *A. thaliana*, ChIP showed that the active rRNA genes that are associated with RNA polymerase I and are hypomethylated on promoter cytosines are also associated with histone H3^{trimethyl}K4. In contrast, the subset of genes with hypermethylated promoter cytosines are associated with H3^{dimethyl}K9 (Lawrence *et al.*, 2004). These observations were extended by analysis of dominant and underdominant rRNA genes in *A. suecica*. Using ChIP, Lawrence *et al.* showed that the silent rRNA genes and NORs derived from *A. thaliana* associate with H3^{dimethyl}K9 but not with H3^{trimethyl}K4 (Lawrence *et al.*, 2004). Likewise, by combining FISH localization of rRNA genes with immunolocalization of modified histones, the highly condensed *A. thaliana* NORs were shown to co-localize with H3^{dimethyl}K9-marked heterochromatin but not with H3^{trimethyl}K4-marked euchromatin during interphase. In contrast, the dominant *A. arenosa*–derived rRNA genes and NORs were found to partially decondense and co-localize both with H3^{dimethyl}K9 and with H3^{trimethyl}K4. Specifically, the condensed regions of the NORs co-localized with H3^{dimethyl}K9 whereas the decondensed, presumably active genes within the NOR co-localized with H3^{trimethyl}K4 (Lawrence *et al.*, 2004). Again, these data are consistent with the interpretation that only a subset of the dominant *A. arenosa* rRNA genes is active (i.e. the fraction enriched in H3^{trimethyl}K4), with the excess genes remaining highly condensed, heterochromatic and associated with histone H3 dimethylated on lysine 9.

7.6 Concerted changes in DNA and histone methylation comprise an on/off switch

As discussed above, differences in DNA methylation, histone acetylation and histone methylation are clearly correlated with the transcriptional activity of rRNA genes in their on or off states. The realization that these modifications are interdependent stemmed from an ability to experimentally flip the switch from off to on and monitor the changes in DNA and histone modifications that ensue. As discussed previously, silenced rRNA genes can be derepressed chemically using aza-dC to block DNA methylation or using TSA to block histone deacetylation (Lawrence *et al.*, 2004). Derepression can also be accomplished genetically by RNAi-mediated knockdown of HDACs *HDT1* (Lawrence *et al.*, 2004) or HDA6 (R. Lawrence and C. S. Pikaard, unpublished data). No matter how one throws the switch from 'off' to 'on', promoter DNA methylation and H3^{dimethyl}K9 association are lost and H3^{trimethyl}K4 and H3^{acetyl}K9 association are gained (Figure 7.2). Because blocking either DNA

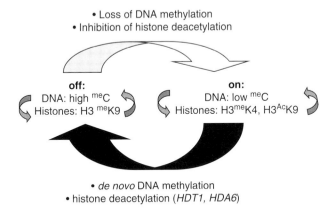

Figure 7.2 Chromatin alterations associated with the epigenetic on-off switch. Silenced ribosomal RNA genes in the off state have promoters displaying cytosine hypermethylation and association with histone H3 dimethylated on lysine 9. In contrast, active rRNA genes have hypomethylated promoters associated with H3 that is trimethylated on lysine 4 and acetylated on lysine 9. Switching states from off to on can be accomplished by blocking cytosine methylation or histone deacetylation, either chemically or through loss-of-function mutations. Conversely, *de novo* DNA methylation and histone deacetylation are presumably required for silencing.

methylation or histone deacetylation has identical consequences for DNA methylation, for H3 methylation and for H3 acetylation, we hypothesize that these chromatin modifications specify one another in a self-reinforcing silencing cycle (Figure 7.3).

The steps in the model depicted in Figure 7.3 are supported by published evidence. DNA methylation can be perpetuated at every round of cell division via maintenance methylation, which imparts methylation to newly replicated daughter DNA strands based on the methylation pattern in the mother strand (Bird, 2002). Cytosine methylation is then the target for MBDs that are known to associate in multi-protein complexes that include one or more HDACs (Feng & Zhang, 2001; Jones *et al.*, 1998; Nan *et al.*, 1998; Zhang *et al.*, 1999) such that cytosine methylation can act as a signal that brings about H3K9 deacetylation. As mentioned previously, K9 acetylation and K9 methylation are mutually exclusive such that K9 deacetylation needs to precede K9 methylation

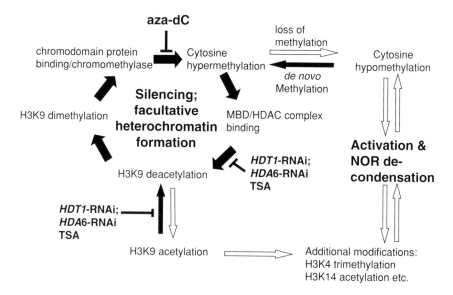

Figure 7.3 A model for how DNA methylation and histone modifications specify one another in the epigenetic switch mechanism that controls rRNA gene transcription. Cytosine methylation patterns can be perpetuated through multiple rounds of cell division by so-called maintenance methylation. Methylated DNA is then recognized by methylcytosine-binding domain proteins (MBDs) that, in turn, associate with histone deacetylases (HDACs). H3K9 deacetylation is a prerequisite for H3K9 dimethylation, which then recruits chromodomain proteins that possess, or associate with, DNA methyltransferases to complete the repression cycle. Blocking cytosine methylation with 5-aza-2' deoxycytosine (aza-dC) or blocking histone deacetylation with trichostatin A (TSA) or by RNAi-mediated knockdown of HDACs HDT1 or HDA6 is sufficient to disrupt the cycle, causing concerted changes in both DNA methylation and histone acetylation/methylation. Histone acetylation/deacetylation and DNA methylation/demethylation are hypothesized to be the likely control points for switching between the silenced and active states.

(Nakayama *et al.*, 2001). For the H3K9 methylation event(s), H3K9 methylases have been identified in numerous eukaryotes (e.g. mammalian SUV39H1 and SUV39H2, *Drosophila* Su(var)3–9, and *Arabidopsis* KYP or SUVH4 (Jackson *et al.*, 2002; Malagnac *et al.*, 2002; O'Carroll *et al.*, 2000; Rea *et al.*, 2000; for reviews see Dobosy & Selker, 2001; Grewal & Moazed, 2003; Richards & Elgin, 2002). In mammals and *Neurospora*, heterochromatin protein 1 (HP1) or related chromodomain-containing proteins are then known to bind to methylated H3K9 (Lachner *et al.*, 2001; Schotta *et al.*, 2002). DNA methyltransferases, in turn, interact with HP1, suggesting a mechanism whereby cytosine methylation can be specified by histone H3K9 methylation (Fuks *et al.*, 2003a; Freitag *et al.*, 2004; Lehnertz *et al.*, 2003). Alternatively, plants have DNA methylases with chromodomains (e.g. chromomethylases CMT1, CMT2 and CMT3 in *Arabidopsis*; only CMT3 is known to be functional) (Henikoff & Comai, 1998) such that an HP1-like intermediary may be unnecessary to link histone methylation to DNA methylation. In either case, H3K9 methylases are known to be required for some or all cytosine methylation in plants, *Neurospora* and mammals because H3K9 methylase mutations cause the loss of DNA methylation (Johnson *et al.*, 2002; Lehnertz *et al.*, 2003; Malagnac *et al.*, 2002; Tamaru *et al.*, 2003). A similar or identical cycle can explain the fact that H3K9 methylation is decreased in *A. thaliana* defective for the maintenance (replication-dependent) DNA methyltransferase *MET1* or the chromatin remodeler *DDM1* (decreased DNA methylation 1) (Gendrel *et al.*, 2002; Tariq *et al.*, 2003).

It is possible that some steps in the pathway depicted in the model can be bypassed. For instance, MBDs associate with histone methylases (Fujita *et al.*, 2003; Fuks *et al.*, 2003b). Likewise, DNA methyltransferases may recruit HDACs without the need for an MBD intermediate because mammalian DNMT1 (a maintenance methyltransferase) and DNMT3a and DNMT3b (*de novo* methyltransferases) co-immunoprecipitate with HDAC1 and HDAC2 (Fuks *et al.*, 2000, 2001). DNMT1 and DNMT3a also interact *in vitro* with the mammalian H3K9 methyltransferase, SUV39H1(Fuks *et al.*, 2003a). Other biochemical evidence has shown that H3K9 deacetylases and methyltransferases can exist in the same complex (Shi *et al.*, 2003). These interactions may also suggest that the chromatin-modifying activities required to establish silencing interact to form multi-functional complexes capable of carrying out concerted reactions that include DNA methylation, histone deacetylation and histone methylation.

7.7 Future studies: identifying genes required for the epigenetic on/off switch

Based on the model depicted in Figure 7.3, we are undertaking a systematic search for genes required for rRNA gene silencing by using transgene-induced

RNAi to knock down genes encoding DNA methyltransferases, methylcyto-sine binding domain proteins, HDACs and histone methylases.

So far, we have completed the screen of the 16 predicted HDAC genes that are expressed in *A. thaliana*. These genes belong to three superfamilies: RPD3/HDA, HD2 (HDTs; unique to plants) and SIR2 (Pandey *et al.*, 2002). HDT1 (Lawrence *et al.*, 2004) and HDA6 (R. Lawrence and C. S. Pikaard, unpublished data) have been found to be required for rRNA gene silencing in nucleolar dominance, as mentioned previously. An interesting feature of HDT1 is that it is localized in the nucleolus (Lawrence *et al.*, 2004), as is maize HD2 (Lusser *et al.*, 1997) and all four members of the HDT family in *Arabidopsis* (Pendle *et al.*, 2004), suggesting that HDT1 may interact directly with the rRNA genes.

An interesting parallel to our finding that HDA6 is required for nucleolar dominance in *A. suecica* is that Probst *et al.* (2004) independently found that HDA6 plays a role in rDNA condensation, acetylation and cytosine methyla-tion in *A. thaliana* in addition to its known roles in transcriptional silencing of transgenes and endogenous silenced *loci* (Aufsatz *et al.*, 2002; Lippman *et al.*, 2003; Murfett *et al.*, 2001). Collectively, these data show that HDA6 is required for nuclear organization of rDNA repeats in both *A. thaliana* and *A. suecica*. It is also noteworthy that yeast RPD3, the founding member of the gene family that includes HDA6, is involved in the conformational changes that make active rRNA genes accessible to psoralen cross-linking (Sandmeier *et al.*, 2002).

Presumably, there are specific HATs required for rRNA gene activation, but nothing is known in plants. However, in mammalian cells, overexpression of HATs CBP, p300 and PCAF has been shown to stimulate rRNA gene tran-scription (Hirschler-Laszkiewicz *et al.*, 2001). Furthermore, CBP has been immunolocalized within the nucleolus, suggesting a possible direct involve-ment in rRNA transcription (Hirschler-Laszkiewicz *et al.*, 2001). In *A. thali-ana*, 12 HAT genes have been identified (Pandey *et al.*, 2002), including five belonging to the CBP family. It will be interesting to ascertain whether these HATs are involved in rRNA gene dosage control, nucleolar dominance and chromatin structure.

We anticipate that current screens to identify the DNA methyltransferases, MBDs and histone methylases required for nucleolar dominance will identify a number of genes involved in rRNA gene silencing. If so, we are keen to know if the proteins that carry out these different functions physically interact with rRNA genes and with each other. Physical interactions among the proteins may help explain why changes in cytosine methylation, histone deacetylation and histone methylation are so tightly coordinated; perhaps these are concerted reactions carried out by a multi-functional enzyme complex. It is also import-ant to determine if overall cytosine methylation density or methylation of specific cytosines is key to rRNA gene silencing. Likewise, how is cytosine methylation lost when rRNA genes switch from off to on? Does this occur through an active process, involving DNA glycosylases such as DEMETER or ROS1 (Choi *et al.*, 2002; Gong *et al.*, 2002), or passively through multiple rounds of replication uncoupled from maintenance methylation?

Recent breakthroughs have finally provided a rationale for why nucleolar dominance occurs, namely as a manifestation of the normal dosage control mechanisms that appear to operate in all eukaryotes, whether they are hybrids or not. However, the big unanswered questions that remain are: how is one parental set of rRNA genes chosen for inactivation? Why not dosage-control both parental sets of rRNA genes, rather than completely inactivate one set? We are actively seeking clues that will lead to answers to these questions.

Acknowledgments

Research in the Pikaard laboratory related to nucleolar dominance and rRNA gene silencing is supported by United States National Institutes of Health grant R01-GM60380 and National Science Foundation grant DBI-9975930. Research in the Neves and Viegas laboratories is supported by Fundação para a Ciência e Tecnologia project POCTI/BCI/38557/2001 to N.N.

References

Ahmad, K. & Henikoff, S. (2002) The histone variant H3.3 marks active chromatin by replication-independent nucleosome assembly. *Molecular Cell*, **9**, 1191–200.

Amado, L., Abranches, R., Neves, N. & Viegas, W. (1997) Development-dependent inheritance of 5-azacytidine-induced epimutations in triticale: analysis of rDNA expression patterns. *Chromosome Research*, **5**, 445–50.

Aufsatz, W., Mette, M. F., Van Der Winden, J., Matzke, M. & Matzke, A. J. (2002) HDA6, a putative histone deacetylase needed to enhance DNA methylation induced by double-stranded RNA. *EMBO Journal*, **21**, 6832–41.

Bannister, A. J., Schneider, R. & Kouzarides, T. (2002) Histone methylation: dynamic or static? *Cell*, **109**, 801–806.

Berger, S. L. (2002) Histone modifications in transcriptional regulation. *Current Opinion in Genetics and Development*, **12**, 142–8.

Bernstein, B. E., Humphrey, E. L., Erlich, R. L., Schneider, R., Bouman, P., Liu, J. S., Kouzarides, T. & Schreiber, S. L. (2002) Methylation of histone H3 Lys 4 in coding regions of active genes. *Proceedings of the National Academy of Sciences of the United States of America*, **99**, 8695–700.

Bestor, T. H. (1998) The host defence function of genomic methylation patterns. *Novartis Foundation Symposium*, **214**, 187–95.

Bird, A. (2002) DNA methylation patterns and epigenetic memory. *Genes & Development*, **16**, 6–21.

Brannan, C. I. & Bartolomei, M. S. (1999) Mechanisms of genomic imprinting. *Current Opinion in Genetics and Development*, **9**, 164–70.

Buzek, J., Riha, K., Siroky, J., Ebert, I., Greilhuber, J. & Vyskot, B. (1998) Histone H4 underacetylation in plant facultative heterochromatin. *Biological Chemistry*, **379**, 1235–41.

Caperta, A. D., Neves, N., Morais-Cecilio, L., Malho, R. & Viegas, W. (2002) Genome restructuring in rye affects the expression, organization and disposition of homologous rDNA loci. *Journal of Cell Science*, **115**, 2839–46.

Caudy, A. A. & Pikaard, C. S. (2002) *Xenopus* ribosomal RNA gene intergenic spacer elements conferring transcriptional enhancement and nucleolar dominance-like competition in oocytes. *Journal of Biological Chemistry*, **277**, 31577–84.

Cavanaugh, A. H., Hirschler-Laszkiewicz, I., Hu, Q., Dundr, M., Smink, T., Misteli, T. & Rothblum, L. I. (2002) Rrn3 phosphorylation is a regulatory checkpoint for ribosome biogenesis. *Journal of Biological Chemistry*, **277**, 27423–32.

Chen, Z. J. & Pikaard, C. S. (1997a) Transcriptional analysis of nucleolar dominance in polyploid plants: biased expression/silencing of progenitor rRNA genes is developmentally regulated in *Brassica*. *Proceedings of the National Academy of Sciences of the United States of America*, **94**, 3442–7.

Chen, Z. J. & Pikaard, C. S. (1997b) Epigenetic silencing of RNA polymerase I transcription: a role for DNA methylation and histone modification in nucleolar dominance. *Genes & Development*, **11**, 2124–36.

Chen, Z. J., Comai, L. & Pikaard, C. S. (1998) Gene dosage and stochastic effects determine the severity and direction of uniparental rRNA gene silencing (nucleolar dominance) in *Arabidopsis* allopolyploids. *Proceedings of the National Academy of Sciences of the United States of America*, **95**, 14891–6.

Choi, Y., Gehring, M., Johnson, L., Hannon, M., Harada, J. J., Goldberg, R. B., Jacobsen, S. E. & Fischer, R. L. (2002) DEMETER, a DNA glycosylase domain protein, is required for endosperm gene imprinting and seed viability in *Arabidopsis*. *Cell*, **110**, 33–42.

Conconi, A., Widmer, R. M., Koller, T. & Sogo, J. M. (1989) Two different chromatin structures coexist in ribosomal RNA genes throughout the cell cycle. *Cell*, **57**, 753–61.

Csordas, A. (1990) On the biological role of histone acetylation. *Biochemistry Journal*, **265**, 23–38.

Cuthbert, G. L., Daujat, S., Snowden, A. W., Erdjument-Bromage, H., Hagiwara, T., Yamada, M., Schneider, R., Gregory, P. D., Tempst, P., Bannister, A. J. & Kouzarides, T. (2004) Histone deimination antagonizes arginine methylation. *Cell*, **118**, 545–53.

Dammann, R., Lucchini, R., Koller, T. & Sogo, J. M. (1993) Chromatin structures and transcription of rDNA in yeast *Saccharomyces cerevisiae*. *Nucleic Acids Research*, **21**, 2331–8.

Dammann, R., Lucchini, R., Koller, T. & Sogo, J. M. (1995) Transcription in the yeast rRNA gene locus: distribution of the active gene copies and chromatin structure of their flanking regulatory sequences. *Molecular and Cellular Biology*, **15**, 5294–303.

Dobosy, J. R. & Selker, E. U. (2001) Emerging connections between DNA methylation and histone acetylation. *Cellular and Molecular Life Sciences*, **58**, 721–7.

Dutnall, R. N. (2003) Cracking the histone code: one, two, three methyls, you're out! *Molecular Cell*, **12**, 3–4.

Felsenfeld, G. & Groudine, M. (2003) Controlling the double helix. *Nature*, **421**, 448–53.

Feng, Q. & Zhang, Y. (2001) The MeCP1 complex represses transcription through preferential binding, remodeling, and deacetylating methylated nucleosomes. *Genes & Development*, **15**, 827–32.

Flavell, R. B. (1986) The structure and control of expression of ribosomal RNA genes. *Oxford Surveys of Plant Molecular and Cell Biology*, **3**, 252–74.

Flavell, R. B. & O'Dell, M. (1979) The genetic control of nucleolus formation in wheat. *Chromosoma*, **71**, 135–52.

Flavell, R. B., O'Dell, M. & Thompson, W. F. (1988) Regulation of cytosine methylation in ribosomal DNA and nucleolus organizer expression in wheat. *Journal of Molecular Biology*, **204**, 523–34.

Freitag, M., Hickey, P. C., Khlafallah, T. K., Read, N. D. & Selker, E. U. (2004) HP1 is essential for DNA methylation in *Neurospora*. *Molecular Cell*, **13**, 427–34.

French, S. L., Osheim, Y. N., Cioci, F., Nomura, M. & Beyer, A. L. (2003) In exponentially growing *Saccharomyces cerevisiae* cells, rRNA synthesis is determined by the summed RNA polymerase I loading rate rather than by the number of active genes. *Molecular and Cellular Biology*, **23**, 1558–68.

Frieman, M., Chen, Z. J., Saez-Vasquez, J., Shen, L. A. & Pikaard, C. S. (1999) RNA polymerase I transcription in a *Brassica* interspecific hybrid and its progenitors: tests of transcription factor involvement in nucleolar dominance. *Genetics*, **152**, 451–60.

Fujita, N., Watanabe, S., Ichimura, T., Tsuruzoe, S., Shinkai, Y., Tachibana, M., Chiba, T. & Nakao, M. (2003) Methyl-CpG binding domain 1 (MBD1) interacts with the Suv39h1-HP1 heterochromatic complex for DNA methylation-based transcriptional repression. *Journal of Biological Chemistry*, **278**, 24132–8.

Fuks, F., Burgers, W. A., Brehm, A., Hughes-Davies, L. & Kouzarides, T. (2000) DNA methyltransferase Dnmt1 associates with histone deacetylase activity. *Nature Genetics*, **24**, 88–91.

Fuks, F., Burgers, W. A., Godin, N., Kasai, M. & Kouzarides, T. (2001) Dnmt3a binds deacetylases and is recruited by a sequence-specific repressor to silence transcription. *EMBO Journal*, **20**, 2536–44.

Fuks, F., Hurd, P. J., Deplus, R. & Kouzarides, T. (2003a) The DNA methyltransferases associate with HP1 and the SUV39H1 histone methyltransferase. *Nucleic Acids Research*, **31**, 2305–12.

Fuks, F., Hurd, P. J., Wolf, D., Nan, X., Bird, A. P., & Kouzarides, T. (2003b) The methyl-CpG-binding protein MeCP2 links DNA methylation to histone methylation. *Journal of Biological Chemistry*, **278**, 4035–40.

Gendrel, A. V., Lippman, Z., Yordan, C., Colot, V. & Martienssen, R. A. (2002) Dependence of heterochromatic histone H3 methylation patterns on the *Arabidopsis* gene DDM1. *Science*, **297**, 1871–3.

Gerbi, S. A. (1985) Evolution of ribosomal DNA. In R. J. McIntyre (ed.) *Molecular Evolutionary Genetics*. Plenum Press, New York, pp. 419–517.

Gong, Z., Morales-Ruiz, T., Ariza, R. R., Roldan-Arjona, T., David, L. & Zhu, J. K. (2002) ROS1, a repressor of transcriptional gene silencing in *Arabidopsis*, encodes a DNA glycosylase/lyase. *Cell*, **111**, 803–14.

González-Melendi, P., Wells, B., Beven, A. & Shaw, P. (2001) Single ribosomal transcription units are linear, compacted Christmas trees in plant nucleoli. *Plant Journal*, **27**, 223–33.

Goodpasture, C. & Bloom, S. E. (1975) Visualization of nucleolar organizer regions in mammalian chromosomes using silver staining. *Chromosoma*, **53**, 37–50.

Grewal, S. I. (2000) Transcriptional silencing in fission yeast. *Journal of Cell Physiology*, **184**, 311–18.

Grewal, S. I. & Moazed, D. (2003) Heterochromatin and epigenetic control of gene expression. *Science*, **301**, 798–802.

Grummt, I. (1999) Regulation of mammalian ribosomal gene transcription by RNA polymerase I. *Progress in Nucleic Acid Research and Molecular Biology*, **62**, 109–54.

Grummt, I. (2003) Life on a planet of its own: regulation of RNA polymerase I transcription in the nucleolus. *Genes & Development*, **17**, 1691–702.

Grummt, I. & Pikaard, C. S. (2003) Epigenetic mechanisms controlling RNA polymerase I transcription. *Nature Reviews. Molecular Cell Biology*, **4**, 641–9.

Grummt, I., Roth, E. & Paule, M. R. (1982) rRNA transcription *in vitro* is species-specific. *Nature*, **296**, 173–4.

Haaf, T., Hayman, D. L. & Schmid, M. (1991) Quantitative determination of rDNA transcription units in vertebrate cells. *Experimental Cell Research*, **193**, 78–86.

Hannan, K. M., Brandenburger, Y., Jenkins, A., Sharkey, K., Cavanaugh, A., Rothblum, L., Moss, T., Poortinga, G., McArthur, G. A., Pearson, R. B. & Hannan, R. D. (2003) mTOR-dependent regulation of ribosomal gene transcription requires S6K1 and is mediated by phosphorylation of the carboxy-terminal activation domain of the nucleolar transcription factor UBF. *Molecular and Cellular Biology*, **23**, 8862–77.

Heard, E., Clerc, P. & Avner, P. (1997) X-chromosome inactivation in mammals. *Annual Review of Genetics*, **31**, 571–610.

Hebbes, T. R., Thorne, A. W. & Crane-Robinson, C. (1988) A direct link between core histone acetylation and transcriptionally active chromatin. *EMBO Journal*, **7**, 1395–402.

Heitz, E. (1931) Nukleolen und chromosomen in der Gattung Vicia. *Planta*, **15**, 495–505.

Henikoff, S. & Comai, L. (1998) A DNA methyltransferase homolog with a chromodomain exists in multiple polymorphic forms in *Arabidopsis*. *Genetics*, **149**, 307–18.

Hernandez-Verdun, D. & Roussel, P. (2003) Regulators of nucleolar functions. *Progress in Cell Cycle Research*, **5**, 301–308.

Hernandez-Verdun, D., Roussel, P. & Gebrane-Younes, J. (2002) Emerging concepts of nucleolar assembly. *Journal of Cell Science*, **115**, 2265–70.

Hirschler-Laszkiewicz, I., Cavanaugh, A., Hu, Q., Catania, J., Avantaggiati, M. L. & Rothblum, L. I. (2001) The role of acetylation in rDNA transcription. *Nucleic Acids Research*, **29**, 4114–24.

Honjo, T. & Reeder, R. H. (1973) Preferential transcription of *Xenopus laevis* ribosomal RNA in interspecies hybrids between *Xenopus laevis* and *Xenopus mulleri*. *Journal of Molecular Biology*, **80**, 217–28.

Houben, A., Belyaev, N. D., Turner, B. M. & Schubert, I. (1996) Differential immunostaining of plant chromosomes by antibodies recognizing acetylated histone H4 variants. *Chromosome Research*, **4**, 191–4.

Houchins, K., O'Dell, M., Flavell, R. B. & Gustafson, J. P. (1997) Cytosine methylation and nucleolar dominance in cereal hybrids. *Molecular Genetics and Genomics*, **255**, 294–301.

Huynh, K. D. & Lee, J. T. (2001) Imprinted X inactivation in eutherians: a model of gametic execution and zygotic relaxation. *Current Opinions in Cell Biology*, **13**, 690–97.

Jackson, D. A., Iborra, F. J., Manders, E. M. M. & Cook, P. R. (1998) Numbers and organisation of RNA polymerases, nascent transcripts and transcription units in HeLa nuclei. *Molecular Biology of the Cell*, **9**, 1523–36.

Jackson, J. P., Lindroth, A. M., Cao, X. & Jacobsen, S. E. (2002) Control of CpNpG DNA methylation by the KRYPTONITE histone H3 methyltransferase. *Nature*, **416**, 556–60.

Jackson, J. P., Johnson, L., Jasencakova, Z., Zhang, X., PerezBurgos, L., Singh, P. B., Cheng, X., Schubert, I., Jenuwein, T. & Jacobsen, S. E. (2004) Dimethylation of histone H3 lysine 9 is a critical mark for DNA methylation and gene silencing in *Arabidopsis thaliana*. *Chromosoma*, **112**, 308–15.

Jacob, S. T. (1995) Regulation of ribosomal gene transcription. *Biochemistry Journal*, **306**, 617–26.

Jenuwein, T. & Allis, C. D. (2001) Translating the histone code. *Science*, **293**, 1074–80.

Jeppesen, P. & Turner, B. M. (1993) The inactive X chromosome in female mammals is distinguished by a lack of histone H4 acetylation, a cytogenetic marker for gene expression. *Cell*, **74**, 281–9.

Jimenez, R., Burgos, M. & Diaz de la Guardia, R. (1988) A study of the Ag-staining significance in mitotic NOR's. *Heredity*, **60**, 125–7.

Johnson, L., Cao, X. & Jacobsen, S. (2002) Interplay between two epigenetic marks. DNA methylation and histone H3 lysine 9 methylation. *Current Biology*, **12**, 1360–67.

Jones, P. L., Veenstra, G. J., Wade, P. A., Vermaak, D., Kass, S. U., Landsberger, N., Strouboulis, J. & Wolffe, A. P. (1998) Methylated DNA and MeCP2 recruit histone deacetylase to repress transcription. *Nature Genetics*, **19**, 187–91.

Jupe, E. R. & Zimmer, E. A. (1993) DNaseI-sensitive and undermethylated rDNA is preferentially expressed in a maize hybrid. *Plant Molecular Biology*, **21**, 805–21.

Kouzarides, T. (2002) Histone methylation in transcriptional control. *Current Opinion in Genetics and Development*, **12**, 198–209.

Kuo, M. H. & Allis, C. D. (1998) Roles of histone acetyltransferases and deacetylases in gene regulation. *BioEssays*, **20**, 615–26.

Lachner, M., O'Carroll, D., Rea, S., Mechtler, K. & Jenuwein, T. (2001) Methylation of histone H3 lysine 9 creates a binding site for HP1 proteins. *Nature*, **410**, 116–20.

Lawrence, R. J., Earley, K., Pontes, O., Silva, M., Chen, Z. J., Neves, N., Viegas, W. & Pikaard, C. S. (2004) A concerted DNA methylation/histone methylation switch regulates rRNA gene dosage control and nucleolar dominance. *Molecular Cell*, **13**, 599–609.

Learned, R. M., Cordes, S. & Tjian, R. (1985) Purification and characterization of a transcription factor that confers promoter specificity to human RNA polymerase I. *Molecular and Cellular Biology*, **5**, 1358–69.

Lee, D. Y., Hayes, J. J., Pruss, D. & Wolffe, A. P. (1993) A positive role for histone acetylation in transcription factor access to nucleosomal DNA. *Cell*, **72**, 73–84.

Lee, J. T. & Jaenisch, R. (1997) The (epi)genetic control of mammalian X-chromosome inactivation. *Current Opinions in Genetics and Development*, **7**, 274–80.

Lehnertz, B., Ueda, Y., Derijck, A. A., Braunschweig, U., Perez-Burgos, L., Kubicek, S., Chen, T., Li, E., Jenuwein, T. & Peters, A. H. (2003) Suv39h-mediated histone H3 lysine 9 methylation directs DNA methylation to major satellite repeats at pericentric heterochromatin. *Current Biology*, **13**, 1192–200.

Leitch, A. R., Mosgoller, W., Shi, M. & Heslop-Harrison, J. S. (1992) Different patterns of rDNA organization at interphase in nuclei of wheat and rye. *Journal of Cell Science*, **101**, 751–7.

Leung, A. K. & Lamond, A. I. (2003) The dynamics of the nucleolus. *Critical Reviews in Eukaryotic Gene Expression*, **13**, 39–54.

Lippman, Z., May, B., Yordan, C., Singer, T. & Martienssen, R. (2003) Distinct mechanisms determine transposon inheritance and methylation via small interfering RNA and histone modification. *Public Library of Science Biology*, **1**, E67.

Litt, M. D., Simpson, M., Gaszner, M., Allis, C. D. & Felsenfeld, G. (2001) Correlation between histone lysine methylation and developmental changes at the chicken beta-globin locus. *Science*, **293**, 2453–5.

Luger, K., Rechsteiner, T. J. & Richmond, T. J. (1999) Preparation of nucleosome core particle from recombinant histones. *Methods in Enzymology*, **304**, 3–19.

Lusser, A., Brosch, G., Loidl, A., Haas, H. & Loidl, P. (1997) Identification of maize histone deacetylase HD2 as an acidic nucleolar phosphoprotein. *Science*, **277**, 88–91.

Malagnac, F., Bartee, L. & Bender, J. (2002) An *Arabidopsis* SET domain protein required for maintenance but not establishment of DNA methylation. *EMBO Journal*, **21**, 6842–52.

Martienssen, R. A. & Colot, V. (2001) DNA methylation and epigenetic inheritance in plants and filamentous fungi. *Science*, **293**, 1070–74.

McClintock, B. (1934) The relationship of a particular chromosomal element to the development of the nucleoli in *Zea mays*. *Zeitschrift fur Zellforschung und Mikroskopische Anatomie*, **21**, 294–328.

McKnight, S. L. & Miller, O. L. (1976) Ultrastructural patterns of RNA synthesis during early embryogenesis of *Drosophila melanogaster*. *Cell*, **8**, 305–19.

Miesfeld, R. & Arnheim, N. (1984) Species-specific rDNA transcription is due to promoter-specific binding factors. *Molecular and Cellular Biology*, **4**, 221–7.

Miesfeld, R., Sollner-Webb, B., Croce, C. & Arnheim, N. (1984) The absence of a human-specific ribosomal DNA transcription factor leads to nucleolar dominance in mouse-human hybrid cells. *Molecular and Cellular Biology*, **4**, 1306–12.

Milkereit, P. & Tschochner, H. (1998) A specialized form of RNA polymerase I, essential for initiation and growth-dependent regulation of rRNA synthesis, is disrupted during transcription. *EMBO Journal*, **17**, 3692–703.

Miller, O. L. & Beatty, B. R. (1969) Visualization of nucleolar genes. *Science*, **164**, 955–7.

Moazed, D. (2001) Common themes in mechanisms of gene silencing. *Molecular Cell*, **8**, 489–98.

Moss, T. & Stefanovsky, V. Y. (2002) At the center of eukaryotic life. *Cell*, **109**, 545–8.

Murfett, J., Wang, X. J., Hagen, G. & Guilfoyle, T. J. (2001) Identification of *Arabidopsis* histone deacetylase HDA6 mutants that affect transgene expression. *Plant Cell*, **13**, 1047–61.

Muscarella, D. E., Vogt, V. M. & Bloom, S. E. (1985) The ribosomal RNA gene cluster in aneuploid chickens: evidence for increased gene dosage and regulation of gene expression. *Journal of Cell Biology*, **101**, 1749–56.

Muscarella, D. E., Vogt, V. M. & Bloom, S. E. (1987) Characterization of ribosomal RNA synthesis in a gene dosage mutant: the relationship of topoisomerase I and chromatin structure to transcriptional activity. *Journal of Cell Biology*, **105**, 1501–13.

Nakayama, J., Rice, J. C., Strahl, B. D., Allis, C. D. & Grewal, S. I. (2001) Role of histone H3 lysine 9 methylation in epigenetic control of heterochromatin assembly. *Science*, **292**, 110–13.

Nan, X., Ng, H. H., Johnson, C. A., Laherty, C. D., Turner, B. M., Eisenman, R. N. & Bird, A. (1998) Transcriptional repression by the methyl-CpG-binding protein MeCP2 involves a histone deacetylase complex. *Nature*, **393**, 386–9.

Navashin, M. (1934) Chromosomal alterations caused by hybridization and their bearing upon certain general genetic problems. *Cytologia*, **5**, 169–203.

Neves, N., Heslop-Harrison, J. S. & Viegas, W. (1995) rRNA gene activity and control of expression mediated by methylation and imprinting during embryo development in wheat x rye hybrids. *Theoretical and Applied Genetics*, **91**, 529–33.

Neves, N., Delgado, M., Silva, M., Caperta, A., Morais-Cecílio, L. & Viegas, W. (2005) Ribosomal DNA heterochromatin in plants. *Cytogenetics and Genome Research*, **109**, 104–111.

Ng, H. H. & Bird, A. (2000) Histone deacetylases: silencers for hire. *Trends in Biochemical Sciences*, **25**, 121–6.

Noma, K., Allis, C. D. & Grewal, S. I. (2001) Transitions in distinct histone H3 methylation patterns at the heterochromatin domain boundaries. *Science*, **293**, 1150–55.

O'Carroll, D., Scherthan, H., Peters, A. H., Opravil, S., Haynes, A. R., Laible, G., Rea, S., Schmid, M., Lebersorger, A., Jerratsch, M., Sattler, L., Mattei, M. G., Denny, P., Brown, S. D., Schweizer, D. & Jenuwein, T. (2000) Isolation and characterization of Suv39h2, a second histone H3 methyltransferase gene that displays testis-specific expression. *Molecular and Cellular Biology*, **20**, 9423–33.

Pandey, R., Muller, A., Napoli, C. A., Selinger, D. A., Pikaard, C. S., Richards, E. J., Bender, J., Mount, D. W. & Jorgensen, R. A. (2002) Analysis of histone acetyltransferase and histone deacetylase families of *Arabidopsis thaliana* suggests functional diversification of chromatin modification among multicellular eukaryotes. *Nucleic Acids Research*, **30**, 5036–55.

Pederson, T. and Politz, J. C. (2000) The nucleolus and the four ribonucleoproteins of translation. *Journal of Cell Biology*, **148**, 1091–1095.

Pelletier, G., Stefanovsky, V. Y., Faubladier, M., Hirschler-Laszkiewicz, I., Savard, J., Rothblum, L. I., Cote, J. & Moss, T. (2000) Competitive recruitment of CBP and Rb-HDAC regulates UBF acetylation and ribosomal transcription. *Molecular Cell*, **6**, 1059–66.

Pendle, A. F., Clark, G. P., Boon, R., Lewandowska, D., Lam, Y. W., Andersen, J., Mann, M., Lamond, A. I., Brown, J. W. & Shaw, P. J. (2004) Proteomic analysis of the *Arabidopsis* nucleolus suggests novel nucleolar functions. *Molecular Biology of the Cell*, **16**, 260–69.

Peters, A. H., Kubicek, S., Mechtler, K., O'Sullivan, R. J., Derijck, A. A., Perez-Burgos, L., Kohlmaier, A., Opravil, S., Tachibana, M., Shinkai, Y., Martens, J. H. & Jenuwein, T. (2003) Partitioning and plasticity of repressive histone methylation states in mammalian chromatin. *Molecular Cell*, **12**, 1577–89.

Peyroche, G., Milkereit, P., Bischler, N., Tschochner, H., Schultz, P., Sentenac, A., Carles, C. & Riva, M. (2000) The recruitment of RNA polymerase I on rDNA is mediated by the interaction of the A43 subunit with Rrn3. *EMBO Journal*, **19**, 5473–82.

Phillips, R. L., Kleese, R. A. & Wang, S. S. (1971) The nucleolus organizer region of maize (*Zea mays* L.): chromosomal site of DNA complementary to ribosomal RNA. *Chromosoma*, **36**, 79–88.

Pikaard, C. S. (2002) Transcription and tyranny in the nucleolus: the organization, activation, dominance and repression of ribosomal RNA genes. In C. R. Somerville & E. M. Meyerowitz (eds) *The Arabidopsis Book*. American Society of Plant Biologists, Rockville, MD.

Pontes, O., Lawrence, R. J., Neves, N., Silva, M., Lee, J. H., Chen, Z. J., Viegas, W. & Pikaard, C. S. (2003) Natural variation in nucleolar dominance reveals the relationship between nucleolus organizer chromatin topology and rRNA gene transcription in *Arabidopsis. Proceedings of the National Academy of Sciences of the United States of America*, **100**, 11418–23.

Probst, A. V., Fagard, M., Proux, F., Mourrain, P., Boutet, S., Earley, K., Lawrence, R. J., Pikaard, C. S., Murfett, J., Furner, I., Vaucheret, H. & Scheid, O. M. (2004) *Arabidopsis* histone deacetylase HDA6 is required for maintenance of transcriptional gene silencing and determines nuclear organization of rDNA repeats. *Plant Cell*, **16**, 1021–34.

Rea, S., Eisenhaber, F., O'Carroll, D., Strahl, B. D., Sun, Z. W., Schmid, M., Opravil, S., Mechtler, K., Ponting, C. P., Allis, C. D. & Jenuwein, T. (2000) Regulation of chromatin structure by site-specific histone H3 methyltransferases. *Nature*, **406**, 593–9.

Reeder, R. H. (1974) Ribosomes from eukaryotes: genetics. In M. Nomura (ed.) *Ribosomes*. Cold Spring Harbor Laboratory Press, Cold Spring Harbor, NY, pp. 489–519.

Reeder, R. H. (1985) Mechanisms of nucleolar dominance in animals and plants. *Journal of Cell Biology*, **101**, 2013–16.

Reeder, R. H. & Roan, J. G. (1984) The mechanism of nucleolar dominance in *Xenopus* hybrids. *Cell*, **38**, 39–44.

Rice, J. C. & Allis, C. D. (2001) Histone methylation versus histone acetylation: new insights into epigenetic regulation. *Current Opinions in Cell Biology*, **13**, 263–73.

Rice, J. C., Briggs, S. D., Ueberheide, B., Barber, C. M., Shabanowitz, J., Hunt, D. F., Shinkai, Y. & Allis, C. D. (2003) Histone methyltransferases direct different degrees of methylation to define distinct chromatin domains. *Molecular Cell*, **12**, 1591–8.

Richards, E. J. (1997) DNA methylation and plant development. *Trends in Genetics*, **13**, 319–23.

Richards, E. J. & Elgin, S. C. (2002) Epigenetic codes for heterochromatin formation and silencing: rounding up the usual suspects. *Cell*, **108**, 489–500.

Richmond, T. J. & Davey, C. A. (2003) The structure of DNA in the nucleosome core. *Nature*, **423**, 145–50.

Ritossa, F. M. & Spiegelman, S. (1965) Localization of DNA complementary to ribosomal RNA in the nucleolus organizer region of *Drosophila melanogaster*. *Proceedings of the National Academy of Sciences of the United States of America*, **53**, 737–45.

Rogers, S. O. & Bendich, A. J. (1987) Ribosomal RNA genes in plants: variability in copy number and in the intergenic spacer. *Plant Molecular Biology*, **9**, 509–20.

Sandmeier, J. J., French, S., Osheim, Y., Cheung, W. L., Gallo, C. M., Beyer, A. L. & Smith, J. S. (2002) RPD3 is required for the inactivation of yeast ribosomal DNA genes in stationary phase. *EMBO Journal*, **21**, 4959–68.

Santoro, R. & Grummt, I. (2001) Molecular mechanisms mediating methylation-dependent silencing of ribosomal gene transcription. *Molecular Cell*, **8**, 719–25.

Sardana, R., O'Dell, M. & Flavell, R. (1993) Correlation between the size of the intergenic regulatory region, the status of cytosine methylation of rRNA genes and nucleolar expression in wheat. *Molecular Genetics and Genomics*, **236**, 155–62.

Scheer, U. & Weisenberger, D. (1994) The nucleolus. *Current Opinions in Cell Biology*, **6**, 354–9.

Schotta, G., Ebert, A., Krauss, V., Fischer, A., Hoffmann, J., Rea, S., Jenuwein, T., Dorn, R. & Reuter, G. (2002) Central role of Drosophila SU(VAR)3–9 in histone H3-K9 methylation and heterochromatic gene silencing. *EMBO Journal*, **21**, 1121–31.

Schwarzacher, H. G. & Mosgoeller, W. (2000) Ribosome biogenesis in man: current views on nucleolar structures and function. *Cytogenetics and Cell Genetics*, **91**, 243–52.

Shaw, P. J. & Jordan, E. G. (1995) The nucleolus. *Annual Review of Cell and Developmental Biology*, **11**, 93–121.

Shaw, P. J., Rawlins, D. & Highett, M. (1993) Nuclear and nucleolar structure in plants. In J. S. Heslop-Harrison & R. B. Flavell (eds) *The Chromosome*. Bios Scientific Publisher, Oxford, pp. 161–71.

Shi, Y., Sawada, J., Sui, G., Affar el, B., Whetstine, J. R., Lan, F., Ogawa, H., Luke, M. P. & Nakatani, Y. (2003) Coordinated histone modifications mediated by a CtBP co-repressor complex. *Nature*, **422**, 735–8.

Shiio, Y. & Eisenman, R. N. (2003) Histone sumoylation is associated with transcriptional repression. *Proceedings of the National Academy of Sciences of the United States of America*, **100**, 13225–30.

Sleutels, F., Barlow, D. P. & Lyle, R. (2000) The uniqueness of the imprinting mechanism. *Current Opinions in Genetics and Development*, **10**, 229–33.

Stefanovsky, V. Y., Pelletier, G., Hannan, R., Gagnon-Kugler, T., Rothblum, L. I. & Moss, T. (2001) An immediate response of ribosomal transcription to growth factor stimulation in mammals is mediated by ERK phosphorylation of UBF. *Molecular Cell*, **8**, 1063–73.

Strahl, B. D. & Allis, C. D. (2000) The language of covalent histone modifications. *Nature*, **403**, 41–5.

Struhl, K. (1998) Histone acetylation and transcriptional regulatory mechanisms. *Genes & Development*, **12**, 599–606.

Tamaru, H. & Selker, E. U. (2001) A histone H3 methyltransferase controls DNA methylation in *Neurospora crassa*. *Nature*, **414**, 277–83.

Tamaru, H., Zhang, X., McMillen, D., Singh, P. B., Nakayama, J., Grewal, S. I., Allis, C. D., Cheng, X. & Selker, E. U. (2003) Trimethylated lysine 9 of histone H3 is a mark for DNA methylation in *Neurospora crassa*. *Nature Genetics*, **34**, 75–9.

Tariq, M., Saze, H., Probst, A. V., Lichota, J., Habu, Y. & Paszkowski, J. (2003) Erasure of CpG methylation in *Arabidopsis* alters patterns of histone H3 methylation in heterochromatin. *Proceedings of the National Academy of Sciences of the United States of America*, **100**, 8823–7.

Thompson, W. F. & Flavell, R. B. (1988) DNase I sensitivity of ribosomal RNA genes in chromatin and nucleolar dominance in wheat. *Journal of Molecular Biology*, **204**, 535–48.

Turner, B. M. (1991) Histone acetylation and control of gene expression. *Journal of Cell Science*, **99**, 13–20.

Turner, B. M. (2000) Histone acetylation and an epigenetic code. *BioEssays*, **22**, 836–45.

Vermaak, D., Ahmad, K. & Henikoff, S. (2003) Maintenance of chromatin states: an open-and-shut case. *Current Opinions in Cell Biology*, **15**, 266–74.

Vettese-Dadey, M., Grant, P. A., Hebbes, T. R., Crane-Robinson, C., Allis, C. D. & Workman, J. L. (1996) Acetylation of histone H4 plays a primary role in enhancing transcription factor binding to nucleosomal DNA *in vitro*. *EMBO Journal*, **15**, 2508–18.

Vogelauer, M., Wu, J., Suka, N. & Grunstein, M. (2000) Global histone acetylation and deacetylation in yeast. *Nature*, **408**, 495–8.

Wallace, H. & Birnstiel, M. L. (1966) Ribosomal cistrons and the nucleolar organizer. *Biochimica et Biophysica Acta*, **114**, 296–310.

Wallace, H. & Langridge, W. H. R. (1971) Differential amphiplasty and the control of ribosomal RNA synthesis. *Heredity*, **27**, 1–13.

Warner, J. R. (1999) The economics of ribosome biosynthesis in yeast. *Trends in Biochemical Sciences*, **24**, 437–40.

Zhang, Y., Ng, H. H., Erdjument-Bromage, H., Tempst, P., Bird, A. & Reinberg, D. (1999) Analysis of the NuRD subunits reveals a histone deacetylase core complex and a connection with DNA methylation. *Genes & Development*, **13**, 1924–35.

8   Virus-induced gene silencing

Tamas Dalmay

8.1   Introduction

The word 'virus' originates from a Latin word meaning 'poison'. This definition is particularly apt, as viruses are responsible for huge yield losses in crops worldwide and there are no pesticides available that can effectively prevent infection. Viruses are small infectious agents that can pass through filters of pore size so small that they hold back all known cellular parasites from the filtrate. A more precise definition is that they are sets of one or more genomic nucleic acid molecules, normally encased in a protective coat, that are able to mediate their own replication only within suitable host cells (Matthews, 1992). Most of the viruses infecting plants have an RNA genome, although a small proportion of plant viruses store their genetic information in DNA.

8.1.1   Transgene-triggered gene silencing targets viruses

For a long time, the only possible weapons against viruses were the application of pesticides that targeted virus vectors or the introduction of natural virus resistance genes into cultivated crop plants. Although these methods can reduce yield losses to a limited extent, they certainly do not solve the problem. The discovery of the Ti plasmid of *Agrobacterium tumefaciens* gave a new hope in the fight against viruses. It provided the technology to introduce genes, other than those derived from close relatives introduced through classical plant breeding methods, into plants. Twenty years ago, Sanford and Johnston (1985) proposed the concept of pathogen-derived resistance (PDR), which took advantage of the new technology of plant transformation. PDR was thought to be manifested in susceptible plants through the expression of pathogen-derived genes that would interfere with the replication and spread of the pathogen. According to this theory, dominant negative forms of pathogen-derived proteins would give the most effective resistance. During the late 1980s and early 1990s a flood of papers described the transformations of several plant species with different viral genes (Fitchen & Beachy, 1993). Analysis of these data suggested two types of mechanism. One required a protein product for resistance and the other did not (Baulcombe, 1996; Beachy, 1997). One set of transgenic plants contributed especially to our understanding of the pathogen-derived protein-independent mechanism. Tobacco plants transformed with a mutated form of the replicase gene of potato virus X (PVX) exhibited extreme resistance against certain PVX strains. The PVX strain from which the transgene was derived showed lower nucleotide sequence homology to the resistance-breaking strains than to strains that were not able to infect the transgenic plants (Longstaff et al., 1993). Careful analysis of these plants revealed that lines accumulating high level of transgene mRNA were susceptible to PVX while the replicase mRNA was not detectable in lines resistant to PVX (Mueller et al., 1995). This observation suggested that this phenomenon is related to posttranscriptional gene silencing (PTGS), which was supported by the fact that the transgenic replicase mRNA was actively transcribed but degraded at the posttranscriptional level. It was proposed that the virus resistance was the result of the PTGS mechanism triggered by the pathogen-derived transgene, which targeted the PVX RNA molecules replicating in the cytoplasm (Mueller et al., 1995). Later it was shown that the PTGS-inducing transgene does not have to be derived from a pathogen. English et al. (1996) demonstrated that tobacco plants exhibiting silencing of the β-glucuronidase (GUS) transgene were resistant to a PVX vector carrying a GUS insert while they were susceptible to PVX or a PVX vector carrying an unrelated insert (i.e. green fluorescent protein – GFP). Thus early data from the Baulcombe laboratory established that viruses are targeted by the silencing machinery in plants where PTGS is triggered by a transgene.

8.1.2   Viruses trigger PTGS

Another line of research suggested that viruses not only are targets of transgene silencing but can themselves trigger silencing of transgenes (Lindbo et al., 1993) and endogenous genes (Kumagai et al., 1995). This phenomenon was later coined 'virus-induced gene silencing' (VIGS) and was characterised by Ruiz et al. (1998). VIGS is discussed in more detail in Section 8.2. Other work led to the discovery that viruses do not need trans- or endogenous genes to induce PTGS. Cauliflower mosaic virus (CaMV)- and tomato black ring virus (TBRV)-infected oilseed rape and tobacco plants, respectively, demonstrated the 'recovery' phenotype (Covey et al., 1997; Ratcliff et al., 1997). In this phenotype, plants exhibited strong symptoms on the inoculated leaves after virus inoculation but these symptoms progressively disappeared in upper leaves. The newly emerging leaves became resistant to CaMV or to TBRV. The resistance was sequence-specific because PVX was able to infect the recovered tissue but recombinant PVX carrying fragments of TBRV was not (Ratcliff et al., 1997). The recovery phenotype has been known for almost 50 years (Harrison, 1958) and PTGS has finally shed light on its underlying mechanism. PTGS also explains, at least partially, another phenomenon that has been known to plant virologists for many years. 'Cross-protection' occurs when the inoculation of plants with a weak virus strain gives resistance to more aggressive strains of the same virus (Sequeira, 1984). Although the mechanism behind this phenomenon is not understood, it has been widely exploited as a means of defence against viral infection. Once Ratcliff *et al.* deciphered the recovery phenotype, they made the link between PTGS and cross-protection, since the latter can also be interpreted as sequence-specific RNA degradation. Ratcliff *et al.* (1999) reported that wild-type plants infected with PVX carrying a *GFP* insert were susceptible to a heterologous virus (tobacco mosaic virus (TMV)) but showed resistance to recombinant TMV, carrying a fragment of *GFP*. All the experiments described above indicated that viruses induce PTGS in wild-type plants but the ultimate proof was provided by Hamilton and Baulcombe (1999) when they demonstrated that PVX-specific small RNAs (later coined short interfering RNAs (siRNAs)) accumulated in PVX-infected *Nicotiana benthamiana* plants. siRNAs are 20- to 21-nt RNA molecules generated from long double-stranded RNA (dsRNA) molecules. An RNaseIII-type enzyme called Dicer recognises and processes dsRNA molecules into short fragments of almost identical length (Bernstein *et al.*, 2001). In *Arabidopsis* there are three dicer-like enzymes (DCL1, DCL2 and DCL3) with specialised functions. The production of viral siRNAs is DCL2-dependent, while DCL1 and DCL3 are required for microRNA (miRNA) and endogenous siRNA biogenesis, respectively (Xie *et al.*, 2004). Initially it was thought that the viral replicative forms (annealed plus and minus strands) are the only substrates of DCL2 but it seems that viral siRNAs can be generated from only one of the strands due to intramolecular secondary structures (Szittya *et al.*, 2002).

8.1.3 Systemic silencing

PTGS shares several similarities with the immune system. It is specific, adaptive and not cell-autonomous. While antibodies can recognise epitopes on the surface of foreign proteins, siRNAs ensure that intruder RNAs are degraded sequence specifically. Although the immune system and PTGS are both adaptive, this is achieved through different mechanisms. Antibody variants are generated by recombination and selected by their affinity to antigens, yet they are always encoded by the host genome. In contrast, the antiviral PTGS sequence specificity determinants (siRNAs) are not encoded by the host genome but are generated from the genetic material of the virus. The first observation that PTGS is not cell-autonomous was by Palauqui *et al.* (1996). They studied cosuppression and targeted the *nitrate reductase* (*Nia*) gene by PTGS through ectopic expression of *Nia*. The first signs of PTGS on these plants were chlorotic spots appearing randomly on leaves. These spots grew as the plants developed until they reached a vein, from which point the upper leaves of the plants showed extensive chlorosis. These plants resembled virus-infected plants and suggested that a diffusible silencing signal moved from the lower leaves to the upper leaves through the stem, just like viruses move from inoculated to systemic leaves. The same group demonstrated the existence of such a signal by grafting cosuppressed rootstocks of tobacco with the scions from non-silenced plants (Palauqui *et al.*, 1997). Efficient transmission of cosupression from rootstock to scion required the expression of a homologous transgene in the scions. Movement of a silencing signal occurred even when the stock and scion were separated by a 30-cm non-transgenic stem (achieved by double-grafting). The PTGS systemic signal was also observed in another experimental system, targeting a *GFP* transgene in *N. benthamiana* plants (Voinnet & Baulcombe, 1997). Delivering ectopic copies of *GFP* into leaves by *A. tumefaciens* infiltration led to spectacular systemic silencing of GFP. Biolistic delivery of a plasmid expressing GUS and GFP confirmed that only a few cells received the constructs and that the silencing signal moved through the phloem and caused uniform silencing of the *GFP* transgene in the whole plant (Voinnet *et al.*, 1998). Recently, systemic silencing was dissected into several steps using transgenic *N. benthamiana* and *Arabidopsis* plants (Himber *et al.*, 2003). The first step is initiated at a single cell level followed by movement of the signal into the 10–15 cells surrounding the initially silenced cell. This short distance movement of the signal does not require homologous transcripts and is the likely explanation of the chlorotic spots on the Nia cosuppressed plants. The second step is the long-distance movement of the signal molecule through the phloem. Surprisingly, the size of the short RNAs involved in these two steps is different. A longer class of short RNAs (24–26 nt), although not required for cell-to-cell spread of silencing, is strongly correlated with the long-distance movement of silencing signals (Hamilton *et al.*, 2002). The third step is the unloading of signals in the upper leaves and then the short-distance movement is repeated, which is SDE1/SGS2 (Dalmay *et al.*, 2000; Mourrain *et al.*, 2000) and SDE3 (Dalmay *et al.*, 2001) independent (Himber *et al.*, 2003; see Chapter 2 for more details on genes involved in PTGS). However, the fourth step, which is the extensive cell-to-cell spread of the signal, requires these genes and involves transitive PTGS (Vaistij *et al.*, 2002). During this step, secondary siRNAs (21 nt but not 24–26 nt in length) are synthesised *de novo* by SDE1/SGS2, SDE3 and probably other factors, which then move into the adjacent 10–15 cells by the same mechanism as in the initially silenced cells in the lower leaves. The successive repetition of trafficking and amplification of 21-nt siRNAs leads to the uniform silencing of the upper leaves (Himber *et al.*, 2003). All the experiments describing and characterising the systemic silencing signal used transgenes because PTGS targeting endogenous genes does not spread systemically (Voinnet *et al.*, 1998). How did this mechanism evolve without transgenes? Substantial evidence has been accumulated that systemic silencing evolved to fight against viral infections. The first clue was the striking similarity between the pattern of the systemic spread of viruses and the PTGS signal (Voinnet *et al.*, 1998).

Further evidence was provided by Voinnet *et al.* (2000) by expressing move-ment-deficient PVX in wild-type *N. benthamiana*. The recombinant PVX was able to move cell-to-cell in the infiltrated leaves but could not spread into systemic leaves. However, the replication of PVX triggered systemic silenc-ing in the upper leaves. The systemic silencing was visualised by inserting endogenous gene fragments into PVX, which led to visible silencing of these genes around the veins in the upper leaves (Voinnet *et al.*, 2000). These results demonstrate that siRNAs generated from viral dsRNAs in the initially infected cells spread into the adjacent 10–15 cells and eventually to upper leaves just as it was observed for transgene-produced siRNAs. These siRNAs are there-fore moving ahead of the virus, programming cells to degrade viral RNA molecules.

8.2   Virus-induced gene silencing

8.2.1   Mechanism of virus-induced gene silencing

The first report, suggesting that viruses not only are targets of transgene-triggered PTGS but also elicit PTGS themselves, studied the expression of transgenic tobacco plants transformed with either a full-length form of the tobacco etch virus (TEV) coat protein or a form truncated at the N terminus (Lindbo *et al.*, 1993). These plants were initially susceptible to TEV infection, and typical systemic symptoms developed. However, 3–5 weeks after a TEV infection was established, transgenic plants recovered from the TEV infection, and new stem and leaf tissue emerged symptom- and virus-free. The recovered leaves could not be infected with TEV but they were susceptible to the closely related potato virus Y (PVY). Interestingly, the transgene transcription rate in recovered and mock inoculated tissue was identical but there was a 12- to 22-fold difference in the accumulation of transgene mRNA. In the recovered leaves the transgene mRNA was degraded at the posttranscriptional level, which resulted in sequence-specific virus resistance. Since transgene mRNA was downregulated only in infected plants it became apparent that in this case the virus, not the transgene, triggered PTGS and initiated the silencing of the transgene (Lindbo *et al.*, 1993). Two years later Kumagai *et al.* (1995) reported that a recombinant TMV virus carrying a fragment of the phytoene desaturase (*PDS*) gene in antisense orientation caused downregulation of the *PDS* gene in tobacco plants. PDS protects chlorophyll from photobleaching and green tissues turn white in the absence of PDS. At that time the mechan-ism of gene expression downregulation was not understood and it was thought that the antisense *PDS* fragment in the plus strand RNA of TMV interfered with the *PDS* mRNA. The fact that both TMV constructs, containing *PDS* fragments in either antisense or sense orientation, led to the same white phenotype was overlooked. It took another few years until it was finally realised that RNA vectors harbouring host gene fragments downregulate the expression of host genes through PTGS. This particular type of PTGS was named virus induced gene silencing or VIGS (Ruiz *et al.*, 1998). In this study fragments of *GFP* and *PDS* were inserted into PVX and wild-type and GFP transgenic *N. benthamiana* plants were inoculated with these recombinant viruses. Both induced silencing of the target genes and VIGS was determined to be posttranscriptional and cytoplasmic because intron sequences were not targeted. However, there were some apparent differences between the VIGS phenotypes triggered by PVX carrying *PDS* or *GFP*. The level of PVX–PDS in tissues exhibiting VIGS of PDS was very similar to the level of wild-type PVX in infected plants. In contrast, accumulation of PVX–GFP was hardly detectable in GFP-silenced tissues. Another difference was the extent of the VIGS phenotype. Silencing of PDS was patchy and occurred only in cells where the virus was replicating. VIGS of GFP, however, was uniform and was initiated in all tissues. Silencing of GFP was maintained even when PVX–GFP was not detected. Based on these results, it was concluded that initiation of VIGS depends on the virus and maintenance of it is virus-independent (Ruiz *et al.*, 1998). The observed difference between targeting an endogenous gene and a transgene by VIGS was confirmed by other studies. Jones *et al.* (1999) reported that the *GFP* transgene became methylated after being targeted by PVX–GFP but the endogenous ribulose-1,5-bisphosphate carboxylase oxygenase (Rubisco) small subunit (rbcS) gene remained unmethylated upon PVX–rbcs infection, albeit the plants exhibited silencing of Rubisco. The maintenance of silencing by transgenes and the fact that silencing is not maintained by endogenous genes was demonstrated using a virus vector different from PVX. *Arabidopsis* plants recover from tobacco rattle virus (TRV) infection and provide a good system to study VIGS. The white phenotype of PDS silencing is elicited by TRV–PDS infection but the leaves that appear after the plants have recovered from viral infection emerge green and do not exhibit the VIGS phenotype. In contrast, newly emerging recovered leaves of GFP transgenic *Arabidopsis* plants maintain GFP silencing following TRV–GFP infection (Dalmay *et al.*, 2000). The exact explanation for these differences between the silencing of endogenous genes and transgenes is not known but two proteins have been identified that are required for the maintenance of transgene silencing: SDE1/SGS2 (Dalmay *et al.*, 2000) and SDE3 (Dalmay *et al.*, 2001). The same proteins are required for secondary siRNA production (transitive PTGS; see Chapter 2 for details; Vaistij *et al.*, 2002). In our experiment, secondary siRNAs were generated only from the targeted transgenes and not from the endogenous genes studied. This suggests that RNAs, which are substrates for the SDE1/SGS2–SDE3 machinery, can maintain silencing in the absence of the inducer by producing secondary siRNAs, which move into other cells and initiate PTGS. Whether all transgenes and none of the endogenous genes are substrate for the maintenance apparatus is a separate issue. Sanders *et al.* (2002) reported that an endogenous *beta-1, 3-glucanase* gene is involved in secondary siRNA production in cosuppressed tobacco plants, suggesting that some endogenous mRNAs are accepted by SDE1/SGS2 and SDE3, can maintain PTGS and therefore can be targeted uniformly by VIGS. Interestingly, viral RNAs show a similar division to transgenes and endogenous genes. Several viruses (i.e. turnip crinkle virus (TCV)) accumulate to the same level in wild-type and *sde1/sgs2* or *sde3* plants but cucumber mosaic virus (CMV) accumulates at a much higher level in the mutant plants (Mourrain *et al.*, 2000; Dalmay *et al.*, 2001). This indicates that TCV RNA is not recognised by SDE1/SGS2 and SDE3 because PTGS works with the same efficiency both in wild-type and in mutant plants but CMV RNA is processed by these genes and probably secondary siRNAs are produced from the viral RNA molecules. These observations suggest that some viruses mimic endogenous genes (i.e. TCV) and other viruses resemble transgenes (i.e. CMV), although we do not know which features of the RNA determine whether or not they are targeted by SDE1/SGS2 and SDE3.

Most plant viruses are RNA viruses and initially they were at the centre of VIGS research. One of the reasons for this is that it is more difficult to create expression vectors from DNA viruses. However, a few VIGS vectors have been successfully generated based on DNA viruses (Turnage *et al.*, 2002). A recent study investigated the genetic requirement of VIGS initiated by a DNA virus vector and found that SDE1/SGS2 and SGS3 are needed for VIGS of endogenous genes but mutations in *SGS1* and *AGO1* (see Chapter 2 for further details on these genes) only delayed the onset of silencing and had a small effect on the VIGS phenotype (Muangsan *et al.*, 2004).

8.2.2 Virus vectors for gene silencing

Eight different virus vectors have been developed for silencing in plants: five RNA viruses and three DNA viruses. The first plant virus vectors were TMV (Donson *et al.*, 1991) and PVX (Chapman *et al.*, 1992), generated just after the first report had been published about the mystical phenomenon of cosuppression (Napoli *et al.*, 1990). Initially the purpose of these vectors was to express proteins in plants and it took several years to realise their potential to induce silencing. TMV was first used for VIGS on *Nicotiana tabacum* and *N. benthamiana*, although at that time the mechanism was not understood (Kumagai *et al.*, 1995). This vector was used in a few studies (Hiriart *et al.*, 2002, 2003) but did not become the first choice for laboratories that wanted to use VIGS as a tool. PVX was the second VIGS vector (Ruiz *et al.*, 1998) and soon became very popular and was used by many different groups. Recently it was shown that the PVX vector can induce silencing in the leaves and tubers of potato, suggesting that it will be useful for large-scale functional analysis of potato expressed sequenced tags (ESTs) (Faivre-Rampant *et al.*, 2004). As TMV and PVX were initially designed as expression vectors both used a duplicated promoter sequence. The second generation of VIGS vectors started with TRV, the first vector constructed specifically for silencing and not for the expression of proteins. The advantages of TRV are that the VIGS phenotype is more uniform than that caused by TMV or PVX and this vector is able to target host mRNAs in the growing points of plants. Two very similar versions of the TRV vector have been generated and were initially used in *N. benthamiana* (Liu *et al.*, 2002a; Ratcliff *et al.*, 2001), although it has since been demonstrated that TRV can also effectively mediate VIGS in cultivated plants like tomato (Liu *et al.*, 2002b) and pepper (Chung *et al.*, 2004) and in wild and cultivated potato species (Brigneti *et al.*, 2004). Due to its host range and improved VIGS phenotype, the TRV vector practically replaced the PVX vector in VIGS experiments on dicotyledonous plants. The first and at present the only VIGS vector for monocotyledonous plants has been developed by Holzberg *et al.* (2002). The tripartite barley stripe mosaic virus (BSMV) carrying a fragment of *PDS* was shown to trigger photobleaching in barley. The latest RNA virus-based VIGS vector takes advantage of the satellite tobacco mosaic virus (STMV) (Gossele *et al.*, 2002). This satellite virus-induced silencing system (SVISS) is a two-component system in which STMV is used as a vector for the delivery of silencing RNA into tobacco plants and TMV is used as a helper virus for providing movement and replication proteins *in trans*. The main advantage of this two-component system is that by uncoupling virus replication components from silencing induction components, the intensity of silencing becomes more pronounced (Gossele *et al.*, 2002). In addition to these RNA virus vectors, three DNA virus-based VIGS vectors have been developed using tomato golden mosaic virus (TGMV) (Kjemtrup *et al.*, 1998), cabbage leaf curl virus (CaLCuV) (Turnage *et al.*, 2002) and DNAbeta satellite DNA found in association with tomato yellow leaf curl China virus (TYLCCNV-Y10) (Tao & Zhou, 2004). Further technical details about the application of these vectors can be found in two recent reviews: Lu *et al.* (2003a) and Muangsan and Robertson (2004).

8.2.3   *Transgenic virus-induced gene silencing*

Transgene-induced silencing systems are convenient because the plants exhibit silencing throughout their life. However, transgenes do not trigger gene silencing consistently; the number of transgenic lines exhibiting silencing varies in an unpredictable way. In contrast, VIGS is very reliable; silencing is induced in every infected plant. However, it is difficult to infect very young or old plants and inoculation of large number of plants can be laborious. There is a hybrid approach, which combines the consistency of VIGS and the convenience of transgenic lines, using amplicon transgenes. Amplicon transgenic plants are transformed with a full-length cDNA copy of a virus downstream of a constitutive promoter. As with the first virus vectors, amplicon plants were originally made to overexpress proteins harboured by the viral construct. However, Angell and Baulcombe (1997) reported that tobacco plants transformed with a PVX–GUS vector expressed GUS at a very low level and that the plants did not exhibit symptoms of PVX. In fact, they found that these amplicon plants showed extreme strain-specific resistance against PVX and suppression of transiently expressed RNA-sharing homology with the transgene. Based on these observations they concluded that amplicon constructs induce gene silencing. One of the important findings was that gene silencing was induced in every single amplicon line. The authors recognised the potential of this technique and demonstrated that it can be used to consistently silence plant genes if the PVX amplicon contains fragment of host genes instead of, or in addition to, *GUS* (Angell & Baulcombe, 1999). Interestingly, after the discovery of viral suppressors (see Section 8.3) the amplicon plus system was generated, which achieved the original goal: very high expression of proteins by blocking gene silencing and allowing high level of virus replication (Mallory *et al.*, 2002). In spite of the usefulness of the amplicon-silencing system, it did not become widely used because soon after its discovery it was shown that inverted repeat constructs are as efficient and consistent as amplicon transgenes but without the difficulty of using viruses (Smith *et al.*, 2000; Waterhouse *et al.*, 1998).

8.2.4   Application of virus-induced gene silencing

The application of VIGS is discussed in two subsections: Section 8.2.4.1 describes how VIGS has been used to identify function of host genes and Section 8.2.4.2 is dedicated to the role of VIGS in deciphering the function of a particular group of plant genes, resistance genes.

8.2.4.1   Identification of gene function
Many groups have used VIGS to investigate the function of a variety of plant genes. Table 8.1 summarises the reports where VIGS was not only demonstrated to work in a certain species by targeting *PDS* or *Rubisco* but in fact used as a tool to gather information about the function of plant genes.

8.2.4.2   Analysing the function of disease resistance genes
Interestingly, VIGS has been used very extensively to investigate a specific group of plant genes. There are many more publications on applying VIGS to plant resistance genes than to all other types of genes. This may be due to the fact that scientists working on resistance genes were more willing to use plant pathogens (i.e. viruses) as tools. Nevertheless, VIGS has made a big impact on the plant–pathogen interaction field and led to the discovery of many new genes involved in different pathways. Table 8.2 summarises the studies on resistance genes using VIGS as a tool.

Table 8.1 Using VIGS to investigate functions of plant genes

Target gene	Host	VIGS vector	Silencing phenotype	Reference
NbCesA	N. benthamiana	PVX	Bulging leaf cells, dwarf plant	Burton et al., 2000
NbPCNA	N. benthamiana	TGMV	Abnormal leaf shape, loss of meristem activity	Peele et al., 2001
NbChlH	N. benthamiana	TMV	Yellow and white leaves	Hiriart et al., 2002
LeCTR1, LeCTR2	Tomato	TRV	Loss of chlorophyll, dwarf plants	Liu et al., 2002b
PS II oxygen evolving complex, 33k subunit	N. benthamiana	TRV	Light green leaves	Abbink et al., 2002
NtTK, NtRpII NtALS, NtPPX, NtPARP NtGln, NtNPK1, NtAct,	N. tabacum	STMV and TMV helper virus	Various visual phenotype	Gossele et al., 2002
NbFtsH	N. benthamiana	PVX	Photobleaching	Saitoh and Terauchi, 2002
NtCYP51	N. benthamiana	PVX	Accumulation of obtusifoliol and growth reduction	Burger et al., 2003
NbPAF, NbRPN9	N. benthamiana	TRV	Delayed programmed cell death	Kim et al., 2003
NtCDPK1, NtRPN3	N. benthamiana	PVX	Abnormal stomatal development	Lee et al., 2003
NaPMT, NaTI	N. attenuate	TRV	Suppressed methyl jasmonate pathway	Saedler and Baldwin, 2004
AtSMO1, AtSMO2	N. benthamiana	TMV	Accumulation of 4,4-dimethylsterols and 4alpha-methylsterols	Darnet and Rahier, 2004
NbIDDS, NbHDS, NbIDI	N. benthamiana	TRV	Albino leaves, disorganised palisade mesophyll, reduced cuticle, fewer plastids and disrupted thylakoid membrane	Page et al., 2004
NbDEK	N. benthamiana	TRV	Arrested organ development	Ahn et al., 2004
NbCHMP1	N. benthamiana	TRV	Altered leaf morphology	Yang et al., 2004

Table 8.2 Studying resistance gene pathways using VIGS

Target gene	Host	VIGS vector	Silencing phenotype	Reference
NbCDPK2	*N. benthamiana*	PVX	Wilting, hypersensitive response (HR)	Romeis *et al.*, 2001
NbSGT1	*N. benthamiana*	PVX	Breakdown of diverse types of disease resistance	Peart *et al.*, 2002a
NbEDS1	*N. benthamiana* transgenic N	PVX	Breakdown of TIR–NBS–LRR but not CC–NBS–LRR protein mediated resistance	Peart *et al.*, 2002b
Nt Rar1, NtEDS1, NtNPR1/NIM1	*N. benthamiana* transgenic N	TRV	Breakdown of N resistance	Liu *et al.*, 2002a
NbCOP9, NbSGT1, NbSKP1	*N. benthamiana* transgenic N	TRV	Breakdown of N resistance	Liu *et al.*, 2002c
NbNPK1	*N. benthamiana* transgenic N	TRV, PVX	Breakdown of N, Bs2 and Rx but not Pto and Cf4 mediated resistance, developmental defects	Jin *et al.*, 2002
NbWIPK, NbSIPK	*N. benthamiana*	PVX	HR	Sharma *et al.*, 2003
P58(IPK)	*N. benthamiana*	PVX	Plant death following viral infection	Bilgin *et al.*, 2003
NtMEK2	*N. benthamiana* transgenic N	PVX	Attenuated N gene–mediated resistance	Jin *et al.*, 2003
NbrbohA, NbrbohB	*N. benthamiana*	PVX	Reduced H_2O_2 accumulation, loss of resistance to Phytophthora, reduction and delay of HR	Yoshioka *et al.*, 2003
4992 random Nb genes	*N. benthamiana*	PVX	Various phenotypes	Lu *et al.*, 2003a, 2003b
21 candidate genes	Tomato	TRV	9 candidate genes were required for Pto-mediated resistance	Ekengren *et al.*, 2003
NbEDS1, NbSGT1	*N. benthamiana*	PVX	Breakdown of Bs4 resistance	Schornack *et al.*, 2004
LePP2Ac1	*N. benthamiana*	TRV	Constitutive expression of PR genes, localised cell death, increased resistance to *P. syringae*	He *et al.*, 2004
CITRX	*N. benthamiana*, tomato	TRV	Accelerated Cf-9/Avr9-triggered HR, enhanced accumulation of H_2O_2, alteration of protein kinase activity, induction of defence-related genes, increased resistance to Cladosporium fulvum	Rivas *et al.*, 2004
NbNTF6/NRK1, NbMEK1/NQK1, NbWRKY1–3, NbMYB1, NbCOI1	*N. benthamiana*	TRV	Compromised N-mediated resistance	Liu *et al.*, 2004a
LevarP	Tomato	TRV	Suppression of *P. syringae*-induced iNOS activation, increased disease symptom severity	Chandok *et al.*, 2004

8.3   Viral suppressors of gene silencing

The first part of this chapter described how viruses induce PTGS and how the silencing machinery targets viral RNA for degradation. The fact that viruses still exist and are passed on from plant to plant suggests that viruses can defend themselves against PTGS. The discovery of a viral defence system came from a series of experiments designed to understand the phenomenon of viral synergism (Vance *et al.*, 1995). Plants co-infected with several viruses often show more severe symptoms than the simple addition of symptoms caused by the viruses on their own. The development of these enhanced symptoms in plants infected with more than one virus is called synergism. Studying a certain virus pair, PVX and PVY, led to the conclusion that the level of PVY is similar in co-infected tissue and in tissue only infected by PVY. However, the level of PVX is much higher in the presence of PVY than in tissues infected only by PVX (Vance *et al.*, 1995). The next set of experiments was designed to find out which PVY proteins are responsible for the enhanced replication of PVX. Inoculating transgenic tobacco plants, expressing each of the PVY proteins, with PVX revealed that a protein called HcPro was sufficient to increase the level of PVX (Pruss *et al.*, 1997). The level of a virus can be increased either by enhancing the virus replication or by protecting it from degradation. Since the *in vitro* replication system that would allow to distinguish between these two possibilities was not available, an indirect approach was taken. It was shown that HcPro can increase the accumulation of non-viral mRNAs targeted by PTGS (Anandalakshmi *et al.*, 1998; Kasschau and Carrington, 1998). Therefore it became clear that HcPro does not enhance virus replication but suppresses PTGS, which allows higher accumulation of viral RNA or other RNAs targeted by PTGS. This was confirmed by another line of experiments where systemic silencing of GFP was suppressed by PVY infection or by inoculation with recombinant PVX virus expressing *HcPro* but not with PVX harbouring a frameshift mutant form of *HcPro* (Brigneti *et al.*, 1998). Interestingly, similar experiments revealed that HcPro is not unique because the 2b protein of CMV was also shown to suppress PTGS (Brigneti *et al.*, 1998). The only common feature of HcPro and 2b is that without these proteins the mutant viruses can infect plants but they accumulate at lower levels and cause milder symptoms. These viral proteins are therefore called pathogenicity determinants. Exploring other pathogenicity determinants from different viruses revealed that almost all viruses studied encode a suppressor of gene silencing (Voinnet *et al.*, 1999). A very interesting feature of viral suppressors of PTGS is that no common sequence or structural motif has been found among the various suppressor proteins. More than a dozen different suppressors have been identified, which are shown in Table 8.3. Further details on viral suppressors of gene silencing can be found in several good reviews (Baulcombe, 2002; Moissiard & Voinnet, 2004; Silhavy & Burgyan, 2004).

8.3.1   Characterisation of P19 and HcPro

The best-understood viral suppressor is the P19 protein encoded by tombus-viruses. In the absence of P19, tombusviruses cannot accumulate in upper leaves because of strong PTGS. Tombusvirus-specific siRNAs accumulate to high levels in wild-type virus–infected plants, suggesting that P19 does not block the generation of these molecules (Silhavy *et al.*, 2002). However, *in vitro* assays showed that P19 can bind to siRNAs, suggesting that the protein inhibits PTGS by blocking the siRNAs to target viral or mRNAs (Silhavy *et al.*, 2002). Further evidence supporting this model is that viral siRNAs were mainly found in P19 complexes in tombusvirus-infected plants and that P19

Table 8.3 Suppressors of PTGS encoded by plant viruses

Virus genus	Virus	Suppressor	Reference
Positive stranded RNA viruses			
Carmovirus	Turnip crinkle virus	P38	Qu *et al.*, 2003
Comovirus	Cowpea mosaic virus	Small coat protein	Liu *et al.*, 2004b
Cucumovirus	Cucumber mosaic virus	2b	Brigneti *et al.*, 1998
Closterovirus	Beet yellows virus	P21	Reed *et al.*, 2003
Hordeivirus	Barley yellow mosaic virus	Gamma-b	Yelina *et al.*, 2002
Pecluvirus	Peanut clump virus	P15	Dunoyer *et al.*, 2002
Poleovirus	Beet western yellow virus	P0	Pfeffer *et al.*, 2002
Potexvirus	Potato virus X	P25	Voinnet *et al.*, 2000
Potyvirus	Potato virus Y	HcPro	Anandalakshmi *et al.*, 1998; Kasschau and Carrington, 1998
Sobemovirus	Rice yellow mosaic virus	P1	Voinnet *et al.*, 1999
Tombusvirus	Tomato bushy stunt virus	P19	Voinnet *et al.*, 1999
Tobamovirus	Tomato mosaic virus	P130	Kubota *et al.*, 2003
Negative strand RNA viruses			
Tospovirus	Tomato spotted wilt virus	NSs	Takeda *et al.*, 2002
Tenuivirus	Rice hoja blanca virus	NS3	Bucher *et al.*, 2003
DNA viruses			
Begomoviruses	African cassava mosaic virus	AC2	Voinnet *et al.*, 1999
	Tomato yellow leaf curl virus	C2	Dong *et al.*, 2003

also suppresses RNA silencing in a heterologous *Drosophila in vitro* system by preventing siRNA incorporation into the RNA-induced silencing complex (RISC) (Lakatos *et al.*, 2004). Interestingly, P19 does not contain a conserved dsRNA-binding motif; however, the crystal structure of P19 showed that it selectively binds to siRNAs based on the length of the duplex region (Vargason *et al.*, 2003; Ye *et al.*, 2003).

The other well-characterised suppressor protein is HcPro. Expression of transgenic HcPro in *Arabidopsis* results in a phenotype very similar to *dcl1* plants (DCL1 cleaves the miRNA precursor molecules). However, while miRNAs are missing from *dcl1* plants, ectopic expression of HcPro does not affect the level of miRNAs (Kasschau *et al.*, 2003). These data suggest that HcPro inhibits miRNA activity by blocking the RISC. This model is supported by the observation that miRNA targets are upregulated both in *dcl1*, *HcPro* transgenic and in turnip mosaic potyvirus–infected plants (Kasschau *et al.*, 2003). The molecular mechanism of HcPro suppression is not known.

8.3.2 Suppressors break pathogen-derived resistance

The discovery of virus-encoded suppressors of PTGS led to concerns about the stability of PTGS-based virus resistance in transgenic plants. Two reports described the breakdown of virus resistance upon CMV infection (Mitter *et al.*, 2003; Simon-Mateo *et al.*, 2003). Both studies used plants resistant to different potyviruses (PVY and plum pox virus) and found that CMV infection led to increased susceptibility. The exact mechanism of breaking PTGS-based virus resistance is not known and two interesting observations suggest that it is more complicated than a simple suppression of PTGS. First, CMV suppression supported only a transient PVY accumulation and did not prevent recovery of the transgenic plants from PVY infection (Mitter *et al.*, 2003). Second, although CMV and tobacco vein mottling virus (TVMV) infection both led to suppression of PTGS of the transgene, only the silencing suppression caused by CMV, but not that originating from TVMV, was able to revert to a PPV-susceptible phenotype (Simon-Mateo *et al.*, 2003). These reports suggest that crop plants should be equipped with resistance function to several viruses to avoid the breakdown of resistance by infection with heterologous viruses.

8.3.3 Application of viral suppressors of gene silencing

The application of viral suppressors is discussed in two subsections: Section 8.3.3.1 describes how they have been used to analyse the silencing machinery and Section 8.3.3.2 illustrates how they can be used to overexpress proteins in plants.

8.3.3.1 *Analysing the silencing machinery*

Viral suppressors are thought to act at distinct steps of the silencing machinery based on the lack of common motifs and on observations that they have different effects on silencing phenotypes (block or do not block the silencing signal, revert or do not revert already silenced tissues, uniform or localised effect, etc.). Dunoyer *et al.* (2004) set up a system in *Arabidopsis* where they compared the effects of five unrelated viral silencing suppressors on the siRNA and miRNA pathways. Although all the suppressors inhibited PTGS, only three of them induced developmental defects, indicating that the two pathways are only partially overlapping. These developmental defects were very similar and correlated with inhibition of miRNA-guided cleavage of endogenous transcripts and not with altered miRNA accumulation per se. Among the suppressors investigated, the P19 protein co-immunoprecipitated with siRNA duplexes and miRNA duplexes corresponding to the primary cleavage products of miRNA precursors. Thus, the study confirmed that P19 blocks PTGS by sequestering both classes of small RNAs (see Section 8.3.1) Finally, the differential effects of the silencing suppressors tested here upon other types of *Arabidopsis* silencing-related small RNAs revealed a great variety of biosynthetic and, presumably, functional pathways for these molecules (Dunoyer *et al.*, 2004). This illustrates that silencing suppressors are valuable tools for investigating the complexity of RNA silencing.

8.3.3.2 *Overexpression of proteins*

Transient gene expression is a fast and flexible technique to express proteins. *A. tumefaciens* can be used for transient expression of genes that have been inserted into the T DNA region of the bacterial Ti plasmid in plants. A bacterial culture is infiltrated into leaves and T DNA transfer is followed by transient expression of the gene of interest in the plant cells. However, the effectiveness of the system is limited because the protein expression fades away after 2–3 days. Voinnet *et al.* (2003) reported that PTGS is the main reason for cessation of protein expression. They describe a system based on the co-expression of P19 that suppresses PTGS in the infiltrated tissues and allows high level of transient expression. Expression of a range of proteins was enhanced 50-fold or more in the presence of P19, and therefore protein purification could be achieved from as little as 100 mg of infiltrated leaf material (Voinnet *et al.*, 2003). The P19-facilitated expression system may be significant in industrial production and as a research tool for isolation and biochemical characterisation of proteins without generating transgenic plants.

Acknowledgement

The author thanks Dr Rachel Rusholme for critical reading of the manuscript.

References

Abbink, T. E., Peart, J. R., Mos, T. N., Baulcombe, D. C., Bol, J. F. & Linthorst, H. J. (2002) Silencing of a gene encoding a protein component of the oxygen-evolving complex of photosystem II enhances virus replication in plants. *Virology*, **295**, 307–19.

Ahn, J. W., Kim, M., Lim, J. H., Kim, G. T. & Pai, H. S. (2004) Phytocalpain controls the proliferation and differentiation fates of cells in plant organ development. *Plant Journal*, **38**, 969–81.

Anandalakshmi, R., Pruss, G. J., Ge, X., Marathe, R., Mallory, A. C., Smith, T. H. & Vance, V. B. (1998) A viral suppressor of gene silencing in plants. *Proceedings of the National Academy of Sciences of the United States of America*, **95**, 13079–84.

Angell, S. M. & Baulcombe, D. C. (1997) Consistent gene silencing in transgenic plants expressing a replicating potato virus X RNA. *EMBO Journal*, **16**, 3675–84.

Angell, S. M. & Baulcombe, D. C. (1999) Technical advance: potato virus X amplicon-mediated silencing of nuclear genes. *Plant Journal*, **20**, 357–62.

Baulcombe, D. C. (1996) Mechanisms of pathogen-derived resistance to viruses in transgenic plants. *Plant Cell*, **8**(10), 1833–1844.

Baulcombe, D. C. (2002) Viral suppression of systemic silencing. *Trends in Microbiology*, **10**, 306–308.

Beachy, R. N. (1997) Mechanisms and applications of pathogen-derived resistance in transgenic plants. *Current Opinion in Biotechnology*, **8**(2), 215–20.

Bernstein, E., Caudy, A. A., Hammond, S. M. & Hannon, G. J. (2001) Role for a bidentate ribonuclease in the initiation step of RNA interference. *Nature*, **409**, 363–6.

Bilgin, D. D., Liu, Y., Schiff, M. & Dinesh-Kumar, S. P. (2003) P58(IPK), a plant ortholog of double-stranded RNA-dependent protein kinase PKR inhibitor, functions in viral pathogenesis. *Developmental Cell*, **4**, 651–61.

Brigneti, G., Voinnet, O., Li, W. X., Ji, L. H., Ding, S. W. & Baulcombe, D. C. (1998) Viral pathogenicity determinants are suppressors of transgene silencing in *Nicotiana benthamiana*. *EMBO Journal*, **17**, 6739–46.

Brigneti, G., Martin-Hernandez, A. M., Jin, H., Chen, J., Baulcombe, D. C., Baker, B. & Jones, J. D. (2004) Virus-induced gene silencing in Solanum species. *Plant Journal*, **39**, 264–72.

Bucher, E., Sijen, T., De Haan, P., Goldbach, R. & Prins, M. (2003) Negative-strand tospoviruses and tenuiviruses carry a gene for a suppressor of gene silencing at analogous genomic positions. *Journal of Virology*, **77**, 1329–36.

Burger, C., Rondet, S., Benveniste, P. & Schaller, H. (2003) Virus-induced silencing of sterol biosynthetic genes: identification of a *Nicotiana tabacum* L. obtusifoliol-14-alpha-demethylase (CYP51) by genetic manipulation of the sterol biosynthetic pathway in *Nicotiana benthamiana*. *Journal of Experimental Botany*, **54**, 1675–83.

Burton, R. A., Gibeaut, D. M., Bacic, A., Findlay, K., Roberts, K., Hamilton, A., Baulcombe, D. C. & Fincher, G. B. (2000) Virus-induced silencing of a plant cellulose synthase gene. *Plant Cell*, **12**, 691–706.

Chandok, M. R., Ekengren, S. K., Martin, G. B. & Klessig, D. F. (2004) Suppression of pathogen-inducible NO synthase (iNOS) activity in tomato increases susceptibility to *Pseudomonas syringae*. *Proceedings of the National Academy of Sciences of the United States of America*, **101**, 8239–44.

Chapman, S., Kavanagh, T. & Baulcombe, D. (1992) Potato virus X as a vector for gene expression in plants. *Plant Journal*, **2**, 549–57.

Chung, E., Seong, E., Kim, Y. C., Chung, E. J., Oh, S. K., Lee, S., Park, J. M., Joung, Y. H. & Choi, D. (2004) A method of high frequency virus-induced gene silencing in chili pepper (*Capsicum annuum* L. cv. Bukang). *Molecular Cell*, **17**, 377–80.

Covey, S. N., Al-Kaff, N. S., Langara, A. & Turner, D. S. (1997) Plants combat infection by gene silencing. *Nature*, **385**, 781–2.

Dalmay, T. D., Hamilton, A. J., Rudd, S., Angell, S. & Baulcombe, D. C. (2000) An RNA-dependent RNA polymerase gene in *Arabidopsis* is required for posttranscriptional gene silencing mediated by a transgene but not by a virus. *Cell*, **101**, 543–53.

Dalmay, T. D., Horsefield, R., Braunstein, T. H. & Baulcombe, D. C. (2001) SDE3 encodes an RNA helicase required for post-transcriptional gene silencing in *Arabidopsis*. *EMBO Journal*, **20**, 2069–78.

Darnet, S. & Rahier, A. (2004) Plant sterol biosynthesis: identification of two distinct families of sterol 4-alpha-methyl oxidases. *Biochemistry Journal*, **378**, 889–98.

Dong, X., van Wezel, R., Stanley, J. & Hong, Y. (2003) Functional characterization of the nuclear localization signal for a suppressor of posttranscriptional gene silencing. *Journal of Virology*, **77**, 7026–33.

Donson, J., Kearney, C. M., Hilf, M. E. & Dawson, W. O. (1991) Systemic expression of a bacterial gene by a tobacco mosaic virus-based vector. *Proceedings of the National Academy of Sciences of the United States of America*, **88**, 7204–208.

Dunoyer, P., Pfeffer, S., Fritsch, C., Hemmer, O., Voinnet, O. & Richards, K. E. (2002) Identification, subcellular localization and some properties of a cysteine-rich suppressor of gene silencing encoded by peanut clump virus. *Plant Journal*, **29**, 555–67.

Dunoyer, P., Lecellier, C. H., Parizotto, E. A., Himber, C. & Voinnet, O. (2004) Probing the microRNA and small interfering RNA pathways with virus-encoded suppressors of RNA silencing. *Plant Cell*, **16**, 1235–50.

Ekengren, S. K., Liu, Y., Schiff, M., Dinesh-Kumar, S. P., Martin, G. B. (2003) Two MAPK cascades, NPR1, and TGA transcription factors play a role in Pto-mediated disease resistance in tomato. *Plant Journal*, **36**, 905–17.

English, J. J., Mueller, E. & Baulcombe, D. C. (1996) Suppression of virus accumulation in transgenic plants exhibiting silencing of nuclear genes. *Plant Cell*, **8**, 179–88.

Faivre-Rampant, O., Gilroy, E. M., Hrubikova, K., Hein, I., Millam, S., Loake, G. J., Birch, P., Taylor, M. & Lacomme, C. (2004) Potato virus X-induced gene silencing in leaves and tubers of potato. *Plant Physiology*, **134**, 1308–16.

Fitchen, J. H. & Beachy, R. N. (1993) Genetically-engineered protection against viruses in transgenic plants. *Annual Review of Microbiology*, **47**, 739–63.

Gossele, V., Fache, I., Meulewaeter, F., Cornelissen, M. & Metzlaff, M. (2002) SVISS – a novel transient gene silencing system for gene function discovery and validation in tobacco plants. *Plant Journal*, **32**, 859–66.

Hamilton, A. J. & Baulcombe, D. C. (1999) A species of small antisense RNA in post-transcriptional gene silencing in plants. *Science*, **286**, 950–52.

Hamilton, A. J., Voinnet, O., Chappell, L. & Baulcombe, D. C. (2002) Two classes of short interfering RNA in RNA silencing. *EMBO Journal*, **21**, 4671–9.

Harrison, B. D. (1958) Relationship between beet ringspot, potato bouquet and tomato black ring viruses. *Journal of General Microbiology*, **18**, 450.

He, X., Anderson, J. C., Pozo Od, O., Gu, Y. Q., Tang, X. & Martin, G. B. (2004) Silencing of subfamily I of protein phosphatase 2A catalytic subunits results in activation of plant defense responses and localized cell death. *Plant Journal*, **38**, 563–77.

Himber, C., Dunoyer, P., Moissiard, G., Ritzenthaler, C. & Voinnet, O. (2003) Transitivity-dependent and -independent cell-to-cell movement of RNA silencing. *EMBO Journal*, **22**, 4523–33.

Hiriart, J. B., Lehto, K., Tyystjarvi, E., Junttila, T. & Aro, E. M. (2002) Suppression of a key gene involved in chlorophyll biosynthesis by means of virus induced gene silencing. *Plant Molecular Biology*, **50**, 213–24.

Hiriart, J. B., Aro, E. M. & Lehto, K. (2003) Dynamics of the VIGS-mediated chimeric silencing of the *Nicotiana benthamiana* ChlH gene and of the tobacco mosaic virus vector. *Molecular Plant–Microbe Interactions*, **16**, 99–106.

Holzberg, S., Brosio, P., Gross, C. & Pogue, G. P. (2002) Barley stripe mosaic virus-induced gene silencing in a monocot plant. *Plant Journal*, **30**, 315–27.

Jin, H., Axtell, M. J., Dahlbeck, D., Ekwenna, O., Zhang, S., Staskawicz, B. & Baker, B. (2002) NPK1, an MEKK1-like mitogen-activated protein kinase kinase kinase, regulates innate immunity and development in plants. *Developmental Cell*, **3**, 291–7.

Jin, H., Liu, Y., Yang, K. Y., Kim, C. Y., Baker, B. & Zhang, S. (2003) Function of a mitogen-activated protein kinase pathway in N gene-mediated resistance in tobacco. *Plant Journal*, **33**, 719–31.

Jones, L., Hamilton, A. J., Voinnet, O., Thomas, C. L., Maule, A. J. & Baulcombe, D. C. (1999) RNA–DNA interactions and DNA methylation in post-transcriptional gene silencing. *Plant Cell*, **11**, 2291–301.

Kasschau, K. D. & Carrington, J. C. (1998) A counterdefensive strategy of plant viruses: suppression of posttranscriptional gene silencing. *Cell*, **95**, 461–70.

Kasschau, K. D., Xie, Z., Allen, E., Llave, C., Chapman, E. J., Krizan, K. A. & Carrington, J. C. (2003) P1/ HC-Pro, a viral suppressor of RNA silencing, interferes with *Arabidopsis* development and miRNA function. *Developmental Cell*, **4**, 205–17.

Kim, M., Yang, K. S., Kim, Y. K., Paek, K. H. & Pai, H. S. (2003) Molecular characterization of NbPAF encoding the alpha6 subunit of the 20S proteasome in *Nicotiana benthamiana*. *Molecular Cell*, **15**, 127–32.

Kjemtrup, S., Sampson, K. S., Peele, C. G., Nguyen, L. V., Conkling, M. A., Thompson, W. F. & Robertson, D. (1998) Gene silencing from plant DNA carried by a Geminivirus. *Plant Journal*, **14**, 91–100.

Kubota, K., Tsuda, S., Tamai, A. & Meshi, T. (2003) Tomato mosaic virus replication protein suppresses virus-targeted posttranscriptional gene silencing. *Journal of Virology*, **77**, 11016–26.

Kumagai, M. H., Donson, J., Della-Cioppa, G., Harvey, D., Hanley, K. & Grill, L. K. (1995) Cytoplasmic inhibition of carotenoid biosynthesis with virus-derived RNA. *Proceedings of the National Academy of Sciences of the United States of America*, **92**, 1679–83.

Lakatos, L., Szittya, G., Silhavy, D. & Burgyan, J. (2004) Molecular mechanism of RNA silencing suppression mediated by p19 protein of tombusviruses. *EMBO Journal*, **23**, 876–84.

Lee, S. S., Cho, H. S., Yoon, G. M., Ahn, J. W., Kim, H. H. & Pai, H. S. (2003) Interaction of NtCDPK1 calcium-dependent protein kinase with NtRpn3 regulatory subunit of the 26S proteasome in *Nicotiana tabacum*. *Plant Journal*, **33**, 825–40.

Lindbo, J. A., Silva-Rosales, L., Proebsting, W. M. & Dougherty, W. G. (1993) Induction of a highly specific antiviral state in transgenic plants: implications for regulation of gene expression and virus resistance. *Plant Cell*, **5**, 1749–59.

Liu, Y., Schiff, M., Marathe, R. & Dinesh-Kumar, S. P. (2002a) Tobacco *Rar1, EDS1* and *NPR1/NIM1* like genes are required for *N*-mediated resistance to tobacco mosaic virus. *Plant Journal*, **30**, 415–29.

Liu, Y., Schiff, M. & Dinesh-Kumar, S. P. (2002b) Virus-induced gene silencing in tomato. *Plant Journal*, **31**, 777–86.

Liu, Y., Schiff, M., Serino, G., Deng, X. W. & Dinesh-Kumar, S. P. (2002c) Role of SCF ubiquitin-ligase and the COP9 signalosome in the N gene-mediated resistance response to tobacco mosaic virus. *Plant Cell*, **14**, 1483–96.

Liu, Y., Schiff, M. & Dinesh-Kumar, S. P. (2004a) Involvement of MEK1 MAPKK, NTF6 MAPK, WRKY/ MYB transcription factors, COI1 and CTR1 in N-mediated resistance to tobacco mosaic virus. *Plant Journal*, **38**, 800–809.

Liu, L., Grainger, J., Canizares, M. C., Angell, S. M. & Lomonossoff, G. P. (2004b) Cowpea mosaic virus RNA-1 acts as an amplicon whose effects can be counteracted by a RNA-2-encoded suppressor of silencing. *Virology*, **323**, 37–48.

Longstaff, M., Brigneti, G., Boccard, F., Chapman, S. N. & Baulcombe, D. C. (1993) Extreme resistance to potato virus X infection in plants expressing a modified component of the putative viral replicase. *EMBO Journal*, **12**, 379–86.

Lu, R., Martin-Hernandez, A. M., Peart, J. R., Malcuit, I. & Baulcombe, D. C. (2003a) Virus induced gene silencing in plants. *Methods*, **30**, 296–303.

Lu, R., Malcuit, I., Moffett, P., Ruiz, M. T., Peart, J., Wu, A. J., Rathjen, J. P., Bendahmane, A., Day, L. & Baulcombe, D. C. (2003b) High throughput virus-induced gene silencing implicates heat shock protein 90 in plant disease resistance. *EMBO Journal*, **22**, 5690–99.

Mallory, A. C., Parks, G., Endres, M. W., Baulcombe, D., Bowman, L. H., Pruss, G. J. & Vance, V. B. (2002) The amplicon-plus system for high-level expression of transgenes in plants. *Nature Biotechnology*, **20**, 622–5.

Matthews, R. E. F. (1992) *Fundamentals of Plant Virology*. Academic Press, San Diego.

Mitter, N., Sulistyowati, E., Dietzgen, R. G. (2003) Cucumber mosaic virus infection transiently breaks dsRNA-induced transgenic immunity to potato virus Y in tobacco. *Molecular Plant–Microbe Interactions*, **16**, 936–44.

Moissiard, G. & Voinnet, O. (2004) Viral suppression of RNA silencing in plants. *Molecular Plant Pathology*, **5**, 71–82.

Mourrain, P., Beclin, C., Elmayan, T., Feuerbach, F., Godon, C., Morel, J. B., Jouette, D., Lacombe, A. M., Nikic, S., Picault, N., Remoue, K., Sanial, M. T. A. & Vaucheret, H. (2000) *Arabidopsis SGS2* and *SGS3* genes are required for posttranscriptional gene silencing and natural virus resistance. *Cell*, **101**, 533–42.

Muangsan, N. & Robertson, D. (2004) Geminivirus vectors for transient gene silencing in plants. *Methods in Molecular Biology*, **265**, 101–15.

Muangsan, N., Beclin, C., Vaucheret, H. & Robertson, D. (2004) Geminivirus VIGS of endogenous genes requires SGS2/SDE1 and SGS3 and defines a new branch in the genetic pathway for silencing in plants. *Plant Journal*, **38**, 1004–14.

Mueller, E., Gilbert, J. E., Davenport, G., Brigneti, G. & Baulcombe, D. C. (1995) Homology-dependent resistance: transgenic virus resistance in plants related to homology-dependent gene silencing. *Plant Journal*, **7**, 1001–13.

Napoli, C., Lemieux, C. & Jorgensen, R. (1990) Introduction of a chimeric chalcone synthase gene into petunia results in reversible co-suppression of homologous genes *in trans*. *Plant Cell*, **2**, 279–89.

Page, J. E., Hause, G., Raschke, M., Gao, W., Schmidt, J., Zenk, M. H. & Kutchan, T. M. (2004) Functional analysis of the final steps of the 1-deoxy-D-xylulose 5-phosphate (DXP) pathway to isoprenoids in plants using virus-induced gene silencing. *Plant Physiology*, **134**, 1401–13.

Palauqui, J. C., Elmayan, T., Deborne, F. D., Crete, P., Charles, C. & Vaucheret, H. (1996) Frequencies, timing, and spatial patterns of co-suppression of nitrate reductase and nitrite reductase in transgenic tobacco plants. *Plant Physiology*, **112**, 1447–56.

Palauqui, J. C., Elmayan, T., Pollien, J. M. & Vaucheret, H. (1997) Systemic acquired silencing: transgene-specific post-transcriptional silencing is transmitted by grafting from silenced stocks to non-silenced scions. *EMBO Journal*, **16**, 4738–45.

Peart, J. R., Lu, R., Sadanandom, A., Malcuit, I., Moffett, P., Brice, D. C., Schauser, L., Jaggard, D. A., Xiao, S., Coleman, M. J., Dow, M., Jones, J. D., Shirasu, K. & Baulcombe, D. C. (2002a) Ubiquitin ligase-associated protein SGT1 is required for host and nonhost disease resistance in plants. *Proceedings of the National Academy of Sciences of the United States of America*, **99**, 10865–9.

Peart, J. R., Cook, G., Feys, B. J., Parker, J. E. & Baulcombe, D. C. (2002b) An EDS1 orthologue is required for N-mediated resistance against tobacco mosaic virus. *Plant Journal*, **29**, 569–79.

Peele, C., Jordan, C. V., Muangsan, N., Turnage, M., Egelkrout, E., Eagle, P., Hanley-Bowdoin, L. & Robertson, D. (2001) Silencing of a meristematic gene using geminivirus-derived vectors. *Plant Journal*, **27**, 357–66.

Pfeffer, S., Dunoyer, P., Heim, F., Richards, K. E., Jonard, G. & Ziegler-Graff, V. (2002) P0 of beet western yellows virus is a suppressor of posttranscriptional gene silencing. *Journal of Virology*, **76**, 6815–24.

Pruss, G., Ge, X., Shi, X. M., Carrington, J. C. & Vance, V. B. (1997) Plant viral synergism: the potyviral genome encodes a broad range pathogenicity enhancer that transactivates replication of heterologous viruses. *Plant Cell*, **9**, 859–68.

Qu, F., Ren, T. & Morris, T. J. (2003) The coat protein of turnip crinkle virus suppresses posttranscriptional gene silencing at an early initiation step. *Journal of Virology*, **77**, 511–22.

Ratcliff, F., Harrison, B. D. & Baulcombe, D. C. (1997) A similarity between viral defense and gene silencing in plants. *Science*, **276**, 1558–60.

Ratcliff, F., MacFarlane, S. & Baulcombe, D. C. (1999) Gene silencing without DNA: RNA-mediated cross protection between viruses. *Plant Cell*, **11**, 1207–15.

Ratcliff, F., Martin-Hernandez, A. M. & Baulcombe, D. C. (2001) Tobacco rattle virus as a vector for analysis of gene function by silencing. *Plant Journal*, **25**, 237–45.

Reed, J. C., Kasschau, K. D., Prokhnevsky, A. I., Gopinath, K., Pogue, G. P., Carrington, J. C. & Dolja, V. V. (2003) Suppressor of RNA silencing encoded by beet yellows virus. *Virology*, **306**, 203–209.

Rivas, S., Rougon-Cardoso, A., Smoker, M., Schauser, L., Yoshioka, H. & Jones, J. D. (2004) CITRX thioredoxin interacts with the tomato Cf-9 resistance protein and negatively regulates defence. *EMBO Journal*, **23**, 2156–65.

Romeis, T., Ludwig, A. A., Martin, R. & Jones, J. D. (2001) Calcium-dependent protein kinase play an essential role in a plant defence response. *EMBO Journal*, **20**, 5556–67.

Ruiz, M. T., Voinnet, O. & Baulcombe, D. C. (1998) Initiation and maintenance of virus-induced gene silencing. *Plant Cell*, **10**, 937–46.

Saedler, R. & Baldwin, I. T. (2004) Virus-induced gene silencing of jasmonate-induced direct defences, nicotine and trypsin proteinase-inhibitors in *Nicotiana attenuate*. *Journal of Experimental Botany*, **55**, 151–7.

Saitoh, H. & Terauchi, R. (2002) Virus induced silencing of FtsH gene in *Nicotiana benthamiana* causes a striking bleached phenotype. *Genes and Genetic Systems*, **77**, 335–40.

Sanders, M., Maddelein, W., Depicker, A., Van Montagu, M., Cornelissen, M. & Jacobs, J. (2002) An active role for endogenous beta-1,3-glucanase genes in transgene-mediated co-suppression in tobacco. *EMBO Journal*, **21**, 5824–32.

Sanford, J. C. & Johnston, S. A. (1985) The concept of parasite-derived resistance-deriving resistance genes from the parasite's own genome. *Journal of Theoretical Biology*, **113**, 395–405.

Schornack, S., Ballvora, A., Gurlebeck, D., Peart, J., Ganal, M., Baker, B., Bonas, U. & Lahaye, T. (2004) The tomato resistance protein Bs4 is a predicted non-nuclear TIR-NB-LRR protein that mediates defense responses to severely truncated derivatives of AvrBs4 and overexpressed AvrBs3. *Plant Journal*, **37**, 46–60.

Sequeira, L. (1984) Cross protection and induced resistance: their potential for plant disease control. *Trends in Biochemical Sciences*, **2**, 26–30.

Sharma, P. C., Ito, A., Shimizu, T., Terauchi, R., Kamoun, S. & Saitoh, H. (2003) Virus-induced silencing of WIPK and SIPK genes reduces resistance to a bacterial pathogen, but has no effect on the INF1-induced hypersensitive response (HR) in *Nicotiana benthamiana*. *Molecular Genetics and Genomics*, **269**, 583–91.

Silhavy, D. & Burgyan, J. (2004) Effects and side-effects of viral RNA silencing suppressors on short RNAs. *Trends in Plant Science*, **9**, 76–83.

Silhavy, D., Molnar, A., Lucioli, A., Szittya, G., Hornyik, C., Tavazza, M. & Burgyan, J. (2002) A viral protein suppresses RNA silencing and binds silencing-generated, 21- to 25-nucleotide double-stranded RNAs. *EMBO Journal*, **21**, 3070–80.

Simon-Mateo, C., Lopez-Moya, J. J., Guo, H. S., Gonzalez, E., Garcia, J. A. (2003) Suppressor activity of potyviral and cucumoviral infections in potyvirus-induced transgene silencing. *Journal of General Virology*, **84**, 2877–83.

Smith, N. A., Singh, S. P., Wang, M. B., Stoutjesdijk, P. A., Green, A. G. & Waterhouse, P. M. (2000) Total silencing by intron-spliced hairpin RNAs. *Nature*, **407**, 319–20.

Szittya, G., Molnar, A., Silhavy, D., Hornyik, C. & Burgyan, J. (2002) Short defective interfering RNAs of tombusviruses are not targeted but trigger posttranscriptional gene silencing against their helper virus. *Plant Cell*, **14**, 359–72.

Takeda, A., Sugiyama, K., Nagano, H., Mori, M., Kaido, M., Mise, K., Tsuda, S. & Okuno, T. (2002) Identification of a novel RNA silencing suppressor, NSs protein of tomato spotted wilt virus. *FEBS Letters*, **532**, 75–9.

Tao, X. & Zhou, X. (2004) A modified viral satellite DNA that suppresses gene expression in plants. *Plant Journal*, **38**, 850–60.

Turnage, M. A., Muangsan, N., Peele, C. G. & Robertson, D. (2002) Geminivirus-based vectors for gene silencing in *Arabidopsis*. *Plant Journal*, **30**, 107–14.

Vaistij, F. E., Jones, L. & Baulcombe, D. C. (2002) Spreading of RNA targeting and DNA methylation in RNA silencing requires transcription of the target gene and a putative RNA-dependent RNA polymerase. *Plant Cell*, **14**, 857–67.

Vance, V. B., Berger, P. H., Carrington, J. C., Hunt, A. G. & Shi, X. M. (1995) 5′ proximal potyviral sequences mediate potato virus X/potyviral synergistic disease in transgenic tobacco. *Virology*, **206**, 583–90.

Vargason, J. M., Szittya, G., Burgyan, J. & Tanaka Hall, T. M. (2003) Size selective recognition of siRNA by an RNA silencing suppressor. *Cell*, **115**, 799–811.

Voinnet, O. & Baulcombe, D. C. (1997) Systemic signalling in gene silencing. *Nature*, **389**, 553.

Voinnet, O., Vain, P., Angell, S. & Baulcombe, D. C. (1998) Systemic spread of sequence-specific transgene RNA degradation is initiated by localised introduction of ectopic promoterless DNA. *Cell*, **95**, 177–87.

Voinnet, O., Pinto, Y. M. & Baulcombe, D. C. (1999) Suppression of gene silencing: a general strategy used by diverse DNA and RNA viruses of plants. *Proceedings of the National Academy of Sciences of the United States of America*, **96**, 14147–52.

Voinnet, O., Lederer, C. & Baulcombe, D. C. (2000) A viral movement protein prevents systemic spread of the gene silencing signal in *Nicotiana benthamiana*. *Cell*, **103**, 157–67.

Voinnet, O., Rivas, S., Mestre, P. & Baulcombe, D. (2003) An enhanced transient expression system in plants based on suppression of gene silencing by the p19 protein of tomato bushy stunt virus. *Plant Journal*, **33**, 949–56.

Waterhouse, P. M., Graham, M. W. & Wang, M. B. (1998) Virus resistance and gene silencing in plants can be induced by simultaneous expression of sense and antisense RNA. *Proceedings of the National Academy of Sciences of the United States of America*, **95**, 13959–64.

Xie, Z., Johansen, L. K., Gustafson, A. M., Kassachau, K. D., Lellis, A. D., Zilberman, D., Jacobsen, S. E. & Carrington, J. C. (2004) Genetic and functional diversification of small RNA pathways in plants. *Public Library of Science Biology*, **2**, 1–11.

Yang, K. S., Jin, U. H., Kim, J., Song, K., Kim, S. J., Hwang, I., Lim, Y. P. & Pai, H. S. (2004) Molecular characterization of NbCHMP1 encoding a homolog of human CHMP1 in *Nicotiana benthamiana*. *Molecular Cell*, **17**, 255–61.

Ye, K., Malinina, L. & Patel, D. J. (2003) Recognition of small interfering RNA by a viral suppressor of RNA silencing. *Nature*, **426**, 874–8.

Yelina, N. E., Savenkov, E. I., Solovyev, A. G., Morozov, S. Y. & Valkonen, J. P. (2002) Long-distance movement, virulence, and RNA silencing suppression controlled by a single protein in hordei- and potyviruses: complementary functions between virus families. *Journal of Virology*, **76**, 12981–91.

Yoshioka, H., Numata, N., Nakajima, K., Katou, S., Kawakita, K., Rowland, O., Jones, J. D. & Doke, N. (2003) *Nicotiana benthamiana* gp91phox homologs NbrbohA and NbrbohB participate in H_2O_2 accumulation and resistance to *Phytophthora infestans*. *Plant Cell*, **15**, 706–18.

9 MicroRNAs: micro-managing the plant genome

Sandra K. Floyd and John L. Bowman

9.1 Abstract

Small non-coding RNAs have been implicated in endogenous gene regulation and in the control of invading genetic entities. Small RNAs derived from double-stranded RNA that may or may not be encoded in the genome are termed short interfering RNAs (siRNAs). Such RNAs act to silence invading nucleic acids, such as viruses and transposons, via RNA cleavage or changes in chromatin structure. These phenomena have been termed posttranscriptional gene silencing (PTGS), virus-induced gene silencing (VIGS) and RNA interference (RNAi), and they function as a cellular or genomic immune system. In contrast, endogenously encoded small RNAs that are processed from stem-loop precursor RNAs and that regulate genes encoded at separate distinct loci are called microRNAs (miRNAs). These two classes of small RNA share similarities in their modes of action, suggesting similar biochemical complexes may mediate their function. However, they differ in both their biogenesis and their target sequences, with endogenous genes encoding miRNAs whose targets are other distinct endogenous genes and with siRNAs, whose targets are invading nucleic acids, being derived from double-stranded RNA due to either aberrant transcriptional patterns within transposons or production during viral infection. This chapter concentrates on the biogenesis, activity and evolution of miRNAs in plants, with reference to other systems when appropriate.

9.2 Discovery of miRNAs

miRNAs were first identified in forward genetic screens in *Caenorhabditis elegans*, when it was discovered that a gene involved in timing in larval development, *lin-4*, did not encode a protein, but rather encoded a small RNA (Lee *et al.*, 1993) that negatively regulates the translation of mRNAs produced from another gene, *lin-14*. *lin-4* was previously shown to be required for repression of *lin-14*, and this repression was mediated by sequences in the 3′ UTR of *lin-14* (Wightman *et al.*, 1993). Complementarity between the 3′ untranslated region of *lin-14* known to be required for proper regulation of *lin-14* and that of the 22 nucleotide *lin-4* led to a model in which pairing between the two RNAs is required for translational repression of the *lin-14* mRNA in a regulatory pathway required for the transition between the first and the second larval stages (Lee *et al.*, 1993; Wightman *et al.*, 1993). Because there was no evidence for sequences similar to *lin-4* outside nematodes, this peculiar mode of gene regulation was thought to be an evolutionary novelty. However, the obscurity of small RNA-mediated eukaryotic gene regulation vanished upon the discovery of a second gene, *let-7*, encoding a small RNA that also regulates developmental timing and is present in all bilateral animals examined (Pasquinelli *et al.*, 2000; Reinhart *et al.*, 2000; Slack *et al.*, 2000). Because of the developmental timing roles of *lin-4* and *let-7*, they were originally referred to as small temporal RNAs (stRNAs), but, as additional small RNAs with diverse roles were identified, they acquired the moniker, miRNAs.

9.3   miRNAs versus siRNAs

Small non-coding RNAs have been implicated in endogenous gene regulation and in the control of invading genetic entities (for reviews see Ambros, 2004; Bartel, 2004; Baulcombe, 2004; He & Hannon, 2004; Matzke *et al.*, 2001; Meister & Tuschl, 2004; Plasterk, 2002; Vaucheret *et al.*, 2001; Waterhouse *et al.*, 2001; Zamore, 2002). Small RNAs derived from double-stranded RNA molecules that may or may not be encoded in the genome are termed siRNAs. Such RNAs act to silence invading nucleic acids, such as viruses and transposons, via RNA cleavage or changes in chromatin structure. The siRNAs are derived from double-stranded RNA either produced due to aberrant transcriptional patterns within transposons or generated during viral infection. In some cases these siRNAs may be amplified by an RNA-dependent RNA polymerase, and may propagate silencing through generations and systemically throughout an organism. These phenomena have been termed PTGS, VIGS, RNAi and transcriptional gene silencing (TGS), and function as a cellular or a genomic immune system. In contrast, endogenously encoded small RNAs that are processed from stem-loop precursor RNAs and that regulate genes encoded at separate distinct loci are called miRNAs. miRNAs target mRNAs derived from distinct endogenous genes, silencing their activity by inhibition of translation, by altering mRNA stability or through changes in chromatin structure. These two classes of small RNA, siRNAs and miRNAs, share similarities in their modes of action, suggesting similar biochemical complexes may mediate their function. Indeed, some siRNAs, termed *trans*-acting siRNAs, derived from both the sense and the antisense strands of a non-coding RNA that is apparently made into double-stranded RNA target distinct endogenously encoded genes (Vazquez *et al.*, 2004b). The generation of these *trans*-acting siRNAs requires some proteins normally required for miRNA biogenesis and other proteins normally required for PTGS, demonstrating that the two pathways can share enzymatic components (Vazquez *et al.*, 2004b). Because the two pathways are likely to be evolutionarily related to (or derived from) one another and because of the similarity in their modes of action, the extent of crosstalk between the siRNA and miRNA pathways needs to be further investigated.

9.4   Biogenesis of miRNAs: pri-miRNA, pre-miRNA, mature miRNAs

miRNAs are derived from longer primary transcripts (pri-miRNAs) that are processed into smaller stem-loop miRNA precursors (pre-miRNAs) by an RNase III referred to as DICER (Meister & Tuschl, 2004). There is strong evidence, in both plants and animals, that RNA polymerase II is the primary, perhaps sole, polymerase generating pri-miRNAs since pri-miRNAs are capped at their 5′ end, are polyadenylated and some contain introns (Auker-man & Sakai, 2003; Parizotto *et al.*, 2004). Pre-miRNAs are then processed to yield mature miRNAs, a reaction also catalyzed by DICER in plants (Papp *et al.*, 2003; Park *et al.*, 2002). In contrast to plants where a single enzyme is thought to perform both processing steps inside the nucleus, in animals, the second processing step occurs in the cytoplasm and is catalyzed by a different enzyme (Bartel, 2004; He & Hannon, 2004; Meister & Tuschl, 2004). In animals, pre-miRNAs are transported from the nucleus to the cytoplasm by Exportin 5, a Ran-GTP-dependent nucleo/cytoplasmic cargo transporter (Lund *et al.*, 2004). In plants, some aspect of nucleo-cytoplasmic transport of miR-NAs may be performed by a similar transporter, HASTY, which, when mutant, has phenotypic defects similar to those associated with alterations in develop-mental genes regulated by miRNAs (Bollman *et al.*, 2003).

Following processing of the pre-miRNA by DICER one strand of the resulting miRNA:miRNA* product, the miRNA strand (the miRNA* strand is likely rapidly degraded), accumulates as a mature miRNA that acts to regulate gene expression. The chosen strand is almost always the strand with the less stable 5′ end as compared with the miRNA* strand (Hutvágner & Zamore, 2002a; Khvorova *et al.*, 2003; Schwarz *et al.*, 2003). The mature miRNAs act within a complex that is assumed to be similar to the RNA-induced silencing complex (RISC), which was first identified as the complex that acts in siRNA-mediated RNAi (Hammond *et al.*, 2000). A core compon-ent of the RISC is a member of the ARGONAUTE protein family and ARGONAUTE proteins bind double-stranded RNA, suggesting that they may be directly associated with small RNAs before (e.g. during selection of the miRNA from the miRNA:miRNA*) or during target recognition (Ham-mond *et al.*, 2001; Martinez *et al.*, 2002; Mourelatos *et al.*, 2002). More recently, based on biochemistry and crystal structure, it has been proposed that Argonaute provides the 'slicer' activity within the RISC involved in siRNA-mediated silencing (Liu *et al.*, 2004; Song *et al.*, 2004). Other genes, such as *HEN1* and *HYL1*, are also required for miRNA biogenesis, but the precise biochemical roles of the encoded proteins are as yet undefined (Boutet *et al.*, 2003; Han *et al.*, 2004; Vazquez *et al.*, 2004a).

While the overall picture of miRNA biogenesis via RNA polymerase II transcription followed by DICER-mediated processing and incorporation into an RISC is well established, the precise roles of genes involved in miRNA, as opposed to siRNA, action are less clear due to redundancy in genes encoding both DICER and Argonaute proteins. Four *DICER-LIKE* (*DCL1–4*) genes are encoded in the *Arabidopsis* genome, and they exhibit distinct as well as redundant activities (Xie *et al.*, 2004). DCL1 appears to be the primary protein involved in miRNA biogenesis while DCL3 is required for the generation of retroelement and transposon siRNAs, and DCL2 is implicated in some aspects of viral siRNA production (Finnegan *et al.*, 2003; Papp *et al.*, 2003; Xie *et al.*, 2004). However, that at least some functions of miRNA165/166 are not altered in the *CARPEL FACTORY* allele of *DCL1* suggests that there may be some functional redundancy with respect to miRNA biogenesis, although this should be tested in the *DCL1* embryo-lethal null allele (Bao *et al.*, 2004). That null *DCL1* alleles are embryo-lethal underscores the importance of miRNA action in plant development (Schauer *et al.*, 2002).

The *Arabidopsis* genome encodes ten ARGONAUTE-like proteins, raising the possibility that several different RISCs with differing activities may be present in plants. While an RISC dedicated to miRNA-, as opposed to siRNA-, mediated activity has not been definitively identified, it is likely that such a complex exists and it may have its own dedicated ARGONAUTE protein family member(s). *ARGONAUTE1* (*AGO1*) is required for the action of many miRNAs, since their accumulation is reduced in *ago1* mutants (Vaucheret *et al.*, 2004). The member of the Argonaute family most closely related to *AGO1* in *Arabidopsis*, *PINHEAD* (*PNH*) (also known as *ZWILLE* and *AGO10*), acts redundantly with *AGO1*, with the double mutant exhibiting embryo lethality (Lynn *et al.*, 1999); but other genes, such as *AGO7*, which exhibit heterochronic developmental defects (Hunter *et al.*, 2003), may also act in an miRNA-mediated pathway. Another member of the Argonaute gene family, *AGO4*, acts in siRNA-mediated DNA methylation (Zilberman *et al.*, 2004). As evidence of potential crosstalk between pathways, *AGO1* also acts in an siRNA-mediated pathway as it is required for PTGS and virus resistance, implying overlapping rather than discrete pathway activities for Argonaute proteins (Morel *et al.*, 2002). There are several scenarios for different RISC-like complexes of varying composition acting in distinct miRNA- and siRNA-mediated pathways. For example, RISCs could be specific to a type of small RNA, such as miRNAs, siRNAs derived from heterochromatin, viral dsRNAs, etc. Alternatively, specific RISCs could be dedicated to specific biochemical activities, such as translational repression, RNA cleavage or RNA-directed DNA methylation, some of which could be dictated by the subcellular location of specific RISCs. Because miRNAs have been postulated to regulate gene expression by all three mechanisms, the different scenarios have different implications for precise roles of RISCs in miRNA-mediated gene regulation. Reinforcing the ambiguity in Argonaute protein function, the single Argonaute protein encoded by the *Schizosaccharomyces pombe* genome can act in distinct protein complexes to mediate either PTGS or TGS, suggesting that Argonaute proteins are not necessarily dedicated to one type of RISC (Sigova *et al.*, 2004).

9.5   miRNA nomenclature

miRNAs are named using the miR prefix followed by a unique number using the normal convention of the organism from which the miRNA is derived (Ambros *et al.*, 2003). The miRNA itself is given an miR name, e.g. miR165. Since many miRNAs are encoded by several loci in the genome, the individual genes are named with a letter suffix, e.g. *MIR165a* and *MIR165b* in *Arabidopsis*. In *Arabidopsis*, mutant versions of these genes would be *mir165a* and *mir165b*. A registry of miRNA names can be found at http://www.sanger. ac.uk/Software/Rfam/mirna/index.shtml (Griffiths-Jones, 2004).

9.6   Modes of gene regulation by miRNAs: translation versus mRNA cleavage versus chromatin

In plants, miRNAs have been postulated to control gene expression by directing cleavage of target mRNAs, inhibiting translation of targeted mRNAs or directing RNA-dependent DNA modifications. Unlike miRNAs in animals, whose complementarity with their targets is usually incomplete, plant miRNAs exhibit more complete complementarity with their putative targets (Reinhart *et al.*, 2002; Rhoades *et al.*, 2002). This has two implications:

(1) it is much simpler to predict the targets of miRNAs based on a computational approach (Rhoades *et al.*, 2002);
(2) plant miRNAs were initially thought to cleave target mRNA, reminiscent of the siRNA system in which siRNAs display complete complementarity with target RNAs, and in contrast to animal miRNAs, some of which have been shown to act at the level of translational suppression.

Consistent with this hypothesis, while most animal miRNAs have imperfect complementarity to sequences in the 3′ UTRs of their targets, one animal miRNA, miR196, displays near-perfect complementarity to its targets and appears to regulate them by cleavage of the mRNAs (Yekta *et al.*, 2004). However, the observation that a plant miRNA, miR172, which exhibits near-perfect complementarity with *AP2*, acts at the level of translational repression in regulating *AP2* (Aukerman & Sakai, 2003; Chen, 2004) calls the assumption into question and suggests that there may not exist universally applicable rules for plant miRNAs.

The detection of cleavage products resulting from miRNA-directed, RISC-mediated endonuclease activity has been a hallmark of miRNA-target mRNA interactions (Jones-Rhoades & Bartel, 2004; Kasschau *et al.*, 2003; Llave *et al.*, 2002; Tang *et al.*, 2003). This, together with the observation that inhibition of the miRNA-mediated pathway by reducing DICER or RISC components (e.g. *dcl1*, *ago1* and *hyl1* mutants) leads to an increase in many predicted target mRNAs, has led to the hypothesis that mRNA cleavage may represent the primary mode of gene regulation for many miRNAs (Kasschau *et al.*, 2003; Vaucheret *et al.*, 2004; Vazquez *et al.*, 2004a). Even in cases where other modes of regulation, such as translational control or RNA-directed DNA methylation, are likely to be more important, cleavage products are readily detected by 5′ RACE experiments, supporting the idea that detection of cleavage products may be diagnostic for miRNA-mediated regulation, but not indicative of the precise mechanism.

One case in which cleavage of target mRNAs does not play a primary role in gene regulation is the translational repression of *AP2* mRNA by miR172 (Aukerman & Sakai 2003; Chen, 2004). In this case, overexpression of *MIR172a* resulted in a dramatic reduction in AP2 protein, but had little effect on the levels of *AP2* mRNA (Aukerman & Sakai, 2003) and overexpression of an miRNA-resistant version of *AP2* mRNA led to an increase in AP2 protein whereas overexpression of wild-type *AP2* mRNA led to increased mRNA levels but not increased AP2 protein levels (Chen, 2004). Taken together, these data strongly suggest that miR172 negatively regulates *AP2* at the level of translation, with the small amount of cleavage product detected a likely by-product (i.e. crosstalk between pathways) of miRNA-mediated regulation.

Another mode of gene regulation, RNA-directed DNA methylation, is exemplified by the interaction between miR165/166 (which differ by only a single nucleotide and are likely to target the same genes) and its targets, *PHABULOSA* (*PHB*) and *PHAVOLUTA* (*PHV*) (Bao *et al.*, 2004). miR165/166 have a 17–18 base pair match with their targets only after their RNA is processed, since the miR-binding site comprises the end of exon 4 and the beginning of exon 5. Therefore, a mechanism by which miR-containing RISCs recognize the processed mRNA is implicated, and tests with plants heterozygous for an miR-resistant allele show that this interaction occurs at the transcription template site. Thus, during transcription, but following splicing, miR165/166 guides an RISC to the DNA, and serves as a signal for DNA modifications (Bao *et al.*, 2004). Because many studies have shown that normal transcription can be carried out on heavily methylated DNA and because mutants deficient in primary components of the DNA methylation machinery often fail to display similar morphological alterations, it is likely that methylation provides a hallmark rather than a mechanism for possible chromatin configuration alterations that are hypothesized to take place. That the RNA-directed DNA methylation was not affected in *ago1* or *dcl1* mutants also calls into question whether all miRNAs are processed in the same manner, and suggests that there may be redundant pathways (Bao *et al.*, 2004).

The multiple modes of gene regulation mediated by miRNAs have several implications. For example, are specific miR:mRNA pairs recognized by specific RISCs? Can specific miR:mRNA pairs be recognized by multiple RISCs with different activities? Do specific cell types predominantly have specific types of RISC activities? The answers to these questions require a better understanding of the activity and expression of specific RISCs. Some types of miRNA-mediated gene regulation may be rapid and potentially transient, such as an inhibition of translation, while others, such as modification of chromatin, may be more permanent in a developmental sense. Both miRNAs and their target mRNAs are standard RNA polymerase II transcribed genes that can be expressed in complex spatial and temporal patterns (Aukerman & Sakai, 2003; Juarez *et al.*, 2004; Parizotto *et al.*, 2004) and components of RISC complexes can be differentially regulated (Lynn *et al.*, 1999; Moussian *et al.*, 1998), creating the potential for precise spatial and temporal control of miRNA-mediated gene regulation.

Most plant miRNAs exhibit extensive sequence complementarity with their putative mRNA targets and may target any region of target mRNAs (Park *et al.*, 2002; Reinhart *et al.*, 2002; Rhoades *et al.*, 2002). This is in contrast with animal miRNAs, which have mismatches with their target mRNAs and usually target sequences in 3′ UTRs (Ambros, 2004). Studies have addressed the specific sequence complementarity requirements for two plant miRNAs, miR165/166 and miR171 (Mallory *et al.*, 2004a; Parizotto *et al.*, 2004). Pairing towards the 5′ end of the miRNA (3′ end of the mRNA; nucleotides 3–9 of the miRNA) was much more important for both *in vitro* and *in vivo* function of the miRNA than pairing towards the 3′ end of the miRNA. That there are similar requirements for pairing in animal miRNAs that translationally regulate targets and miR165/166, which acts at least in part at the level of RNA-mediated DNA methylation, indicates that pairing requirements are not specific to mode of action (Mallory *et al.*, 2004a). When cleavage of the message occurs, it is predominantly between the nucleotides pairing with residues 10 and 11 of the miRNA (Floyd & Bowman, 2004; Jones-Rhoades & Bartel, 2004; Kasschau *et al.*, 2003; Llave *et al.*, 2002; Mallory *et al.*, 2004a, b; Palatnik *et al.*, 2003; Vazquez *et al.*, 2004a; Xie *et al.*, 2003). While miRNAs that act at the level of mRNA cleavage (miR171) and RNA-mediated

DNA meythlation (miR165/166) have similar requirements for pairing towards the 5′ end of the miRNA, differences are seen in their pairing requirements at the site of cleavage (Mallory *et al.*, 2004a; Parizotto *et al.*, 2004). In the case of miR171 nucleotide, pairing at the cleavage site is critical for activity of the miRNA *in vivo* while in the case of miR165/166 pairing at the cleavage site was less critical for *in vivo* activity of the miRNA (Mallory *et al.*, 2004a; Parizotto *et al.*, 2004).

9.7   miRNAs and their targets

Many of the miRNAs initially identified in plants are predicted to target transcription factors that are involved in developmental processes, suggesting that miRNA-mediated regulation may be an efficient mechanism for clearing transcripts from specific daughter cell lineages following determinative cell division events (Bartel & Bartel, 2003; Park *et al.*, 2002; Reinhart *et al.*, 2002; Rhoades *et al.*, 2002). Notable exceptions to the preponderance of transcription factors amongst potential targets were two genes involved in the biogenesis of miRNAs themselves, *DCL1* and *AGO1*, suggesting feedback regulation controls the production or activity of miRNAs (Vaucheret *et al.*, 2004; Xie *et al.*, 2003). Subsequently, by screening for additional miRNAs by computational methods or by observing plants that are grown under conditions of stress, additional miRNAs that target more diverse functions have been identified, including some miRNAs that are induced by environmental conditions (Adai *et al.*, 2005; Bonnet *et al.*, 2004; Jones-Rhoades & Bartel, 2004; Sunkar & Zhu, 2004; Wang *et al.*, 2004). A list of known miRNAs of *Arabidopsis*, published by All Hallows Eve (2004), is presented in Table 9.1, along with their potential targets. While all miRNAs listed have been shown to be expressed *in planta*, only a small fraction, those described below in detail, have been tested for function *in vivo*.

In a landmark study, Reinhart *et al.* (2002) identified 16 miRNA gene families by cloning small, 20–24 nucleotide, RNAs that are encoded in the genome and could potentially be processed from stem-loop precursors by DCL1. This study demonstrated that plants have miRNAs and they are similar to those originally identified in animals in many respects. Subsequently, additional miRNAs were discovered by similar techniques (Park *et al.*, 2002) and more recently, by using RNA isolated from plants grown under cold or drought stress, Sunkar and Zhu (2004) identified additional miRNAs, at least some of which were shown to be induced by stress conditions. These studies also demonstrated that there are a large number of small RNA molecules in plants cells that are not miRNAs, but may be siRNAs or RNAs with other biological functions.

In 2004, four different groups used computational and bioinformatics methods to identify additional putative miRNAs in *Arabidopsis* (Adai *et al.*,

2004; Bonnet *et al.*, 2004; Jones-Rhoades & Bartel, 2004; Wang *et al.*, 2004). Three of the studies used conservation between the *Arabidopsis* and *Oryza* genomes and the conserved stem-loop structure of pre-miRNAs as criteria for identifying potential genes (Bonnet *et al.*, 2004; Jones-Rhoades & Bartel, 2004; Wang *et al.*, 2004). The set of resulting putative miRNAs was refined by eliminating repetitive DNA and using slightly differing algorithms, leading the three groups to present different partially overlapping sets of putative miRNAs (Table 9.1). In contrast, Adai *et al.* (2004) started with target sequences (those sequences thought to be transcribed into mRNA) and

Table 9.1 List of *Arabidopsis* miRNAs (*circa* All Hallow's Eve (2004))

Name	Loci	Target protein class (loci)	Detection	Reference
miR156	12	Squamosa promoter-binding TF (11)	E, C	a,b, m, r, v
miR157	4	Squamosa promoter-binding TF (9)	E, C	a, b, m, u
miR158	2	Unknown protein (1)	E	a, b, u
miR159	3	MYB TF (8)	E, C, F	a, b, c, m, r, u, x
miR160	3	ARF TF (3)	E, C	a, b, m, r, v
miR161	1	PPR repeat proteins (9)	E, C	a, b, m
miR162	2	DCL-LIKE1 (1)	E, C, F	a, b, d, m, r
miR163	1	?	E	a, b
miR164	3	NAC TF (6)	E, C, F	a, b, e–f, m, r, u–v
miR165	2	Class III HD-ZIP TF (5)	E, C, F	a, b, g–m
miR166	9	Class III HD-ZIP TF (5)	E, C, F	a, b, g–m, r
miR167	4	ARF TF (2)	E, C	a, b, m, r, u, v
miR168	2	AGO1 (1)	E, C, F	a, b, m, r
miR169	14	CCAAT-binding factor HAP2-like (7)	E	a, b, r, t, u
miR170	1	SCARECROW-like TF (3)	E, C	a, b, m–o, r, u
miR171	4	SCARECROW-like TF (3)	E, C, F	a, b, m–o, r–u
miR172	5	AP2 TF (5)	E, C, F	p–q, r, t–w
miR173	1	Unknown protein (1)	E	w
miR319/JAW	3	TCP TF (5)	E, C, F	c
miR390	2	7 different proteins	E	s, u
miR393	2	TIR1 F-box proteins (4)	E, C	r, s, t, u
miR394	2	F-box protein (1); LIM (1); AAA-ATPase (1)	E, C	r, u
miR395	6	ATP sulfurylases (3), sulfate transporter (1)	E, C	r, u
miR396	2	GRL TF (7)	E, C	r, t
miR397	2	Laccases (3)	E, C	r, s

Table 9.1 *Continued*

Name	Loci	Target protein class (loci)	Detection	Reference
miR398	3	Copper superoxide dismutases (2); cytochrome C oxidase (1)	E, C	r, s, u
miR399	6	Ubiquitin conjugation enzyme (1)	E	r, s, u
miR400	1	PPR proteins (11)	E	s
miR401	1	Unknown protein (3)		s
miR402	1	DNA glycosylase (1)	E	s
miR403	1	Unknown protein (1)	E	s
miR404	1	LRR-TM protein kinase (1)		s
miR405	3	Unknown protein (1)		s
miR406	1	Spliceosomal proteins (2)		s
miR407	1	5 different proteins (5)	E	s
miR408	1	Peptide chain release factor (2)	E	s
miR413	1	?	E	t
miR414	1	?	E	t
miR415	1	?	E	t
miR416	1	?	E	t
miR417	1	?	E	t
miR418	1	?	E	t
miR419	1	?	E	t
miR420	1	?	E	t

E, expression by Northern; C, cleavage or reduction in amount of target mRNA; F, functionally characterized.
Source: (a) Reinhart *et al.*, 2002; (b) Rhoades *et al.*, 2002; (c) Palatnik *et al.*, 2003; (d) Xie *et al.*, 2003; (e) Mallory *et al.*, 2004b; (f) Laufs *et al.*, 2004; (g) Tang *et al.*, 2003; (h) Emery *et al.*, 2003; (i) Juarez *et al.*, 2004; (j) Mallory *et al.*, 2004a; (k) Zhong & Ye, 2004; (l) Kidner & Martienssen, 2004; (m) Vaucheret *et al.*, 2004; (n) Llave *et al.*, 2002; (o) Parizotto *et al.*, 2004; (p) Aukerman & Sakai, 2003; (q) Chen, 2004; (r) Jones-Rhoades & Bartel, 2004; (s) Sunkar & Zhu, 2004; (t) Wang *et al.*, 2004; (u) Adai *et al.*, 2004; (v) Kasschau *et al.*, 2003; (w) Park *et al.*, 2002; (x) Achard *et al.*, 2004.

searched for potential miRNAs based on known miRNA precursor structures, and these data can be accessed from a website (http://bioinformatics.icmb.utexas.edu/mirna) so that one can examine the potential miRNAs targeting any gene of interest. This study also produced a set of putative miRNAs that partially overlapped with the other studies. Since the former studies rely on sequence conservation between *Arabidopsis* and *Oryza*, those miRNAs that may be unique to the respective taxa were not identified. All studies identified additional loci that could encode previously known miRNAs as well as new putative miRNAs, and the new classes of miRNAs were verified by Northern (RNA gel blot) analyses, demonstrating that they are expressed. miR395, which can target ATP sulfurylases, is induced by low sulfur growth conditions, suggesting that additional miRNA genes may be identified by screening in different environmental conditions. Due to the nature of the algorithms used, some miRNAs may have been overlooked if they contained introns between the miRNA and its complement within the pri-miRNA, and given that miRNAs can act at the level of RNA-mediated DNA methylation, miRNAs targeting intronic sequences cannot be excluded as a possibility.

How conserved are the *Arabidopsis* miRNAs? While most *Arabidopsis* miRNAs are shared with *Oryza sativa*, some such as miR158, miR161, miR163 and miR173 have not been detected in *Oryza*, suggesting that they may be more restricted in their phylogenetic distribution. Searching EST databases, Jones-Rhoades and Bartel (2004) showed that at least nine miRNA families (156, 160, 166, 167, 393, 395, 396, 397 and 398) have conserved complementary sites in potential targets in gymnosperms, with miR171 having a conserved target site in *Ceratopteris richardii* and a *Physcomitrella patens* EST encoding a precursor of miR159/319. These data, and the finding that complementary sites for miR165/166 are conserved in Class III homeodomain/leucine zipper (HD-Zip) genes throughout land plants, indicate that many of the miRNA gene families in *Arabidopsis* have a long evolutionary history (Floyd & Bowman, 2004; see Section 9.9).

Several features of the *Arabidopsis* genome's miRNA content are striking. First, while there are only 43 (or fewer, see below) different types of miRNAs, they are encoded at well over 100 loci due to multiple loci encoding the same miRNA for the majority of the miRNAs. Second, there are sets of closely related miRNAs that likely have the same or similar targets: miRNA165/166 targets Class III HD-Zip genes and miRNA170/171 likely target three SCARECROW-like genes. Third, a single miRNA may have a single or many targets. For example, miRNA156 may target 11 different SQUAMOSA promoter-binding (SPB) transcription factor genes, while miR168 appears to target only *AGO1*. Fourth, some miRNAs have the potential to target multiple types of genes. For example, miR398 could target genes encoding both copper superoxide dismutases and cytochrome C oxidase, and miR395 can target both ATP sulfurylases and a sulfate transporter to influence sulfur metabolism. Fifth, while the initial set of cloned miRNAs primarily targeted transcription factors, computational studies later identified many types of potential targets for miRNA genes. Finally, it must be stressed not all the miRNAs listed have been validated beyond preliminary expression experiments, which may not be conclusive due to the large numbers of small RNAs present in plant cells (Reinhart *et al.*, 2002; Sunkar & Zhu, 2004). Thus, the miRNA gene content of *Arabidopsis* resembles the protein-coding gene content in having extensive redundancy. What are the implications of redundant sets of miRNAs? Since each miRNA has a unique promoter, each miRNA could be expressed in a distinct temporal and spatial pattern, providing a complex and precise mechanism to regulate gene activity of the target genes.

9.8   Functional characterization of miRNAs – case studies

While dozens of miRNAs have been identified in the *Arabidopsis* genome, only a handful have been analyzed by what Baulcombe (2004) refers to as the gold standard: expression of a modified miRNA-resistant target with silent mutations in the putative miRNA complementary region. Brief descriptions of those cases for which this experimental validation had been published by All Hallow Eve (2004) are given below, highlighting developmental functions and modes of action for each miRNA studied.

9.8.1   *miR165/166 and Class III HD-Zip genes*

Two miRNAs (miR165/166) target the five Class III HD-Zip gene family members, *PHB*, *PHV*, *REVOLUTA* (*REV*), *ATHB8* and *ATHB15*, present in the *Arabidopsis* genome (Reinhart *et al.*, 2002; Rhoades *et al.*, 2002). Since miR165 (encoded by two genomic loci) and miR166 (encoded by nine loci) differ only by a single nucleotide, it may be that both miRNAs can target all five Class III HD-Zip gene family members. Class III HD-Zip proteins are characterized by an amino-terminal HD-ZIP followed by a region exhibiting sequence similarity to mammalian sterol/lipid-binding domain (START domain) (McConnell *et al.*, 2001; Pontig & Aravind, 1999; Sessa *et al.*, 1998). *PHB*, *PHV* and *REV* are involved in establishment of adaxial identity of leaves, in the development of the apical meristem and the vascular bundles (Emery *et al.*, 2003; McConnell & Barton, 1998; McConnell *et al.*, 2001; Otsuga *et al.*, 2001; Talbert *et al.*, 1995; Zhong & Ye, 2004). While single loss-of-function mutations exhibit minor, or no apparent, defects in development, *phb phv rev* plants lack an apical meristem, with the apical part of the plant consisting of a single radialized cotyledon (Emery *et al.*, 2003). In contrast, dominant gain-of-function mutations result in an abaxial to adaxial conversion of tissue types in leaves (*phb* and *phv* alleles) and radialization of normally asymmetric vascular bundles (*rev* alleles) (Emery *et al.*, 2003; McConnell & Barton, 1998; McConnell *et al.*, 2001; Zhong & Ye, 2004).

All identified gain-of-function mutations in *PHB*, *PHV* and *REV* occur in the START domain, in the sequence complementary to that of miRNA165 and miRNA166. A biochemical approach taken by Tang *et al.* (2003) demonstrated that *PHV* and *PHB* mRNAs are cleaved at the sequence complementary to miR165, while dominant gain-of-function *phv* mRNA remains intact when incubated with a wheat germ extract. Efficient cleavage of the mutant mRNA can be achieved only when a synthetic miRNA with restored complementarity to the mutant sequence is added to the extract. In addition, cleavage products can be detected *in vivo* for all the Class III HD-Zip gene family members (Emery *et al.*, 2003; Mallory *et al.*, 2004a; Zhong & Ye, 2004). Fulfilling

Baulcombe's (2004) gold standard, *Arabidopsis* plants transformed with *REV* or *PHB* cDNAs carrying mutations altering the miRNA165/166 complementary site but not the translated REV or PHB proteins exhibit a phenotype similar to that of the gain-of-function mutants, confirming the *in vivo* action of the predicted miRNA-mediated regulation (Emery *et al.*, 2003; Mallory *et al.*, 2004a). In addition, constitutive expression of wild-type versions of *PHB* or *REV* do not result in an aberrant phenotype, suggesting that these transgenes are negatively regulated in tissues where the genes are not normally active (Emery *et al.*, 2003; Mallory *et al.*, 2004a; McConnell *et al.*, 2001). Thus, the dominant gain-of-function alleles are hypothesized to be due to a loss of miRNA-mediated regulation, and recent studies by Bao *et al.* (2004) suggest that the primary mode of miRNA-mediated regulation is likely not via target mRNA cleavage.

miRNA165/166-dependent RNA-directed DNA methylation occurs in a template-specific manner at the *PHB* and *PHV* loci during transcription and splicing of the *PHB* and *PHV* mRNAs, suggesting that chromatin architectural changes may underlie miR165/166 regulation of Class III HD-Zip genes (Bao *et al.*, 2004). The DNA methylation phenomenon cannot be equally detected in all cell types. Inflorescence meristems of *ap1 cal* plants as well as wild-type siliques exhibit very low levels of methylation in contrast to differentiated tissues that exhibit higher levels of methylation (Bao *et al.*, 2004). This observation could be explained either by a lack of miR165/166 expression in these tissues or by a lack of one or more of the specific RISC-mediating complexes. Thus, rather than a rapid developmentally induced clearing of transcripts, miR165/166 may direct more stable epigenetic changes by influencing chromatin remodeling at the *PHB* and *PHV* loci. Based on the roles of these genes in meristem formation and/or maintenance, one hypothesis is that as cells are displaced from meristems and begin to terminally differentiate, Class III HD-Zip loci are permanently repressed via chromatin remodeling.

Juarez *et al.* (2004) extended functional analyses of the role of miR165/166 in repressing Class III HD-Zip gene expression to maize with the identification of *Rld1* as a Class III HD-Zip gene. Dominant gain-of-function alleles of *Rld1* map to the miR165/166 complementary site within the *Rld1* mRNA, suggesting that they interrupt miRNA-mediated regulation of *rld1*. Interestingly, all *Rld1* mutant alleles map to a single nucleotide towards the 5' end of the miRNA complementary site, in contrast to the semi-dominant *PHB*, *PHV* and *REV* alleles that map towards the 3' end. This could be one possible explanation for the weaker phenotypic defects of *Rld1* mutants as compared with *rev*, *phv* and *phb* gain-of-function mutants. Similar to *PHB* in *phb-1d* mutants, *Rld1* is ectopically expressed in *Rld1* mutants, indicating the importance of negative regulation by miR166 (Juarez *et al.*, 2004; McConnell *et al.*, 2001). The regulation of Class III HD-Zip genes extends to all land plants as the miRNA-binding site is conserved from all gene family members identified in mosses, liverworts, hornworts, lycophytes, ferns, gymnosperms and angiosperms (Floyd & Bowman, 2004). Cleavage products, as a hallmark of miRNA regulation, were detected in all the taxa tested, and a pre-miRNA cloned from *Selaginella*, a lycophyte, is very similar in structure and sequence to miR166b from *Arabidopsis*, implicating a role for miRNA regulation of Class III HD-Zip genes throughout land plants.

Both Juarez *et al.* (2004) and Kidner and Martienssen (2004) speculate that miRNAs may act as mobile signals during the establishment of polarity in developing leaves, an attractive hypothesis for a couple of reasons. First, based on classical dissection experiments, a signal emanating from the apical meristem was proposed to be essential for establishing polarity in developing leaves (Sussex, 1955). In one scenario, the signal induces adaxial fates in cells in close proximity to the meristem, while abaxial fate would be the default in cells of the leaf primordium at a distance from the meristem. The biochemical nature of the hypothetical signal remains elusive, although a potential ligand perceived by the START domain of Class III HD-Zip proteins remains an attractive candidate (McConnell *et al.*, 2001). Second, PTGS and VIGS can act systemically in plants, implying the movement of a silencing signal, possibly via the vasculature, although the silencing signals are excluded from the apical meristem (Beclin *et al.*, 1998; Voinnet *et al.*, 1998). While the systemic silencing signal has not been conclusively identified, that it acts in a sequence-specific manner suggests that it includes a nucleic acid, e.g. siRNAs. To counteract VIGS, viruses have evolved viral suppressors of gene silencing, which appear to target the production or mobility of the silencing signal and small miRNAs (Dunoyer *et al.*, 2004; Hamilton *et al.*, 2002; Llave *et al.*, 2000; Mallory *et al.*, 2001, 2002; Voinnet *et al.*, 2000). Furthermore, constitutive expression of the white clover mosaic virus protein TGB1, which is related to viral suppressers of gene silencing, in *Nicotiana benthamiana* results in a phenotype similar to that of *ago1* mutants of *Arabidopsis*, with radially symmetric leaves that express a marker of abaxial cells, *NtYABBY*, in a sporadic and inconsistent manner, suggesting that polarity establishment is disrupted at an early stage (Foster *et al.*, 2002). In these transgenic plants, viruses are able to move into the shoot apex, a situation not observed in wild-type plants, indicating that communication pathways, likely via plasmodesmata, are altered in these plants (Foster *et al.*, 2002). Thus, disruption of communication pathways used in viral movement and systemic silencing also alters polarity establishment in leaves, suggesting a possible link with the Class III HD-Zip genes that are required for establishment of leaf polarity.

Consistent with the proposed role of miR165/166 in negatively regulating Class III HD-Zip gene expression, the distribution of miR165/166 detected in both *Arabidopsis* and maize is largely complementary to that of Class III HD-Zip expression. Class III HD-Zip genes are expressed in meristems, adaxially in lateral organs, and in the vascular cambium and differentiating xylem (Baima *et al.*, 1995; Emery *et al.*, 2003; Kang *et al.*, 2003; McConnell *et al.*, 2001; Ohashi-Ito & Fukuda, 2003; Otsuga *et al.*, 2001). In contrast, miR165/166 have been detected abaxially in lateral organs and in developing phloem (Juarez *et al.*, 2004; Kidner & Martienssen, 2004). The exception to this is expression of miR166 in cells underlying leaf primordia, prompting the speculation that this represents a source of miR166 that could move into the leaf and phloem (Juarez *et al.*, 2004). While an attractive hypothesis, there is no compelling reason to propose miRNA movement at present. Two arguments against miRNA movement include the precise spatial pattern of miR172, which exactly mirrors that of *AP2*, one of its target genes (Chen, 2004; see below). Movement of miR172 would need to be controlled precisely such that the boundary between the second and third whorls of the flower is not blurred. Perhaps more compelling experimental evidence against movement is the precise coincidence of transcription and activity of miR171 (Parizotto *et al.*, 2004; see below). An alternative hypothesis is that the pattern of miR165/166 expression observed reflects its transcriptional regulation. A simple test of this hypothesis would be to assay the transcriptional activity of miR165/166 promoters and drive the miRNAs with heterologous promoters to assay cell autonomy in their action. Differential regulation of each of the loci encoding miRNA165/166 would provide greater developmental flexibility in the regulation of Class III HD-Zip gene expression. In this regard, miR165/166 expression, abaxially in leaves and in the phloem, is reminiscent of that of the KANADI genes, which have also been proposed to negatively regulate Class III HD-Zip gene expression, and could potentially act upstream of, or in conjunction with, miR165/166 (Emery *et al.*, 2003; Eshed *et al.*, 2001, 2004; Kerstetter *et al.*, 2001). Regardless, that each miRNA locus could be independently spatially and temporally regulated adds another layer of complexity in the interactions between miRNA and target mRNA interactions.

9.8.2   *miR319/JAW and TCP genes*

Due to a high level of redundancy amongst miRNA gene family members, loss-of-function alleles have not been detected through standard forward genetics. However, using an activation-tagging approach, Palatnik *et al.* (2003) identified a gain-of-function allele of miR319 (also known as miRJAW) in which leaf lamina formation was disrupted due to uneven growth. This phenotype is similar to that of loss-of-function alleles of *CINCINNATA* of *Antirrhinum majus*, where growth of leaf lamina occurs too much at the edges relative to the middle (Nath *et al.*, 2003). *CINCINNATA* encodes a transcription factor of the TCP family, and is expressed at and preceding the region of mitotic arrest seen in leaves as they differentiate from their tips towards the bases, leading Nath *et al.* (2003) to propose that CINCINNATA may control cell cycle arrest regulation and disruptions of this process result in uneven growth patterns in the leaf. miR319/JAW-complementary sequences are present in five TCP gene family members in *Arabidopsis*, and these genes are downregulated in gain-of-function miR319/JAW mutants (Palatnik *et al.*, 2003). Furthermore, expression of miRNA-resistant versions of *TCP4* and *TCP2* under their endogenous promoters in transgenic *Arabidopsis* led to aberrant phenotypes, with the miRNA-resistant version of *TCP4* being seedling-lethal, confirming that these genes are miRNA-regulated *in vivo* (Palatnik *et al.* 2003). In contrast, constitutive expression of wild-type versions of *TCP2* or *TCP4* leads to little or no aberrant phenotype. Based on the detection of cleavage products and reduction in target mRNA levels, miR319/JAW probably acts to reduce mRNA stability.

The sequences of miR159 and miR319/JAW differ by only three nucleotides, two of which are at the 3′ end and one is near the center of the miRNAs (Palatnik *et al.*, 2003; Reinhart *et al.*, 2002). However, they appear to have different specificities in their targets, with miR159 targeting a set of MYB-encoding mRNAs and miR319/JAW targeting TCP-encoding mRNAs (Achard *et al.*, 2004; Palatnik *et al.*, 2003). This was shown experimentally with miR319/JAW gain-of-function alleles lacking any effect on the expression of the predicted MYB targets of miR159. It may be that the single change in the middle of the region deemed most important for miRNA activity, as shown for miR165/166 and miR171, may be determinate with regard to substrate specificity in this case (Mallory *et al.*, 2004a; Parizotto *et al.*, 2004).

9.8.3 *miR159 and MYB genes*

miR159 is a posttranscriptional regulator of a subset of MYB transcription factors termed GAMYBs due to their role in gibberellic acid (GA) responses (Achard *et al.*, 2004; Palatnik *et al.*, 2003). Constitutive expression of an miRNA-resistant version of MYB33 causes leaves to be smaller and upwardly curled, validating this GAMYB gene as an miRNA target (Palatnik *et al.*, 2003). Constitutive ectopic expression of miR159a leads to a delay in the transition to flowering and male sterility due to aberrant anther development, two phenotypes previously associated with GA responses, and the levels of miR159 are reduced in mutants lacking either GA or GA signal transduction and are induced by exogenously applied GA (Achard *et al.*, 2004). Since the putative targets of miR159 are thought to promote GA responses, miR159 may act in a negative feedback loop to dampen the activity of GA by acting as a homeostatic regulator of GAMYB function (Achard *et al.*, 2004).

9.8.4 *miR164 and CUC-like NAC genes*

Plants harboring loss-of-function alleles of both *CUC1* and *CUC2* exhibit fused lateral organs, fusion of cotyledons and a failure in apical meristem formation (Aida *et al.*, 1997). Consistent with miR164 negatively regulating *CUC1* and *CUC2*, constitutive ectopic expression of miR164 phenocopies a *cuc1 cuc2* double mutant (Laufs *et al.*, 2004; Mallory *et al.*, 2004b). Additional phenotypes are seen as well, likely due to the negative regulation of additional targets by miR164. Conversely, expression of an miRNA-resistant version of either *CUC1* or *CUC2* under control of their endogenous promoters results in a complementary phenotype, with large gaps between the organs produced from the shoot apical meristem (Laufs *et al.*, 2004; Mallory *et al.*, 2004b). However, in contrast to the TCP and Class III HD-Zip genes described above, constitutive expression of a wild-type allele of *CUC1* causes a dramatic aberrant phenotype, with ectopic meristems developing from the adaxial sides of cotyledons and leaves (Takada *et al.*, 2001). This indicates that miR164 is unlikely to be expressed widely throughout the plant, but rather more locally, refining CUC activity in the shoot apical meristem at the point of lateral organ production. In this sense, miR164 may be a candidate for acting to clear transcripts from cells rapidly following cell division, resulting in different fates for daughter cells as proposed by Bartel and Bartel (2003).

9.8.5   miR172 and AP2 and related genes

Thus far, miR172 is the only confirmed plant miRNA that regulates at least some of its targets, e.g. *AP2*, via repression of translation (Aukerman & Sakai, 2003; Chen, 2004). Aukerman and Sakai (2003) cloned *MIR172a* via an activation-tagging approach, with the overexpression of *MIR172a* resulting in early flowering and floral defects reminiscent of *apetala2* (*ap2*) mutants in which the sepals and petals are replaced by carpels and stamens, respectively. The potential targets of miR172 in *Arabidopsis* include *AP2* and four closely genes (Aukerman & Sakai, 2003; Chen, 2004; Jones-Rhoades & Bartel, 2004; Rhoades *et al.*, 2002). Based on the early flowering phenotype of gain-of-function alleles of *MIR172a* (called *EAT* in Aukerman & Sakai, 2003) and late flowering phenotypes caused by overexpression of one of its *AP2*-like targets (called *TOE1* in Aukerman & Sakai, 2003), miR172 is proposed to regulate the transition to flowering in *Arabidopsis*. Furthermore, a double mutant, *toe1 toe2*, in which two targets of miR172 are compromised, is early flowering (Aukerman & Sakai, 2003). Consistent with these *AP2*-like genes regulating phase changes in plants, a homolog, *GLOSSY15*, that is also potentially miRNA-regulated is involved in the juvenile to adult transition in maize (Moose & Sisco, 1996). Such targets of miRNA regulation could explain the phase change defects described in some mutants, such as *hasty* and *zippy*, and in genes involved in miRNA biogenesis or function (Bollman *et al.*, 2003; Hunter *et al.*, 2003).

   miRNA-resistant alleles of *AP2* result in phenotypes similar to *agamous* (*ag*) mutants that have stamens and carpels replaced by additional whorls of petals and sepals (Chen, 2004). This phenotype is consistent with models of floral organ identity establishment in which *AP2* and *AG* activities are antagonistic (Bowman *et al.*, 1991). Consistent with the phenotypic defects observed, miR172 is expressed in the inner two whorls of the flower, the whorls in which *AG* is active and *AP2* largely inactive (Chen, 2004). miR172 is also expressed in stage 2 flower meristems (Chen, 2004), and likely elsewhere in the plant before the transition to flowering. The discovery of posttranscriptional regulation of *AP2* resolves the contradiction of *AP2* mRNA being present in the inner whorls of the flower, but AP2 activity is largely lacking in these tissues and that constitutive expression of wild-type allele of *AP2* lacks a conspicuous aberrant phenotype (Chen, 2004; Jofuku *et al.*, 1994). Again, activation of miR172 in the flower meristem could be a mechanism for cell fate specification, by clearing *AP2* transcripts rapidly (Bartel & Bartel, 2003). In this respect, it will be of interest to determine whether at least one locus-encoding miR172 is positively regulated by *AG* in the inner two whorls of the flower. As with other miRNA:mRNA pairs, which have five miR172 loci and five potential target loci, differential regulation of the miR172 loci could provide a sensitive mechanism to control the levels of *AP2/TOE* gene activity, which may be particularly relevant considering the many inputs, both intrinsic and environmental, that influence the transition to flowering.

9.8.6 *miR170/171 and HAM-like GRAS genes*

In *Arabidopsis*, both miR171 and miR170 can potentially target three members of the GRAS family of transcription factors (Llave *et al.*, 2002; Parizotto *et al.*, 2004; Rhoades *et al.*, 2002). Based on phylogenetic analyses, the three target genes (sometimes referred to as *SCARECROW-LIKE6* (*SCL6*), *SCL22*, and *SCL27*) are most closely related to the *Petunia* gene, *HAIRY MERISTEM* (*HAM*), suggesting that they may be involved in a similar developmental role in *Arabidopsis* (Bolle, 2004; Stuurman *et al.*, 2002; Tian *et al.*, 2004). In *Petunia*, *HAM* is required for the maintenance of the shoot apical meristem, and as it is expressed beneath the meristem, it appears to act in a non-cell-autonomous manner (Stuurman *et al.*, 2002). This is of interest as another member of the GRAS family, *SHORT ROOT*, also acts non-cell-autonomously, with the protein moving between cell layers to radial pattern tissues in the root (Gallagher *et al.*, 2004; Nakajima *et al.*, 2001). While the developmental roles of miR170/171 and the *HAM*-like genes in *Arabidopsis* have not been reported, studies of this miRNA:mRNA pair have shed some light on miRNA functions in plants.

Detection of target mRNA cleavage products in wild-type plants and an increase in target mRNA levels in mutants impaired in miRNA biogenesis or function suggest that miR170/171 acts by reducing target mRNA stability (Llave *et al.*, 2002; Parizotto *et al.*, 2004; Vaucheret *et al.*, 2004; Vazquez *et al.*, 2004a). Parizotto *et al.* (2004) demonstrated that the overall structure of the miRNA precursor, rather than the miRNA sequence itself, is critical for proper processing by DCL1 by replacing the miRNA sequence and its complement in miR171 with sequences that could target GFP mRNA sequences. This artificial miRNA efficiently targeted GFP mRNA for endonucleolytic cleavage, indicating that as long as the free-energy requirements and complementary pairing requirements are taken into account, artificial miRNAs could theoretically be designed to target any mRNA sequence of interest (Parizotto *et al.*, 2004). Parizotto *et al.* (2004) then utilized a 'sensor' gene, a GFP gene transcriptionally fused with an miR171 target sequence, to assay miR171 activity *in vivo*. This was done in an *sde1* mutant background as the sensor gene was silenced via an siRNA pathway in *SDE1* plants. That the transcriptional activity conferred by an *MIR171* promoter and the activity of miR171 as assayed with the sensor gene nearly perfectly coincided supported the idea that miRNAs could act to ensure clearance of transcripts following cell division and contribute to cell fate specification (Parizotto *et al.*, 2004). Furthermore, miR171 activity and *MIR171* promoter activity precisely coincide strongly against movement of miR171.

9.8.7   *miR168 and ARGONAUTE1 and miR162 and DICER-LIKE1*

Two genes involved in miRNA biogenesis and/or function are themselves regulated by miRNAs (Vaucheret *et al.*, 2004; Xie *et al.*, 2003). This regulation could be providing a negative feedback loop such that the levels of miRNAs and their activity are moderated. Alternatively, miRNA regulation of these genes could provide cell type specificity in the production or activity of miRNAs, possibly restricting them from certain cell or tissue types. Knowledge of the spatial and temporal distribution of the miRNAs and their target genes are required to resolve these questions.

Loss-of-function mutations in *AGO1* are pleiotropic with defects in leaf and meristem development and they are exacerbated by mutations in *PNH*, with the double mutant being embryo/seedling-lethal, again highlighting the importance of miRNA-mediated gene expression in plant development (Bohmert *et al.*, 1998; Lynn *et al.*, 1999; McConnell & Barton, 1995 Moussian *et al.*, 1998). Redundancy between *AGO1* and *PNH* complicates the picture as while *AGO1* appears to be uniformly expressed, *PNH* is expressed in a tissue-specific manner, suggesting differences in RISC activity in different tissues (Lynn *et al.*, 1999; Moussian *et al.*, 1998). These Argonaute family proteins, by analogy, are likely to have roles in RISCs, either providing 'slicer' activity or some other biochemical activity, e.g. translational repression (Liu *et al.*, 2004; Song *et al.*, 2004). That *AGO1* is involved in miRNA function in *Arabidopsis* is supported by increases in mRNA targets of known miRNAs in *ago1* loss-of-function mutants (Vaucheret *et al.*, 2004). Misregulation of many different targets in *ago1* mutants is likely the cause of the pleiotropic phenotype that has been difficult to interpret. For example, *ago1* leaves have been described as either abaxialized or adaxialized, implicating miR65/166 regulation of the Class III HD-Zip genes as contributing to the phenotypic defects (Bohmert *et al.*, 1998; Kidner & Martienssen, 2004; Lynn *et al.*, 1999). However, that miR165/166-mediated RNA-directed DNA methylation of *PHB* still occurs in *ago1* mutants calls these interpretations into question (Bao *et al.*, 2004).

That the regulation of *AGO1* by miR168 is important is confirmed by the severe developmental defects observed in mutants of *AGO1* that are impaired in miR168-mediated cleavage (Vaucheret *et al.*, 2004). These miRNA-resist-ant mutants are not phenotypically similar to *ago1* null mutants, suggesting that miR168 is limited in its expression pattern (Vaucheret *et al.*, 2004). Consistent with this idea, ectopic overexpression of *AGO1* results in aberrant phenotype that could be interpreted as converse to the loss-of-function alleles (Bohmert *et al.*, 1998). Furthermore, *AGO1* also acts in an siRNA-mediated pathway as it is required for PTGS and virus resistance, implying overlapping rather than discrete pathway activities for this Argonaute protein (Morel *et al.*, 2002). Thus, components of RISCs display developmentally controlled activ-ity and may also be promiscuous in the pathways in which they act, making interpretations of their phenotypes difficult.

DCL1 appears to be the primary DICER-like RNase involved in the bio-genesis of miRNAs in *Arabidopsis* and null mutations result in embryo lethality, highlighting the importance of miRNA-mediated gene regulation (Finnegan *et al.*, 2003; Papp *et al.*, 2003; Schauer *et al.*, 2002; Xie *et al.*, 2004). The increase in *DCL1* mRNA in *dcl1* hypomorphic alleles, and other mutants compromised in miRNA production, confirms that miR162 provides negative feedback regulation of *DCL1* (Xie *et al.*, 2003). The target site of miR162 in *DCL1* mRNA spans exons 12 and 13 of *DCL1*, and Xie *et al.* (2003) propose that miRNA-mediated regulation of *DCL1* mRNA occurs in the nucleus. The precise role of this negative feedback regulation may be elucidated by construction of miRNA-resistant *DCL1* alleles.

9.8.8   Summary

Can any generalization be made for these few cases in which plant miRNA-mediated gene regulation has been investigated in some detail? First, both miRNAs and their targets can be both spatially and temporally regulated, providing ample fodder to precisely regulate target gene function. Second, that an miRNA-resistant version of *AGO1* has severe phenotypic defects indicates that feedback regulation of miRNA production and/or function is important. Third, it is difficult to predict, *a priori*, the mode of action of miRNA-mediated regulation. Fourth, that loss of miRNA production results in embryo lethality underscores the importance that these 40+ families of miRNAs have in plant development. This may not be too surprising as loss of activity of any 40 random genes may result in a similar phenotype. Fifth, miRNAs may not be mobile, in contrast to the systemic signaling observed with siRNAs in PTGS and VIGS. How might this regulation be achieved: through different RISC activities or something more intrinsic differentiating mobile signals and miRNAs, such as length? Finally, due to the diverse nature of miRNA targets, developmental and physiological processes, in addition to rapid clearing of transcripts during cell fate decisions, are likely to be mediated through miRNAs.

9.9   Evolution of miRNA-mediated gene regulation

9.9.1   Within the plant kingdom

From our discussion above it should be clear that the importance of miRNA regulation in plant growth and development has been clearly established, although much remains to be learned. Identification of miRNAs and experimental work demonstrating their role in plant development has largely been restricted to the model organism, *Arabidopsis thaliana*, and to a lesser extent *Zea mays*. Most miRNAs identified in *Arabidopsis* have homologs in rice, indicating that they at least date back to the common ancestor of monocots and eudicots, with nearly complete miRNA and target sequence conservation (Bartel & Bartel, 2003; Jones-Rhoades & Bartel, 2004). The exceptions are miR158, miR161, miR163 and miR173, which are known from *Arabidopsis*, but are not found in *Oryza*. Very little is known about the targets and functions of these miRNAs. It is possible that they have been lost in rice or in monocots; it is also possible these miRNAs have relatively recent origins within the eudicot clade. However, most miRNAs identified in *Arabidopsis* can be traced back to an origin that predates the divergence of monocots and eudicots.

It is less clear whether or not most of the miRNAs identified in flowering plants have an origin more ancient than that of angiosperms. *Arabidopsis APETALA2* (*AP2*) is known to be regulated at the level of translation by miR172. Although no other experimental evidence is available, the nucleotides corresponding to the binding site in *AP2* homologs from *Z. mays* (*GLOSSY15*) and two conifers *Pinus* (*PtAP2L1*; *PtAP2L2*) and *Picea* (*PaAP2L1*) are highly conserved (Figure 9.1A). This indicates that the origin of miRNA regulation of *AP2* mRNAs may trace back to at least the last common ancestor of seed plants. Jones-Rhoades and Bartel (2004) also report that conserved binding sites for several other miRNA families (156, 160, 167, 393, 395, 396, 397, 398) were found in coding sequences of gymnosperm homologs. Jones-Rhoades and Bartel (2004) report finding a potential *MIR159* EST from the moss *Physcomitrella*, suggesting an ancient origin of that

A

```
miR172              3' tacgucguaguaguucuaaga 5'

AP2         UUGACAAAUGCUGCAGCAUCAUCAGGAUUCUCUCCUCAUCA
GLOSSY15    AGCGCCGCCGCUGCAGCAUCAUCAGGAUUCCACUGUGGCA
PtAP2L1     AUGAUUGAAAGUGCAGCAUCAUCAGGAUUCUCACCCCAAAU
PtAP2L2     UCAAUGUUGGCAGCAGCAUCAUCAGGAUUCUCACCCCAAGC
AhAP2       UUUGCAACCGCUGCAGCAUCAUCAGGAUUCCACCACAACU
PaAP2L1     AUGAUUGAAAGUGCAGCAUCAUCAGGAUUCUCACCCCAAAU
                       * * * * * * * * * * * *  * * * * *
```

B

```
miR166              3' ccccuuacuucggaccaggcu 5'

PHB         GGUUCAGAUGAUUGGGAUGAAGCCUGGUCCGGAUUCUAUUGG
PHV         GGUCCAGAUGAUUGGGAUGAAGCCUGGUCCGGAUUCUAUUGG
REV         GGUUCAGAUGCCUGGGAUGAAGCCUGGUCCGGAUUCGGUUGG
ATHB8       GGUCCAAAUGCCUGGGAUGAAGCCUGGUCCGGAUUCCAUAGG
ATHB15      GGUUCAGAUGCCUGGAAUGAAGCCUGGUCCGGAUUCCAUUGG
PmC3HDZIP1  GGUCCAGAUGCCUGGGAUGAAGCCUGGUCCGGAUUCGAUUGG
TgC3HDZIP1  GGUCCAGAUGCCUGGGAUGAAGCCUGGUCCGGAUUCGAUUGG
TgC3HDZIP2  GAUCCAGAUGCCUGGGAUGAAGCCUGGUCCGGAUGCUAUUGG
CrC3HDZIP1  GAUUCAGAUGCCUGGAAUGAAGCCUGGUCCUGACUCUAUUGG
SkC3HDZIP1  GAUCCAGAUUCCUGGGAUGAAGCCUGGUCCGGAUUCAAUUAG
SkC3HDZIP2  GAUCCACAUGCCUGGGAUGAAGCCUGGUCCGGAUUCCGUUGG
PpC3HDZIP1  GAUACAGUUACCUGGUAUGAAGCCUGGUCCGGAUGCCAUUGG
PpC3HDZIP2  GAUUCAGUUACCUGGUAUGAAGCCUGGUCCGGAUGCCAUUGG
PpHB10      GAUUCAGUUACCUGGUAUGAAGCCUGGUCCGGAUGCCAUUGG
MpC3HDZIP1  GAUCCAGAUGCCCGGGAUGAAGCCUGGUCCGGACUCGAUUGG
PcC3HDZIP1  GAUCCAAAUGCCUGGGAUGAAGCCUGGUCCGGAAUCUAUUGG
                       * *  * * * * * * * * * * * * * * *
```

Figure 9.1 (A) Alignment of miRNA172 with aligned nucleotides including its binding site in seed plant *AP2*-coding sequences. *AP2*, *Arabidopsis*; *GLOSSY15*, *Zea mays*; Pt, *Pinus thunbergii*; Ah, *Malus domestica*; Pa, *Picea abies* (Shigyo & Ito, 2004). (B) Alignment of miR166 with aligned nucleotides including its binding site in Class III HD-Zip coding sequences. *PHB*; *PHV*; *REV*; *ATHB8*; *ATHB15*; *Arabidopsis*; Pm, *Pseudotsuga menziesii*; Tg, *Taxus globosa*; Cr, *Ceratopteris richardi*; Sk, *Selaginella kraussiana*; Pp, *Physcomitrella patens*; Mp, *Marchantia polymorpha*; Pc, *Phaeoceros carolinianus* (Floyd & Bowman, 2004). Watson–Crick mismatches of the mRNA to the miRNA are indicated by bold type. Asterisks below the alignment indicate nucleotides that are the same in all sequences (within the miRNA-binding site).

miRNA regulation of TCP or MYB mRNAs, although target sequences have not yet been identified. Regulation of *SCARECROW-LIKE* (*SCL*) mRNAs by miR171 is likely to predate the divergence of ferns and seed plants since an *SCL* homolog from *Ceratopteris richardii* encodes a complementary site.

The only known case of miRNA regulation shown to occur in all land plants, thus tracing back at least to the common ancestor of extant embryophytes, is regulation of Class III HD-Zip mRNAs by miR165/166. Floyd and Bowman (2004) found that the target sequence for miR165/166 was highly conserved at the nucleotide level in mRNA sequences of Class III HD-Zip homologs from representatives of all land plant lineages (Figure 9.1B). PCR evidence of cleavage of messenger RNAs was shown for bryophytes, ferns and vascular plants. Furthermore, a putative pri-miRNA encoding miR166 was cloned from *Selaginella kraussiana*. It is likely that several more cases of miRNA regulation will eventually be found to have arisen early in land plant evolution. To date, the complete phylogenetic history of most plant miRNA targets is simply not known and the lack of genomic sequence makes it difficult to search for homologous miRNA genes and targets in non-flowering plants.

9.9.2   miRNAs in plants versus metazoans

Because miRNA regulation has been found to be common and important in both plants and metazoans, it has been postulated that this form of regulation is ancient and predated the divergence of plant and animal lineages (Baulcombe, 2004; Hutvagner & Zamore, 2002b; Lim *et al.*, 2003a; Park *et al.*, 2002; Reinhart *et al.*, 2002; Rhoades *et al.*, 2002). The assumption has been made that since miRNAs have not been found in yeast (Reinhart & Bartel, 2002), miRNA regulation may have been lost in that lineage (Lim *et al.*, 2003b). Contributing to this assumption is the finding that there are many miRNAs that have homologs in vertebrates and invertebrates, indicating an ancient origin in metazoan evolution, including *let-7* (first identified in *C. elegans*), and several other miRNAs shared between *C. elegans* and humans (Lagos-Quintana *et al.*, 2001; Lau *et al.*, 2001; Lim *et al.*, 2003b; Pasquinelli *et al.*, 2000, 2003). However, no miRNA has been identified in animal lineages outside of the bilaterian clade (Pasquinelli *et al.*, 2000, 2003). As in plants, it is more difficult to search for evidence of miRNA regulation in more basal animal lineages. However, at present miRNA regulation in animals cannot be traced back to the metazoan ancestor. A summary of the current ideas on the phylogenetic relationships of eukaryotes is shown in Figure 9.2. Animal and plant clades are separated by several nodes. If miRNAs are not found in the intervening lineages, many of which have not yet been examined experimentally, a common origin of miRNA regulation for plants and animals would entail a single gain and several losses in intervening lineages (seven in Figure 9.2, but more in reality because not all lineages are shown) for a total of eight evolutionary steps. In contrast, independent origins within embryophytes and metazoans would entail two gains and only two evolutionary steps. Thus, the assumption that miRNA regulation in plants was inherited from a

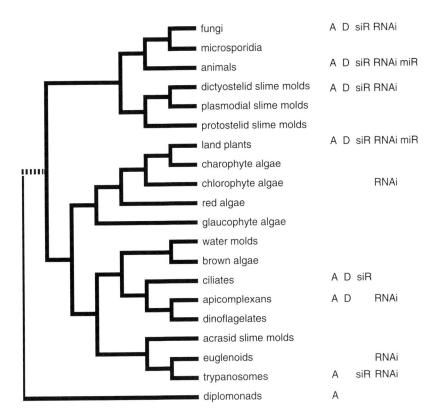

Figure 9.2 Phylogeny of eukaryotes showing distribution of proteins and processes of RNA-mediated regulation (Catalanotto *et al.*, 2004; Koblenz *et al.*, 2003; Martens *et al.*, 2002; Ntefidou *et al.*, 2003; Ullu *et al.*, 2004). Phylogeny redrawn based on Baldauf (2000, 2003). The hatched branch indicates that placement of that branch is not certain. To the right we have indicated in which taxa Argonaute proteins (A), Dicer proteins (D), siRNAs (siR), RNAi and miRNAs (miR) are known to occur.

common eukaryotic ancestor is not well supported by the known phylogenetic distribution of miRNAs.

What would seem to further call into question the idea of a common origin of miRNA regulation in plants and animals is the remarkable fact that out of 207 miRNAs identified in humans not one is shared with plants (Bartel, 2004). Since most of the miRNA targets are mRNAs from genes that are either plant- or animal-specific, it is clear that numerous cases of miRNA regulation have evolved within animal and plant lineages following the origin of genes unique to each lineage. However, one might expect that for genes that predate the divergence of plants and animals, given the apparent high levels of miRNA and target sequence conservation, there would be miRNA-regulated genes shared between plants and animals if they inherited the mechanism from a common ancestor.

There are a few miRNA targets in plants that have known homologs in most eukaryotes, including animals. These include ATP sulfurylases (miR395), *ARGONAUTE1* (miR168) and *DICER-LIKE1* (miR162) (Table 9.1). Interestingly, Dicer and Argonaute genes encode components of the processing machinery of small RNA-mediated gene regulation, including miRNA regulation (see discussion above). ATP sulfurylases catalyze the first reaction in sulfate metabolism (Leustek, 1996).

The miRNA395 family was discovered in a recent computational search for miRNAs and their targets in plants (Jones-Rhoades & Bartel, 2004). The targets of miRNA395 were predicted to be ATP sulfurylase mRNAs. There are four different ATP sulfurylases in *Arabidopsis*: *APS1*, *APS2*, *APS3* and *APS4*. Jones-Rhoades and Bartel (2004) showed that *APS4* mRNAs are cleaved *in vivo* at the putative binding site for miR395. They also predicted that miR395 would target mRNAs of *APS3* and *APS4* as well as a conifer homolog from *Cryptomeria* but not *Arabidopsis APS2*. ATP sulfurylase sequences are known from numerous eukaryotes including the flowering plants, animals and fungi and are highly conserved at the amino acid level (Hatzfeld *et al.*, 2000; Leustek, 1996; Murillo & Leustek, 1995). We obtained full or partial cDNA sequences from Genbank for ATP sulfurlyases from *Arabidopsis*, rice, potato, soybean, onion, the conifer *Cryptomeria*, the red alga *Porphyra*, *Pemicillium*, yeast, the worm *Urechis* and humans. A partial *Physcomitrella* cDNA contig was also obtained from a BLAST search of the PHYSCOBASE DNA database (Nishiyama *et al.*, 2003). The nucleotides were aligned (using an amino acid alignment as a guide) in the region of the predicted binding site of miR395 (Figure 9.3A). We recognize two groups of coding sequences, one group with three or fewer nucleotide mismatches within the miRNA complementary site, including the *Arabidopsis* sequences that are predicted to be targets of miR395, and one group with five or more mismatches, including *APS2* that was not predicted to be a target of miR395.

The group of putative miR395 targets includes sequences from other flowering plants as well as the gymnosperm *Cryptomeria*. The 'mismatch' group includes sequences from flowering plants, the moss *Physcomitrella* and all fungal and animal sequences. Phylogenetic analysis of inferred amino acid sequences for plant, fungal and animal ATP sulfurylases indicates that all the plant sequences form a clade and that the fungal and animal sequences together form a clade (Figure 9.3B). Within the plant clade, the moss sequence is sister to the flowering plant sequences. The flowering plant clade includes two clades that each contain monocot and eudicot sequences, one of which has sequences that are not predicted to be a target of miR395 and one of which, except for one gene, has three or fewer mismatches to miR395 and includes predicted *Arabidopsis* targets. This suggests that miRNA regulation of ATP sulfurylases

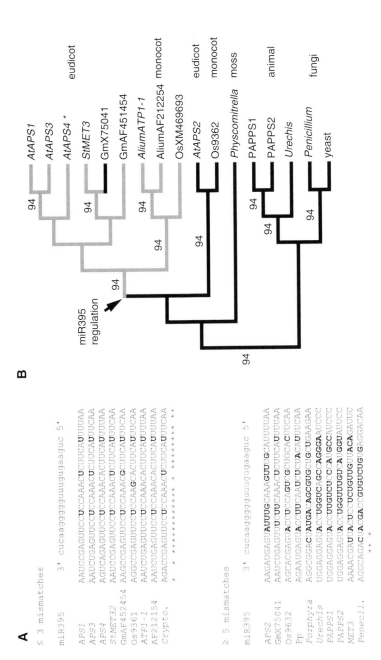

Figure 9.3 (A) Alignment of miRNA395 with aligned nucleotides including its binding site in plants, animals and fungal ATP sulfurylase mRNA sequences. Watson–Crick mismatches of the mRNA to the miRNA are indicated by bold type. Asterisks below the alignment indicate nucleotides that are the same in all sequences. One group includes mRNA sequences with three or fewer mismatches to miR395 including the sequences known and predicted by Jones-Rhoades and Bartel (2004) to be a target of miR395. The second group includes mRNA sequences with five or more mismatches including *APS2*, which was not predicted to be a target. (B) Parsimony analysis of ATP sulfurylase amino acid sequences. Bootstrap values greater than 50% shown on branches. Black branches indicate sequences that are not predicted to be targets of miR395, gray branches indicate sequences that are predicted to be targets of miR395. The asterisk indicates the only sequence for which there is experimental evidence of targeting by miR395. At, *Arabidopsis*; St, *Solanum tuberosum*; Gm, *Glycine max*; Os, *Oryza sativa*; PAPPS1, 2, human genes.

evolved within the land plant clade, following a duplication event that preceded the divergence of monocots and eudicots, but occurred after the divergence of mosses and vascular plants. One of the duplicated ATP sulfurylases became the target of miR395 and this has been maintained in all descendent genes, except perhaps one of the soybean genes. Thus, although sequence and function of ATP sulfurylases is ancient and conserved, miRNA regulation of ATP sulfurylases has evolved uniquely within the land plant clade.

Dicer homologs have been identified from diverse eukaryotes (Figure 9.2) and are probably present in all eukaryotes. Dicer proteins are known to be involved in the processing of double-stranded RNAs in gene-silencing pathways in plants, animals, fungi and slime molds (He & Hannon, 2004; Martens *et al.*, 2002; Matzke *et al.*, 2004; Meister & Tuschl, 2004; Park *et al.*, 2002). Phylogenetic analysis suggests that all four *Arabidopsis* Dicer genes are more closely related to each other than they are to animal and fungal Dicer genes (He & Hannon, 2004). miRNA regulation (by miR162) is limited to *Arabidopsis DCL1* (Xie *et al.*, 2004). Thus again, although plants and animals inherited a Dicer gene from a common ancestor, miRNA regulation of a *DCL1* likely arose within the land plant lineage.

The Argonaute family of proteins is also ancient, with a homolog in archaebacteria (Carmell *et al.*, 2002) and homologs known from all major eukaryotic lineages (Figure 9.2) (Fagard *et al.*, 2000; Koblenz *et al.*, 2003; Song *et al.*, 2004). While plant and metazoan lineages have multiple Argonaute family members, only a single Argonaute homolog has been identified in most other eukaryotes. Phylogenetic relationships of Argonaute sequences from different organisms have been variously resolved in different phylogenetic analyses (Carmell *et al.*, 2002; Hunter *et al.*, 2003; Sasaki *et al.*, 2003), none of which agree well with accepted hypotheses of relationship among animals, fungi and plants (see Figure 9.2). However, plant and animal genomes encode different numbers of Argonaute gene family members (e.g. 10 in *Arabidopsis*, over 20 in *C. elegans*, 8 in humans) (Meister & Tuschl, 2004), suggesting diversification within individual lineages. *Arabidopsis AGO1* is the only Argonaute that is targeted by an miRNA (miR168) (Rhoades *et al.*, 2002; Vaucheret *et al.*, 2004), which again suggests that miRNA regulation evolved once within flowering plants, after the divergence of plants and animals and diversification of Argonaute genes within the embryophyte lineage. The miRNA-mediated regulation of *AGO1* in Arabidopsis is likely to be a relatively recent event since the sister gene in *Arabidopsis*, *PNH*, is not regulated by miR168. It may be of biological importance that the two genes *DCL1* and *AGO1* that have evolved miRNA-mediated regulation may be the two members of their respective gene families to be most critical for the biogenesis and action of miRNAs in *Arabidopsis*.

There is abundant evidence that miRNA regulation of specific genes has evolved independently within plant and animal lineages. This should lead us to examine the details of miRNA regulation in plants and animals and ask if there are differences that are consistent with two independent origins of miRNA biogenesis and processing? In animals, the pri-miRNA is processed into a pre-miRNA stem-loop structure in the nucleus by the enzyme Drosha and then exported to the cytoplasm for processing into an miRNA duplex by Dicer. In *Arabidopsis*, both steps occur in the nucleus catalyzed by *DCL1* and there is no known homolog of Drosha (Bartel, 2004; He & Hannon, 2004; Meister & Tuschl, 2004). The fact that animal pre-miRNA hairpins are exported from the nucleus before processing may explain why the sizes of animal pre-miRNA hairpins appear to be constrained to around 70 nt in length whereas *Arabidopsis* hairpins are variable and can be much larger (Reinhart *et al.*, 2002). In plants, miRNAs have extensive complementarity to target sequences and many are known to mediate cleavage of mRNAs whereas animal miRNAs do not exhibit near-perfect complementarity to their target sequences and most lead to disruption of translation rather than mRNA degradation (He & Hannon, 2004; Jones-Rhoades & Bartel, 2004; Mallory *et al.*, 2004a; Rhoades *et al.*, 2002). Plant miRNA target sites are located in coding regions of mRNAs whereas animal miRNAs target 3' UTRs (He & Hannon, 2004). Several different animal miRNAs have been shown to target the same regulatory motifs that occur in the 3' UTRs of unrelated genes (Lai, 2002). No common binding motifs have been identified in plant miRNA-binding sites. Thus there are differences in plant and animal miRNA processing and function that could reflect independent evolutionary origins.

Despite numerous differences, the basic machinery of miRNA processing and targeting (described above) in plants is similar to that of animals. Both plants and animals require Dicer proteins for the processing of the pri-miRNA into pre-miRNA stem-loop structures (Meister & Tuschl, 2004) and Argonaute proteins as essential components of RISCs that interact with miRNAs and mRNAs to either disrupt translation, cleave the mRNA or interact with DNA. As mentioned above, Dicer and Argonaute proteins are ancient, are known to be involved in pathways of RNA-mediated gene silencing in many eukaryotic lineages and are likely to be ubiquitous in eukaryotes.(Figure 9.2). There are three pathways of gene silencing involving small RNA molecules that have been identified in eukaryotes – cytoplasmic siRNA silencing, silencing of endogenous mRNAs by miRNAs and transcriptional suppression due to siRNA-guided, sequence-specific DNA methylation (Baulcombe, 2004; Cogoni, 2001; Matzke *et al.*, 2004; Meister & Tuschl, 2004). It has become evident that there is a high level of conservation of the mRNA degradation machinery among different gene-silencing pathways (Fagard *et al.*, 2000; Hutvagner & Zamore, 2002b; Okamura *et al.*, 2004) and it has been proposed that the siRNA-mediated gene-silencing mechanisms were derived from an ancestral mechanism that functioned to protect the genome from invasive nucleic acids (Catalanotto *et al.*, 2004; Cogoni, 2001; Fagard *et al.*, 2000;

Liu *et al.*, 2004; Ullu *et al.*, 2004). Therefore, it is not an unreasonable scenario that miRNA regulation of gene expression evolved independently in land plants and metazoans, from a shared small RNA gene-silencing mechanism already present in all eukaryotes. Metazoans and land plants are the only two lineages to evolve true three-dimensional tissues (as opposed to modified filamentous growth as in fungi and brown algae) requiring complex developmental process to pattern and direct growth and differentiation. In both groups, miRNAs are involved in numerous processes essential for normal development. The origin of miRNA regulation may have played important roles in the origin of multi-cellularity in both groups (Bartel, 2004; Reinhart *et al.*, 2002).

There is still much that is not known about miRNA regulation in eukaryotes. It is clear that miRNAs play a major role in plant development. We are only beginning to know the full extent of this role. Many plant miRNAs have only just been identified and they remain to be characterized in terms of their mode of action and developmental roles. Furthermore, the first evidence of miRNA involvement in transcriptional silencing in plants has been reported (Bao *et al.*, 2004). The extent of miRNA involvement in chromatin remodeling is unknown and remains to be determined. We do not know the extent of miRNA regulation throughout the land plant clade. It will be essential to determine the history of miRNA targets and miRNA genes in order to understand the potential role miRNA regulation has had on the evolution of plant development. Understanding all these processes in plants is essential for comparison with animals since miRNA regulation in plants is distinct from that in animals. Comparison of miRNA-mediated gene regulation will allow us to gain a better understanding of how plant and animal miRNAs fit into the larger context of small RNA-directed processes and the role of small RNAs in the biology and evolution of eukaryotes.

Acknowledgments

Work in my laboratory is supported by the US National Science Foundation. We thank Cameron Johnson for help with Table 9.1. We apologize to our colleagues if we have omitted any pertinent references in this rapidly advancing field.

References

Achard, P., Herr, A., Baulcombe, D. C. & Harberd, N. P. (2004) Modulation of floral development by a gibberellin-regulated microRNA. *Development*, **131**, 3357–65.
Adai, A., Johnson, C., Mlotshwa, S., Archer-Evans, S., Manocha, V., Vance, V. & Sundaresan V. (2005) Computational prediction of miRNAs in *Arabidopsis thaliana*. *Genome Research*, **15**(1), 78–9.

Aida, M., Ishida, T., Fukaki, H., Fujisawa, H. & Tasaka, M. (1997) Genes involved in organ separation in *Arabidopsis*: an analysis of the *cup-shaped cotyledon* mutant. *Plant Cell*, **9**, 841–57.

Ambros, V. (2004) The functions of animal microRNAs. *Nature*, **431**, 350–55.

Ambros, V., Bartel, B., Bartel, D. P., Burge, C. B., Carrington, J. C., Chen, X., Dreyfuss, G., Eddy, S. R., Griffiths-Jones, S., Marshall, M., Matzke, M., Ruvkun, G. & Tuschl, T. (2003) A uniform system for microRNA annotation. *RNA*, **9**, 277–9.

Auckerman, M. J. & Sakai, H. (2003) Regulation of flowering time and floral organ identity by a microRNA and its *APETALA2*-like target genes. *Plant Cell*, **15**, 2730–41.

Baima, S., Nobili, F., Sessa, G., Lucchetti, S., Ruberti, I. & Morelli, G. (1995) The expression of the *Athb-8* homeobox gene is restricted to provascular cells in *Arabidopsis thaliana*. *Development*, **121**, 4171–82.

Bao, N., Lye, K. -W. & Barton, M. K. (2004) Mutations at a microRNA complementary site in processed *PHABULOSA* and *PHAVOLUTA* mRNAs are associated with decreased methylation of the template chromosome. *Developmental Cell*, **7**, 653–62.

Bartel, D. P. (2004) MicroRNAs: genomics, biogenesis, mechanism, and function. *Cell*, **116**, 281–97.

Bartel, B. & Bartel, D. (2003) At the root of plant development. *Plant Physiology*, **132**, 709–17.

Baldauf, S. L. (2003) The deep roots of eukaryotes. *Science*, **300**, 1703–706.

Baldauf, S. L., Roger, A. J., Wenk-Siefert, I. & Doolittle, W. F. (2000) A kingdom-level phylogeny of eukaryotes based on combined protein data. *Science*, **290**, 972–7.

Baulcombe, D. C. (2004) RNA silencing in plants. *Nature*, **431**, 356–63.

Beclin, C., Berthome, R., Palauqui, J. C., Tepfer, M. & Vaucheret, H. (1998) Infection of tobacco or *Arabidopsis* plants by CMV counteracts systemic post-transcriptional silencing of nonviral (*trans*)-genes. *Virology*, **252**, 313–17.

Bohmert, K., Camus, I., Bellini, C., Bouchez, D., Caboche, M. & Benning, C. (1998) *AGO1* defines a novel locus of *Arabidopsis* controlling leaf development. *EMBO Journal*, **17**, 170–80.

Bolle, C. (2004) The role of GRAS proteins in plant signal transduction and development. *Planta*, **218**, 683–92.

Bollman, K. M., Aukerman, M. J., Park, M., Hunter, C., Berardini, T. Z. & Poethig, R. S. (2003) *Hasty*, the *Arabidopsis* ortholog of exportin 5/MSN5, regulates phase change and morphogenesis. *Development*, **130**, 1493–504.

Bonnet, E., Wuyts, J., Rouze, P. & Van de Peer, Y. (2004) Detection of 91 potential conserved plant microRNAs in *Arabidopsis thaliana* and *Oryza sativa* identifies important target genes. *Proceedings of the National Academy of Sciences of the United States of America*, **101**, 11511–16.

Boutet, S., Vazquez, F., Liu, J., Beclin, C., Fagard, M., Gratias, A., Morel, J. B., Crete, P., Chen, X. & Vaucheret, H. (2003) *Arabidopsis HEN1*: a genetic link between endogenous miRNA controlling development and siRNA controlling transgene silencing and virus resistance. *Current Biology*, **13**, 843–8.

Bowman, J. L., Smyth, D. R. & Meyerowitz, E. M. (1991) Genetic interactions among floral homeotic genes of *Arabidopsis*. *Development*, **112**, 1–20.

Carmell, M. A., Xuan, Z., Zhang, M. Q. & Hannon, G. J. (2002) The Argonaute family: tentacles that reach into RNAi, developmental control, stem cell maintenance, and tumorigenesis. *Genes & Development*, **16**, 2733–42.

Catalanotto, C., Pallotta, M., ReFalo, P., Sachs, M. S., Vayssie, L., Macino, G. & Cogoni, C. (2004) Redundancy of the two Dicer genes in transgene-induced posttranscriptional gene silencing in *Neurospora crassa*. *Molecular and Cellular Biology*, **24**, 2536–45.

Chen, X. (2004) A microRNA as a translational repressor of *APETALA2* in *Arabidopsis* flower development. *Science*, **303**, 2022–5.

Cogoni, C. (2001) Homology-dependent gene silencing mechanisms in fungi. *Annual Review of Microbiology*, **55**, 381–406.

Dunoyer, P., Lecellier, C. -H., Parizotto, E. A., Himber, C. & Vionnet, O. (2004) Probing the microRNA and small interfering RNA pathways with virus-encoded suppressors of RNA silencing. *Plant Cell*, **16**, 1235–50.

Emery, J. F., Floyd, S. K., Alvarez, J., Eshed, Y., Hawker, N. P., Izhaki, A., Baum, S. F. & Bowman, J. L. (2003) Radial patterning of *Arabidopsis* shoots by Class III HD-ZIP and KANADI genes. *Current Biology*, **13**, 1768–74.

Eshed, Y., Baum, S. F., Perea, J. V. & Bowman J. L. (2001) Establishment of polarity in lateral organs of plants. *Current Biology*, **11**, 1251–60.

Eshed, Y., Izhaki, A., Baum, S. F., Floyd, S. K. & Bowman, J. L. (2004) Asymmetric leaf development and blade expansion in *Arabidopsis* are mediated by KANADI and YABBY activities. *Development*, **131**, 2997–3006.

Fagard, M., Boutet, S., Morel, J. -B., Bellini, C. & Vaucheret, H. (2000) AGO1, QDE-2, and RDE-1 are related proteins required for post-transcriptional gene silencing in plants, quelling in fungi, and RNA interference in animals. *Proceedings of the National Academy of Sciences of the United States of America*, **97**, 11650–54.

Finnegan, E. J., Margis, R. & Waterhouse, P. M. (2003) Posttranscriptional gene silencing is not compromised in the *Arabidopsis CARPEL FACTORY* (*DICER-LIKE1*) mutant, a homolog of Dicer-1 from *Drosophila*. *Current Biology*, **13**, 236–40.

Floyd, S. K. & Bowman, J. L. (2004) Ancient microRNA regulation of gene expression in land plants. *Nature*, **428**, 485–6.

Foster, T. M., Lough, T. J., Emerson, S. J., Lee, R. H., Bowman, J. L., Forster, R. L. S. & Lucas, W. J. (2002) Reversible establishment of organ polarity as a probe for signaling in plant development. *Plant Cell*, **14**, 1497–508.

Gallagher, K. L., Paquette, A. J., Nakajima, K. & Benfey, P. N. (2004) Mechanisms regulating *SHORT-ROOT* intercellular movement. *Current Biology*, **14**, 1847–51.

Griffiths-Jones, S. (2004) The microRNA Registry. *NAR*, **32**, Database Issue, D109–11.

Hamilton, A., Voinnet, O., Chappell, L. & Baulcombe, D. (2002) Two classes of short interfering RNA in RNA silencing. *EMBO Journal*, **21**, 4671–9.

Hammond, S. M., Bernstein, E., Beach, D. & Hannon, G. J. (2000) An RNA-directed nuclease mediates post-transcriptional gene silencing in *Drosophila* cells. *Nature*, **404**, 293–6.

Hammond, S. M., Boettcher, S., Caudy, A. A., Kobayashi, R. & Hannon, G. J. (2001) Argonaute2, a link between genetic and biochemical analyses of RNAi. *Science*, **293**, 1146–50.

Han, M. H., Goud, S., Song, L. & Fedoroff, N. (2004) The *Arabidopsis* double-stranded RNA-binding protein *HYL1* plays a role in microRNA-mediated gene regulation. *Proceedings of the National Academy of Sciences of the United States of America*, **101**, 1093–8.

Hatzfeld, Y., Lee, M., Leustek, T. & Saito, K. (2000) Functional characterization of a gene encoding a fourth ATP sulfurylase isoform from *Arabidopsis*. *Gene*, **248**, 51–8.

He, L. & Hannon, G. J. (2004) MicroRNAs: small RNAs with a big role in gene regulation. *Nature Reviews. Genetics*, **5**, 522–31.

Hunter, C., Sun, H. & Poethig, R. S. (2003) The *Arabidopsis* heterochronic gene *ZIPPY* is an ARGONAUTE family member. *Current Biology*, **13**, 1734–9.

Hutvágner, G. & Zamore, P. D. (2002a) A microRNA in a multiple-turnover RNAi enzyme complex. *Science*, **297**, 2056–60.

Hutvágner, G. & Zamore, P. D. (2002b) RNai: nature abhors a double-strand. *Current Opinion in Genetics and Development*, **12**, 225–32.

Jofuku, K. D., den Boer, B. G., Van Montagu, M. & Okamuro, J. K. (1994) Control of *Arabidopsis* flower and seed development by the homeotic gene *APETALA2*. *Plant Cell*, **6**, 1211–25.

Jones-Rhoades, M. W. & Bartel, D. P. (2004) Computational identification of plant microRNAs and their targets, including a stress-induced miRNA. *Molecular Cell*, **14**, 787–99.

Juarez, M. T., Kui, J. S., Thomas, J., Heller, B. A. & Timmermans, M. C. P. (2004) MicroRNA-mediated repression of *rolled leaf* specifies maize leaf polarity. *Nature*, **428**, 84–8.

Kang, J., Tang, J., Donnelly, P. & Dengler, N. (2003) Primary vascular pattern and expression of ATHB-8 in shoots of *Arabidopsis*. *New Phytologist*, **158**, 443–54.

Kasschau, K. D., Xie, Z., Allen, E., Llave, C., Chapman, E. J., Krizan, K. A. & Carrington, J. C. (2003) P1/HC-Pro, a viral suppressor of RNA silencing, interferes with *Arabidopsis* development and miRNA function. *Developmental Cell*, **4**, 205–17.

Kerstetter, R. A., Bollman, K., Taylor, R. A., Bomblies, K. & Poethig, R. S. (2001) *KANADI* controls organ polarity in *Arabidopsis*. *Nature*, **411**, 706–709.

Khvorova, A., Reynolds, A. & Jayasena, S. D. (2003) Functional siRNAs and miRNAs exhibit strand bias. *Cell*, **115**, 209–16.

Kidner, C. A. & Martienssen, R. A. (2004) Spatially restricted microRNA directs leaf polarity through *ARGONAUTE1*. *Nature*, **428**, 81–4.

Koblenz, B., Schoppmeier, J., Grunow, A. & Lechtreck, K. F. (2003) Centrin deficiency in *Chlamydomonas* causes defects in basal body replication, segregation and maturation. *Journal of Cell Science*, **116**, 2635–46.

Lagos-Quintana, M., Rauhut, R., Lendeckel, W. & Tuschl, T. (2001) Identification of novel genes coding for small expressed RNAs. *Science*, **294**, 853–8.

Lai, E. C. (2002) MicroRNAs are complementary to the 3' UTR sequence motifs that mediate negative post-transcriptional regulation. *Nature Genetics*, **30**, 363–4.

Lau, M. C., Lim, L. P., Weinstein, E. G. & Bartel, D. P. (2001) An abundant class of tiny RNAs with probable regulatory roles in *Caenorhabditis elegans*. *Science*, **294**, 858–62.

Laufs, P., Peaucelle, A., Morin, H. & Traas, J. (2004) MicroRNA regulation of the *CUC* genes is required for boundary size control in *Arabidopsis* meristems. *Development*, **131**, 4311–22.

Lee, R. C., Feimbaum, R. L. & Ambros, V. (1993) The *C. elegans* heterochronic gene *lin-4* encodes small RNAs with antisense complementarity to *lin-14*. *Cell*, **75**, 843–54.

Leustek, T. (1996) Molecular genetics of sulfate assimilation in plants. *Physiologia Plantarum*, **97**, 411–19.

Lim, L. P., Glasner, M. E., Yekta, S., Burge, C. B. & Bartel, D. P. (2003a) Vertebrate microRNA genes. *Science*, **299**, 1540.

Lim, L. P., Lau, N. C., Weinstein, E., Abdelhakim, A., Yekta, S., Rhoades, M. W., Burge, C. B. & Bartel, D. P. (2003b) The microRNAs of *Caenorhabditis elegans*. *Genes & Development*, **17**, 991–1008.

Liu, J., Carmell, M. A., Rivas, F. V., Marsden, C. G., Thomson, J. M., Song, J. -J., Hammond, S. M., Joshua-Tor, L. & Hannon, G. J. (2004) Argonaute2 is the catalytic engine of mammalian RNAi. *Science*, **305**, 1437–41.

Llave, C., Kasschau, K. D. & Carrington, J. C. (2000) Virus-encoded suppressor of posttranscriptional gene silencing targets a maintenance step in the silencing pathway. *Proceedings of the National Academy of Sciences of the United States of America*, **97**, 13401–406.

Llave, C., Xie, Z., Kasschau, K. D. & Carrington, J. C. (2002) Cleavage of scarecrow-like mRNA targets directed by a class of *Arabidopsis* miRNA. *Science*, **297**, 2053–6.

Lund, E., Guttinger, S., Calado, A., Dahlberg, J. E. & Kutay, U. (2004) Nuclear export of miRNA precursors. *Science*, **303**, 95–8.

Lynn, K., Fernandez, A., Aida, M., Sedbrook, J., Tasaka, M., Masson, P. & Barton, M. K. (1999) The *PINHEAD/ZWILLE* gene acts pleiotropically in *Arabidopsis* development and has overlapping functions with the *ARGONAUTE1* gene. *Development*, **126**, 1–13.

Mallory, A. C., Ely, L., Smith, T. H., Marathe, R., Anandalakshmi, R., Fagard, M., Vaucheret, H., Pruss, G., Bowman, L. & Vance, V. B. (2001) HC-Pro suppression of transgene silencing eliminates the small RNAs but not transgene methylation or the mobile signal. *Plant Cell*, **13**, 571–83.

Mallory, A. C., Reinhart, B. J., Bartel, D., Vance, V. B. & Bowman, L. H. (2002) A viral suppressor of RNA silencing differentially regulates the accumulation of short interfering RNAs and micro-RNAs in tobacco. *Proceedings of the National Academy of Sciences of the United States of America*, **99**, 15228–33.

Mallory, A. C., Reinhart, B. J., Jones-Rhoades, M. W., Tang, G., Zamore, P. D., Barton, M. K. & Bartel, D. P. (2004a) MicroRNA control of *PHABULOSA* in leaf development: importance of pairing to the micro-RNA 5' region. *EMBO Journal*, **23**, 3356–64.

Mallory, A. C., Dugas, D. V., Bartel, D. P. & Bartel, B. (2004b) MicroRNA regulation of NAC-domain targets is required for proper formation and separation of adjacent embryonic, vegetative, and floral organs. *Current Biology*, **14**, 1035–46.

Martens, H., Novotny, J., Oberstrass, J., Steck, T. L., Postlethwait, P. & Nellen, W. (2002) RNAi in *Dictyostelium*: the role of RNA-directed RNA polymerases and double-stranded RNase. *Molecular Biology of the Cell*, **13**, 445–53.

Martinez, J., Patkaniowska, A., Urlaub, H., Luhrmann, R. & Tuschl, T. (2002) Single-stranded antisense siRNAs guide target RNA cleavage in RNAi. *Cell*, **110**, 563–74.

Matzke, M., Matzke, A. J. & Kooter, J. M. (2001) RNA: guiding gene silencing. *Science*, **293**, 1080–83.

Matzke, M., Aufsatz, W., Kanno, T., Daxinger, L., Papp, I., Mette, M. F. & Matzke, A. J. M. (2004) Genetic analysis of RNA-mediated transcriptional gene silencing. *Biochimica et Biophysica Acta*, **1677**, 129–41.

McConnell, J. R. & Barton, M. K. (1995) Effect of mutations in the *PINHEAD* gene of *Arabidopsis* on the formation of shoot apical meristems. *Developmental Genetics*, **16**, 358–66.

McConnell, J. R. & Barton, M. K. (1998) Leaf polarity and meristem formation in *Arabidopsis*. *Development*, **125**, 2935–42.

McConnell, J. R., Emery, J. F., Eshed, Y., Bao, N., Bowman, J. & Barton, M. K. (2001) Role of *PHABULOSA* and *PHAVOLUTA* in determining radial patterning in shoots. *Nature*, **411**, 709–13.

Meister, G. & Tuschl, T. (2004) Mechanisms of gene silencing by double-stranded RNA. *Nature*, **431**, 343–9.

Moose, S. P. & Sisco, P. H. (1996) *Glossy15*, an *APETALA2*-like gene from maize that regulates leaf epidermal cell identity. *Genes & Development*, **10**, 3018–27.

Morel, J. B., Godon, C., Mourrain, P., Beclin, C., Boutet, S., Feuerbach, F., Proux, F. & Vaucheret, H. (2002) Fertile hypomorphic *ARGONAUTE* (*ago1*) mutants impaired in post-transcriptional gene silencing and virus resistance. *Plant Cell*, **14**, 629–39.

Mourelatos, Z., Dostie, J., Paushkin, S., Sharma, A., Charroux, B., Abel, L., Rappsilber, J., Mann, M. & Dreyfuss G. (2002) miRNPs: a novel class of ribonucleoproteins containing numerous microRNAs. *Genes & Development*, **16**, 720–28.

Moussian, B., Schoof, H., Haecker, A., Jurgens, G. & Laux, T. (1998) Role of the *ZWILLE* gene in the regulation of central shoot meristem cell fate during *Arabidopsis* embryogenesis. *EMBO Journal*, **17**, 1799–809.

Murillo, M. & Leustek, T. (1995) Adenosine-5'-triphosphate-sulfurlyase from *Arabidopsis thaliana* and *Escherichia coli* are functionally equivalent but structurally and kinetically divergent: nucleotide sequence of two adenosine-5'-triphosphate-sulfurlyase cDNAs from *Arabidopsis thaliana* and analysis of a recombinant enzyme. *Archives of Biochemistry and Biophysics*, **323**, 195–204.

Nakajima, K., Sena, G., Nawy, T. & Benfey, P. N. (2001) Intercellular movement of the putative transcription factor *SHR* in root patterning. *Nature*, **413**, 307–11.

Nath, U., Crawford, B. C., Carpenter, R. & Coen, E. (2003) Genetic control of surface curvature. *Science*, **299**, 1404–407.

Nishiyama, T., Fujita, T., Shin-I, T., Seki, M., Nishide, H., Uchiyama, I., Kamiya, A., Carninci, P., Hayashizaki, Y., Shinozaki, K., *et al.* (2003) Comparative genomics of *Physcomitrella patens* gametophytic transcriptome and *Arabidopsis thaliana*: implication for land plant evolution. *Proceedings of the National Academy of Sciences of the United States of America*, **100**, 8007–12.

Ntefidou, M., Iseki, M., Watanabe, M., Lebert, M. & Hader, D. -P. (2003) Photoactivated adenylyl cyclase controls phototaxis in the flagellate *Euglena gracilis*. *Plant Physiology*, **133**, 1517–21.

Ohashi-Ito, K. & Fukuda, H. (2003) HD-zip III homeobox genes that include a novel member, *ZeHB-13* (*Zinnia*)/*ATHB-15* (*Arabidopsis*), are involved in procambium and xylem cell differentiation. *Plant Cell Physiology*, **44**, 1350–58.

Okamura, K., Ishizuka, A., Siomi, H. & Siomi, M. C. (2004) Distinct roles for Argonaute proteins in small RNA-directed RNA cleavage pathways. *Genes & Development*, **18**, 1655–66.

Otsuga, D., DeGuzman, B., Prigge, M. J., Drews, G. N. & Clark, S. E. (2001) *REVOLUTA* regulates meristem initiation at lateral positions. *Plant Journal*, **25**, 223–36.

Palatnik, J. F., Allen, E., Wu, X., Schommer, C., Schwab, R., Carrington, J. C. & Weigel, D. (2003) Control of leaf morphogenesis by microRNAs. *Nature*, **425**, 257–63.

Papp, I., Mette, M. F., Aufsatz, W., Daxinger, L., Schauer, S. E., Ray, A., van der Winden, J., Matzke, M. & Matzke, A. J. M. (2003) Evidence for nuclear processing of plant microRNA and short interfering RNA precursors. *Plant Physiology*, **132**, 1382–90.

Parizotto, E. A., Dunoyer, P., Rahm, N., Himber, C. & Voinnet, O. (2004) *In vivo* investigation of the transcription, processing, endonucleolytic activity, and functional relevance of the spatial distribution of a plant miRNA. *Genes & Development*, **18**, 2237–42.

Park, W., Li, J., Song, R., Messing, J. & Chen, X. (2002) *CARPEL FACTORY*, a Dicer homolog, and *HEN1*, a novel protein, act in microRNA metabolism in *Arabidopsis thaliana*. *Current Biology*, **12**, 1484–95.

Pasquinelli, A. E., Reinhart, B. J., Slack, F., Martindale, M. Q., Kuroda, M. I., Maller, B., Hayward, D. C., Ball, E. E., Degnan, B., Muller, P., Spring, J., Srinivasan, A., Fishman, M., Finnerty, J., Corbo, J., Levine, M., Leahy, P., Davidson, E. & Ruvkun, G. (2000) Conservation of the sequence and temporal expression of *let-7* heterochronic regulatory RNA. *Nature*, **408**, 86–9.

Pasquinelli, A. E., McCoy, A., Jimenez, E., Salo, E., Ruvkin, G., Martindale, M. Q. & Baguña, J. (2003) Expression of the 22 nucleotide *let-7* heterochronic RNA throughout the Metazoa: a role in life history evolution? *Evolution and Development*, **5**, 372–8.

Plasterk, R. H. (2002) RNA silencing: the genome's immune system. *Science*, **296**, 1263–5.

Pontig, C. P. & Aravind, L. (1999) START: a lipid-binding domain in StAR, HD-ZIP and signalling proteins. *Trends in Biochemical Sciences*, **24**, 130–32.

Reinhart, B. J. & Bartel, D. P. (2002) Small RNAs correspond to centromere heterochromatic repeats. *Science*, **297**, 1831.

Reinhart, B. J., Slack, F. J., Basson, M., Pasquinelli, A. E., Bettinger, J. C., Rougvie, A. E., Horvitz, H. R. & Ruvkun, G. (2000) The 21-nucleotide *let-7* RNA regulates developmental timing in *Caenorhabditis elegans*. *Nature*, **403**, 901–906.

Reinhart, B. J., Weinstein, E. G., Rhoades, M. W., Bartel, B. & Bartel, D. P. (2002) MicroRNAs in plants. *Genes & Development*, **16**, 1616–26.

Rhoades, M. W., Reinhart, B. J., Lim, L. P., Burge, C. B., Bartel, B. & Bartel, D. P. (2002) Prediction of plant microRNA targets. *Cell*, **110**, 513–20.

Sasaki, T., Shiohama, A., Minoshima, S. & Shimizu, N. (2003) Identification of eight members of the Argonaute family in the human genome. *Genomics*, **82**, 323–30.

Schauer, S. E., Jacobsen, S. E., Meinke, D. W. & Ray, A. (2002) *DICER-LIKE1*: blind men and elephants in *Arabidopsis* development. *Trends in Plant Science*, **7**, 487–91.

Schwarz, D. S., Hutvágner, G., Du, T., Xu, Z., Aronin, N. & Zamore, P. D. (2003) Asymmetry in the assembly of the RNAi enzyme complex. *Cell*, **115**, 199–208.

Sessa, G., Steindler, C., Morelli, G. & Ruberti, I. (1998) The *Arabidopsis Athb-8, -9* and *-14* genes are members of a small gene family coding for highly related HD-ZIP proteins. *Plant Molecular Biology*, **38**, 609–22.

Shigyo, M. & Ito, M. (2004) Analysis of gymnosperm two-AP2-domain-containing genes. *Development Genes and Evolution*, **214**, 105–14.

Sigova, A., Rhind, N. & Zamore, P. D. (2004) A single Argonaute protein mediates both transcriptional and post-transcriptional silencing in *Schizosaccharomyces pombe*. *Genes & Development*, **18**, 2359–67.

Slack, F. J., Basson, M., Liu, Z., Ambros, V., Horvitz, H. R. & Ruvkun, G. (2000) The *lin-41* RBCC gene acts in the *C. elegans* heterochronic pathway between the *let-7* regulatory RNA and the *lin-29* transcription factor. *Molecular Cell*, **5**, 659–69.

Song, J. -J., Smith, S. K., Hannon, G. J. & Joshua-Tor, L. (2004) Crystal structure of Argonaute and its implications for RISC slicer activity. *Science*, **305**, 1434–7.

Stuurman, J., Jaggi, F. & Kuhlemeier, C. (2002) Shoot meristem maintenance is controlled by a GRAS-gene mediated signal from differentiating cells. *Genes & Development*, **16**, 2213–18.

Sunkar, R. & Zhu, J. K. (2004) Novel and stress-regulated microRNAs and other small RNAs from *Arabidopsis*. *Plant Cell*, **16**, 2001–19.

Sussex, I. M. (1955) Morphogenesis in *Solanum tuberosum* L.: experimental investigation of leaf dorsoventrality and orientation in the juvenile shoot. *Phytomorphology*, **5**, 286–300.

Takada, S., Hibara, K., Ishida, T. & Tasaka, M. (2001) The *CUP-SHAPED COTYLEDON1* gene of *Arabidopsis* regulates shoot apical meristem formation. *Development*, **128**, 1127–35.

Talbert, P. B., Adler, H. T., Parks, D. W. & Comai, L. (1995) The *REVOLUTA* gene is necessary for apical meristem development and for limiting cell divisions in the leaves and stems of *Arabidopsis thaliana*. *Development*, **121**, 2723–35.

Tang, G., Reinhart, B. J., Bartel, D. P. & Zamore, P. D. (2003) A biochemical framework for RNA silencing in plants. *Genes & Development*, **17**, 49–63.

Tian, C., Wan, P., Sun, S., Li, J. & Chen, M. (2004) Genome-wide analysis of the GRAS gene family in rice and *Arabidopsis*. *Plant Molecular Biology*, **54**, 519–32.

Ullu, E., Tschudi, C. & Chakraborty, T. (2004) RNA interference in protozoan parasites. *Cellular Microbiology*, **6**, 509–19.

Vaucheret, H., Beclin, C. & Fagard, M. (2001) Post-transcriptional gene silencing in plants. *Journal of Cell Science*, **114**, 3083–91.

Vaucheret, H., Vazquez, F., Crátá, P. & Bartel, D. P. (2004) The action of *ARGONAUTE1* in the miRNA pathway and its regulation by the miRNA pathway are crucial for plant development. *Genes & Development*, **18**, 1187–97.

Vazquez, F., Gasciolli, V., Crete, P. & Vaucheret, H. (2004a) The nuclear dsRNA binding protein *HYL1* is required for microRNA accumulation and plant development, but not posttranscriptional transgene silencing. *Current Biology*, **14**, 346–51.

Vazquez, F., Vaucheret, H., Rajagopalan, R., Lepers, C., Gasciolli, V., Mallory, A. C., Hilbert, J. L., Bartel, D. P. & Crete, P. (2004b) Endogenous *trans*-acting siRNAs regulate the accumulation of *Arabidopsis* mRNAs. *Molecular Cell*, **16**, 69–79.

Voinnet, O., Vain, P., Angell, S. & Baulcombe, D. C. (1998) Systemic spread of sequence-specific transgene RNA degradation in plants is initiated by localized introduction of ectopic promoterless DNA. *Cell*, **95**, 177–87.

Voinnet, O., Lederer, C. & Baulcombe, D. C. (2000) A viral movement protein prevents spread of the gene silencing signal in *Nicotiana benthamiana*. *Cell*, **103**, 157–67.

Wang, X. J., Reyes, J. L., Chua, N. H. & Gaasterland, T. (2004) Prediction and identification of *Arabidopsis thaliana* microRNAs and their mRNA targets. *Genome Biology*, **5**, R65.

Waterhouse, P. M., Wang, M. B. & Lough, T. (2001) Gene silencing as an adaptive defence against viruses. *Nature*, **411**, 834–42.

Wightman, B., Ha, I. & Ruvkun, G. (1993) Post-transcriptional regulation of the heterochronic gene *lin-14* by *lin-4* mediates temporal pattern formation in *C. elegans*. *Cell*, **75**, 855–62.

Xie, Z., Kasschau, K. D. & Carrington, J. C. (2003) Negative feedback regulation of *Dicer-Like1* in *Arabidopsis* by microRNA-guided mRNA degradation. *Current Biology*, **13**, 784–9.

Xie, Z., Johansen, L. K., Gustafson, A. M., Kasschau, K. D., Lellis, A. D., Zilberman, D., Jacobsen, S. E. & Carrington, J. C. (2004) Genetic and functional diversification of small RNA pathways in plants. *Public Library of Science Biology*, **2**, E104.

Yekta, S., Shih, I. H. & Bartel, D. P. (2004) MicroRNA-directed cleavage of *HOXB8* mRNA. *Science*, **304**, 594–6.

Zamore, P. D. (2002) Ancient pathways programmed by small RNAs. *Science*, **296**, 1265–9.

Zhong, R. & Ye, Z. H. (2004) *Amphivasal vascular bundle 1*, a gain-of-function mutation of the *IFL1/REV* gene, is associated with alterations in the polarity of leaves, stems and carpels. *Plant Cell Physiology*, **45**, 369–85.

Zilberman, D., Cao, X., Johansen, L. K., Xie, Z., Carrington, J. C. & Jacobsen, S. E. (2004) Role of *Arabidopsis ARGONAUTE4* in RNA-directed DNA methylation triggered by inverted repeats. *Current Biology*, **14**, 1214–20.

Index

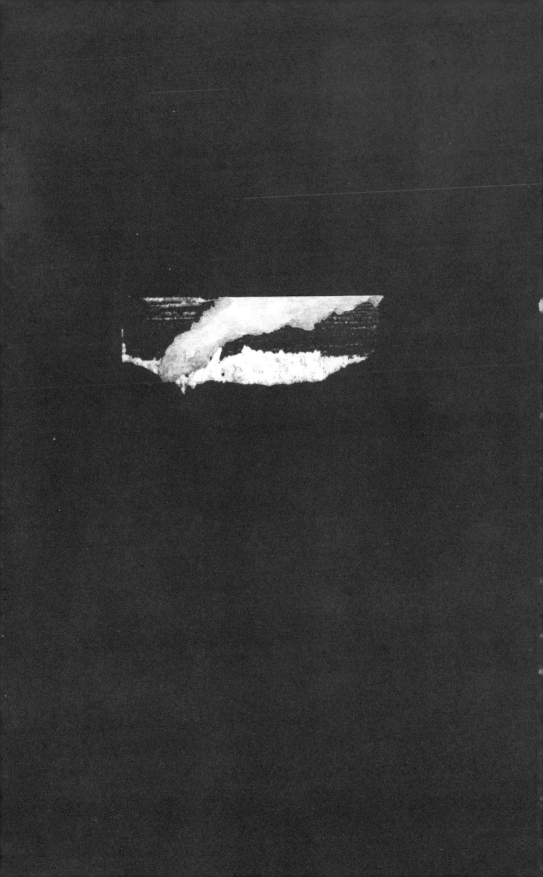